AF786008

45

Springer Series in Chemical Physics

Edited by Fritz P. Schäfer

Springer Series in Chemical Physics

Editors: Vitalii I. Goldanskii Fritz P. Schäfer J. Peter Toennies

Volume 40 **High-Resolution Spectroscopy of Transient Molecules**
By E. Hirota

Volume 41 **High Resolution Spectral Atlas of Nitrogen Dioxide 559–597 nm**
By K. Uehara and H. Sasada

Volume 42 **Antennas and Reaction Centers of Photosynthetic Bacteria**
Structure, Interactions, and Dynamics
Editor: M. E. Michel-Beyerle

Volume 43 **The Atom-Atom Potential Method. Applications to Organic Molecular Solids**
By A. J. Pertsin and A. I. Kitaigorodsky

Volume 44 **Secondary Ion Mass Spectrometry SIMS V**
Editors: A. Benninghoven, R. J. Colton, D. S. Simons, and H. W. Werner

Volume 45 **Thermotropic Liquid Crystals, Fundamentals**
By G. Vertogen and W. H. de Jeu

Volume 46 **Ultrafast Phenomena V**
Editors: G. R. Fleming and A. E. Siegman

Volume 47 **Complex Chemical Reaction Systems**
Editors: J. Warnatz and W. Jäger

Volumes 1-39 are listed on the back inside cover

G. Vertogen W. H. de Jeu

Thermotropic Liquid Crystals, Fundamentals

With 93 Figures

Springer-Verlag Berlin Heidelberg New York
London Paris Tokyo

Professor Dr. Ger Vertogen

Institute for Theoretical Physics, Catholic University, Toernooiveld
NL-6525 ED Nijmegen, The Netherlands

Professor Dr. Wim H. de Jeu

Open University, P.O. Box 2960, NL-6401 DL Heerlen and
FOM-Institute for Atomic and Molecular Physics, Kruislaan 407
NL-1098 SJ Amsterdam, The Netherlands

Series Editors

Professor Vitalii I. Goldanskii

Institute of Chemical Physics
Academy of Sciences, Kosygin Street 3
Moscow V-334, USSR

Professor Dr. Fritz Peter Schäfer

Max-Planck-Institut für
Biophysikalische Chemie
D-3400 Göttingen-Nikolausberg, FRG

Professor Dr. J. Peter Toennies

Max-Planck-Institut für Strömungsforschung
Böttingerstraße 6–8
D-3400 Göttingen, FRG

ISBN-13: 978-3-642-83135-5 e-ISBN-13: 978-3-642-83133-1

DOI: 10.1007/978-3-642-83133-1

Library of Congress Cataloging-in-Publication Data. Vertogen, G. (Gerrit), 1940- Thermotropic liquid crystals, fundamentals. (Springer series in chemical physics ; v. 45) Bibliography: p. Includes index. 1. Liquid crystals. I. Jeu, W. H. de (Wilhelmus Hendrikus), 1943-. I. Title. II. Series. QD923.V47 1987 548'.9 87-14673

This work is subject to copyright. All rights are reserved, whether the whole or part of the material is concerned, specifically the rights of translation, reprinting, reuse of illustrations, recitation, broadcasting, reproduction on microfilms or in other ways, and storage in data banks. Duplication of this publication or parts thereof is only permitted under the provisions of the German Copyright Law of September 9, 1965, in its version of June 24, 1985, and a copyright fee must always be paid. Violations fall under the prosecution act of the German Copyright Law.

© Springer-Verlag Berlin Heidelberg 1988

Softcover reprint of the hardcover 1st edition 1988

The use of registered names, trademarks, etc. in this publication does not imply, even in the absence of a specific statement, that such names are exempt from the relevant protective laws and regulations and therefore free for general use.

2153/3150-543210

I libri non sono fatti per crederci,
ma per essere sottoposti a indagine
(Guglielmo da Baskervilla)

Preface

The motto of this book is quoted from *"Il Nome della Rosa"* by Umberto Eco. The English translation reads *"Books are not made to be believed, but to be subjected to inquiry"*. The veracity of this statement, which not only holds for books but also for scientific papers in general, seems trivial. Unfortunately, however, the reality often turns out differently. We hope that our book will be subjected to a serious inquiry and that it may lead to many fruitful discussions. We would certainly appreciate receiving readers' reactions on possible faults or mistaken starting points.

In writing the book we aimed at a unified and critical account of the fundamental aspects of liquid crystals. Instead of attempting to compile the latest results and cover the field of liquid crystals in the widest possible sense, we concentrate on and discuss the assumptions made in developing theories – and hence in analyzing experimental data – bearing in mind Einstein's dictum *"Erst die Theorie entscheidet darüber, was beobachtet werden kann"* (Only the theory decides on what can be observed). In order to facilitate the study of the book we have avoided intimidating mathematics. The only unfamiliar mathematical tool may be tensor analysis, which we felt was essential for a clear presentation. For the reader's convenience, however, we have incorporated a simple introduction to this subject.

The book has been divided into four parts. Part I about liquid crystals in general is quite descriptive in character. Part II deals with the macroscopic continuum theory of liquid crystals. In particular we present a systematic development of the theory starting from a tensorial point of view, thus emphasizing the symmetry of the various phases. Part III is mainly devoted to a discussion of those experimental techniques that provide relevant microscopic information on the orientational behaviour of liquid crystals. Much attention is paid to the interpretation of experimental data. Part IV discusses the theory of the various phases and their attendant phase transitions, both from a Landau and from a molecular-statistical point of view. The merits of the molecular approach are critically examined. For the sake of clarity we have simplified the existing models as far as possible, thus hoping to facilitate a better understanding of the theory.

Finally we wish to express our sincere appreciation to M.J.R. Straatman for her invaluable support in typing and organizing this text. We also wish to thank J.F.M. Wieland for making the drawings. We greatly benefited from the comments of S.J. Picken, who read the entire manuscript and improved on our English. Finally our thanks go to C.A.M. Govers, E.F. Gramsbergen, B.C.H. Krutzen and L. Longa for valuable discussions and comments on parts of the book. It is clear, however, that we remain entirely responsible for the final contents.

October 1987 *G. Vertogen, W.H. de Jeu*

Contents

Part I MESOMORPHIC BEHAVIOUR

Part II CONTINUUM THEORY

Part I

Mesomorphic Behaviour

1. Introduction

In this chapter we shall begin by describing the basic features that distinguish the liquid crystalline phases from the other condensed phases. The essential characteristic is the presence of orientational order of the non-spherical molecules, while in addition the positional order of the centres of mass is either absent (nematic phases) or reduced (smectic and columnar phases). Next the types of molecules leading to these phases are discussed, and typical examples will be given. Finally we shall review some of the experimental methods that are commonly used to test the liquid crystalline properties of a compound.

1.1 Mesomorphic Behaviour

The *liquid crystalline* phase (or better phases) can be observed in certain organic compounds. These are phases in between the solid crystalline phase and the isotropic liquid phase. A liquid crystal can flow like an ordinary liquid, however other properties, such as birefringence for example, are reminiscent of the crystalline phase. This combination explains the name liquid crystal, which is in fact a *contradictio in terminis*. Other names in use are *mesophase* (meaning intermediate phase) and *mesomorphic phase.*

In order to understand the nature of liquid crystals, one should recall the structure of a crystal. Here the centres of mass of the molecules are located on a three-dimensional periodic lattice, i.e. the centres of mass have a long-range order. Consequently one finds x-ray diffraction patterns showing sharp Bragg reflections. On the other hand, in isotropic liquids only short-range order between the molecular centres of mass is present. Here the centres of mass have, in principle, three translational degrees of freedom. The x-ray diffraction patterns now only show broad, diffuse reflections, corresponding to this short-range structure. Dealing with anisotropic molecules, however, positional order alone does not suffice to characterize the crystalline structure. Long-range order of the orientations of the molecules will be present as well. In many of these solid substances both types of order disappear simultaneously at the melting point. The remaining substances can be divided into two classes characterized by their intermediate phases.

- *Plastic crystalline phases,* in which the positional order is still present, but the orientational order has disappeared or is strongly reduced.
- *Liquid crystalline phases,* characterized by orientational order, while the positional order is reduced or has even completely disappeared.

Roughly speaking, two classes of liquid crystals can be distinguished, *thermotropic* and *lyotropic* liquid crystalline phases. Single component systems that show mesomorphic behaviour in a definite temperature range are called thermotropic. Lyotropic liquid crystalline phases on the other hand show mesomorphic behaviour in solution. The temperature range in which these lyotropic liquid crystals exist is mainly determined by the amount of solvent. The constituent units of lyotropic liquid crystals are either anisotropic molecules or associated groups of many molecules (micelles). If the constituent entities or molecules are elongated (rod-like), the long axes are on average parallel to each other. Due to thermal fluctuations this orientational order is not ideal. If the molecules (or associated groups) have a disc-like structure, the orientational order arises because the normals to the discs are oriented, on average, parallel to each other. This monograph deals mainly with the classical thermotropic liquid crystals formed by simple elongated molecules. A short introduction to the more "exotic" liquid crystals will be given in Chap. 4.

The main subdivision of thermotropic liquid crystals of elongated molecules is into *nematic* and *smectic* mesophases (see Fig. 1.1). In the nematic phase the centres of mass of the molecules have three translational degrees of freedom, and are thus distributed at random. In this sense one can truly speak of an anisotropic liquid. Smectic phases are characterized in addition by a positional order in at least one dimension. The centres of the molecules are, on average, arranged in equidistant planes. Though there may be considerable deviations from the equilibrium positions, one often talks in a somewhat loose way of a layer structure. There exist many types of smectic phases, indicated as S_A, S_B, ... , S_I. They differ in: (i) the orientation of the preferred direction of the molecules with respect to the layer normal

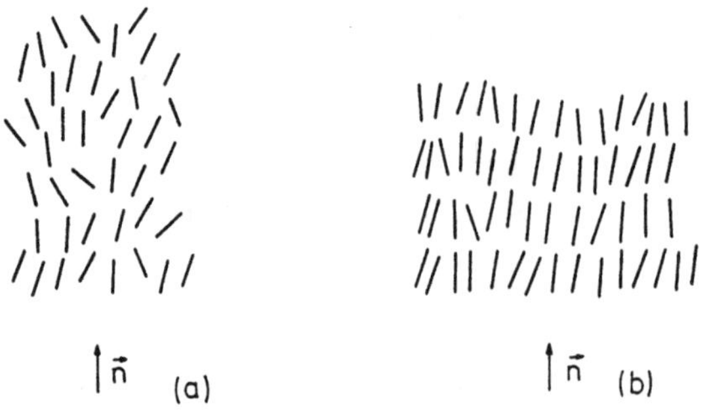

Fig. 1.1. Schematic picture of the nematic **(a)** and smectic A **(b)** phase

(orthogonal and tilted smectic phases); (ii) the organization of the centres of the molecules within the layers. When the preferred direction is tilted with respect to the layer normal there is a long-range correlation of the tilt direction of the various layers, and the smectic phase becomes biaxial. A more detailed description follows in Chap. 3.

When a substance showing a thermotropic liquid crystalline phase is heated, the solid changes at the melting point into a rather turbid liquid. The fluidity may be high for a nematic phase and relatively low for some of the smectic phases. When observed between crossed polarizers, this fluid phase is found to be strongly birefringent. Upon further heating another transition point is reached where the turbid liquid becomes isotropic and consequently optically clear (clearing point T_c, see Fig. 1.2). The melting point and the clearing point define the temperature range in which the mesophase is thermodynamically stable. Both phase transitions are first order, as indicated by the occurrence of a latent heat and a discontinuous change in the density. At the clearing point, however, these quantities are usually an order of magnitude smaller than those at the melting point. This implies that the clearing point, in contrast to the melting point, can hardly be supercooled. A compound that possesses a mesophase is called a mesogenic compound. Here the term liquid crystal will be used for a mesogenic compound in its liquid crystalline phase(s).

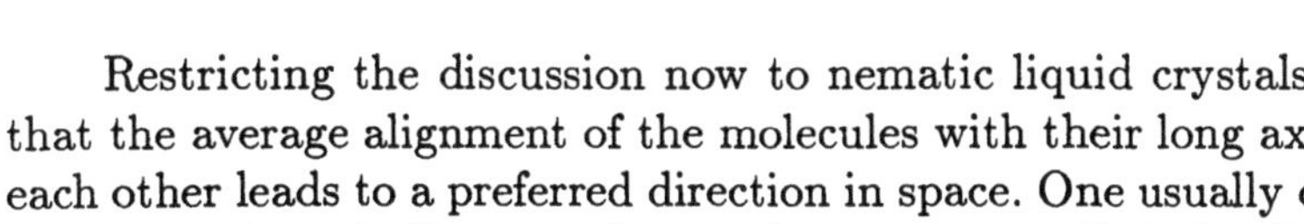

Fig. 1.2. The liquid crystalline phases on a temperature scale

Restricting the discussion now to nematic liquid crystals one can say that the average alignment of the molecules with their long axes parallel to each other leads to a preferred direction in space. One usually describes this local direction of alignment by a unit vector $\boldsymbol{n}$, called the director, which gives at each point in a sample the direction of the preferred axis. The states described by $\boldsymbol{n}$ and $-\boldsymbol{n}$ appear to be indistinguishable, i.e. as far as has been observed until now the polarity of the constituent molecules does not lead to a macroscopic effect. In an actual sample the orientation of $\boldsymbol{n}$ is imposed by the boundary conditions and/or by external fields. Without special measures the boundary conditions will be irregular. This leads to variations in the director pattern, and thus to differences in birefringence. With a polarizing microscope one can observe a characteristic pattern, or texture, in thin liquid crystalline layers. These textures can be used for characterization, as some features depend on the symmetry (nematic, type of smectic) of the phase under consideration.

For the characterization of the orientational order two aspects have to be taken into account: the local preferred direction $\boldsymbol{n}(\boldsymbol{r})$ and the amount of ordering, i.e. the distribution of the long molecular axes around $\boldsymbol{n}$. In order to construct an order parameter suppose one can define a unit vector $\boldsymbol{a}(\boldsymbol{r})$, describing the orientation of the long axis of a molecule at the site $\boldsymbol{r}$. In that case one should imagine the molecules to be rodlike entities. Now, consider the thermal averages of various tensors, that are composed of $\boldsymbol{a}$, over a small but macroscopic volume around the point $\boldsymbol{r}$. As the average $\langle \boldsymbol{a} \cdot \boldsymbol{a} \rangle$ is a constant by definition, it is clear that a scalar order parameter (tensor order parameter of rank zero) is out of the question. Next, one might try a vector order parameter $\langle \boldsymbol{a} \rangle$, analogous to the magnetization in a ferromagnet. However, a non-zero value of $\langle \boldsymbol{a} \rangle$ would violate the equivalence of $\boldsymbol{n}$ and $-\boldsymbol{n}$. It describes a ferroelectric nematic state which has not been observed so far. Accordingly, in order to avoid contradictions, the reader should be on the alert and use the concept of director very carefully. For a proper description one must turn to the next possibility, which is a second-rank tensor. The elements of the tensor order parameter $\tilde{\boldsymbol{S}}$ are defined by

$$S_{\alpha\beta} = \langle a_\alpha a_\beta \rangle - \tfrac{1}{3}\delta_{\alpha\beta} \quad , \quad \alpha, \beta = x, y, z \quad . \tag{1.1}$$

Here x, y, z is a laboratory-fixed frame, while $\delta_{\alpha\beta}$ is the Kronecker tensor which equals one if $\alpha = \beta$ and is zero otherwise. The latter term is added to make $S_{\alpha\beta} = 0$ in the isotropic phase, where $\langle a_\alpha^2 \rangle = \frac{1}{3}$. The tensor order parameter $\tilde{\boldsymbol{S}}$ is a symmetric traceless tensor and thus has in general five different elements. By choosing a suitable coordinate system such a tensor can be brought into diagonal form. In that case either two of the diagonal elements are equal (uniaxial phase) or all three are different (biaxial phase). For the uniaxial case (1.1) can be written as

$$S_{\alpha\beta} = S(n_\alpha n_\beta - \tfrac{1}{3}\delta_{\alpha\beta}) \quad , \quad \alpha, \beta = x, y, z \quad . \tag{1.2}$$

Choosing $\boldsymbol{n}$ along the z axis, the three non-zero diagonal elements are

$$S_{zz} = \tfrac{2}{3}S \quad , \quad S_{xx} = S_{yy} = -\tfrac{1}{3}S \quad . \tag{1.3}$$

The scalar S is a measure of the alignment of the long axes of the molecules. The average orientation of the molecules can be specified by a distribution function $f(\theta)$, where $f(\theta)2\pi \sin\theta \, d\theta$ is the fraction of the molecules on a cone making an angle between θ and $\theta + d\theta$ with $\boldsymbol{n}$. The average value of a quantity $X(\theta)$ over the orientations of all molecules is then given by

$$\langle X \rangle = 2\pi \int_0^\pi f(\theta) X(\theta) \sin\theta \, d\theta \tag{1.4}$$

and S is defined as

$$S = \langle \tfrac{3}{2} \cos^2\theta - \tfrac{1}{2} \rangle \quad . \tag{1.5}$$

If the distribution of the long molecular axes is random as in the isotropic phase, then $\langle \cos^2\theta \rangle = \frac{1}{3}$ and $S = 0$. The value $S = 1$ corresponds to the case of perfectly aligned molecules. In practice, S varies from values around 0.3–0.4 at T_c to about 0.8 at much lower temperatures. Hence the average value of θ may be as large as 40°; the deviations from perfect alignment are considerable. It follows from the properties of liquid crystals that the distribution function will be strongly peaked around the directions $\theta = 0$ and $\theta = \pi$ (the director), while for $\theta = \frac{1}{2}\pi$ a minimum has to be expected. Because $\theta = 0$ and $\theta = \pi$ are equivalent, it holds that $f(\theta) = f(\pi - \theta)$. Consequently, if $X(\theta)$ in (1.4) changes sign when θ is replaced by $\pi - \theta$, its average value will be zero. This applies, for example, to $\cos\theta$ for which $\langle \cos\theta \rangle = 0$. A more general definition of the order parameter will be given in Sect. 5.2, while the orientational distribution and its relation with the anisotropy of various physical properties will be fully discussed in Part III.

When discussing the degree of order, it has been assumed so far that the director is oriented along the z axis everywhere. Unless special measures are taken this will not be true for a real liquid crystal. However, the variations of $\boldsymbol{n}$ are only important over distances much larger than the molecular dimensions, and hence a nematic can still be taken to be locally uniaxial. S then specifies the average degree of orientation with respect to the local director. In other words, the space dependence of $\boldsymbol{n}$ and the temperature dependence of S can be treated separately. This property is the basis of the continuum theory of liquid crystals, which is the subject of Part II.

So far it has been implicitly assumed that the molecules constituting the liquid crystalline phases have mirror symmetry. In the case of optically active molecules a nematic phase is found only for a racemic mixture of right- and left-handed species. As soon as one species of molecules is in the majority a *chiral nematic* mesophase results (abbreviated as N^*). A simple picture of this phase is sketched in Fig. 1.3. In addition to the long-range orientational order there is a spatial variation of the director leading to a helical structure. If a series of planes perpendicular to the helix axis is considered, then each plane shows orientational order as in the nematic phase. However, the local preferred direction of alignment of the molecules

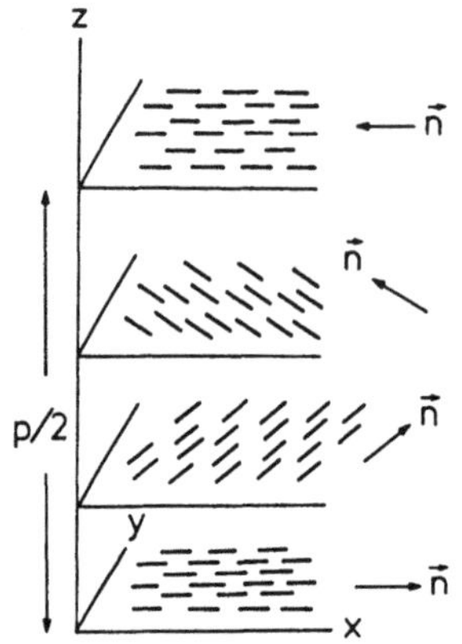

Fig. 1.3. Schematic picture of the chiral nematic phase. For the sake of clarity perfect alignment is assumed

is slightly rotated in adjacent planes. The helical structure can be described by an intrinsically non-constant director field (see Fig. 1.3):

$$n_x = \cos\phi \quad , \quad n_y = \sin\phi \quad , \quad n_z = 0 \quad . \tag{1.6}$$

The helix axis has been taken parallel to the z axis with $\phi = t_0 z +$ constant. A full rotation of $\boldsymbol{n}$ is completed over a distance $p = 2\pi/|t_0|$, called the pitch. The sign of t_0 distinguishes between left-handed and right-handed chiral nematics. Because $\boldsymbol{n}$ and $-\boldsymbol{n}$ are still equivalent the repetition period is $p/2$. In practice, p can vary between, say 200 nm and infinity, which means that for molecules, adjacent in the z direction, the difference in the preferred direction remains extremely small. The local order is thus essentially the same as in nematics. Many derivatives of cholesterol belong to this class of liquid crystals, which explains the historical name *cholesteric* as an alternative to chiral nematic. Because of the great similarity to the nematic phase the word chiral nematic should be preferred. The existence of the helicoidal structure gives rise to interesting optical properties that will be discussed in Sect. 7.4. Optical activity of the molecules influences some of the smectic states as well. The tilted smectic phases change into chiral smectic phases with a helicoidal ordering of the tilt directions.

One of the interesting aspects of liquid crystals is their sensitivity even to weak external stimuli (i.e. boundary conditions, external fields), due to the cooperative effect of the long-range orientational order. This property can be used to obtain samples with a uniform director pattern, thus allowing the direct determination of the anisotropy of various physical properties. For example, if for a uniaxial phase like a nematic a light beam is incident along the director, the ordinary refractive index n_o will be measured. If an incident light beam is perpendicular to the director and is polarized along $\boldsymbol{n}$ the other extreme situation is found, giving the extraordinary refractive index n_e. Indicating the directions parallel and perpendicular to $\boldsymbol{n}$ by the subscripts $\|$ and $\perp$, respectively, one finds an anisotropy $\Delta n = n_e - n_o = n_{\|} - n_{\perp}$. Liquid crystals can be strongly birefringent with Δn reaching values up to 0.5. Similarly an anisotropy will be measured for other properties like the electric or magnetic susceptibility. This gives the possibility of manipulating the orientation of $\boldsymbol{n}$ with external fields. The corresponding change in birefringence leads to the important applications of liquid crystals in display devices.

1.2 Mesogenic Compounds

In Table 1.1 some examples of nematogenic compounds are presented. The various phases and transition temperatures (in °C) are given in a linear notation, where K stands for the crystalline state, N for the nematic phase and I for the isotropic liquid. The first compound, p,p'-dimethoxyazoxybenzene or p-azoxyanisole (PAA) is the *drosophila* (fruitfly) of liquid crystal research.

Table 1.1. Some examples of nematic liquid crystals

p-azoxyanisole, PAA

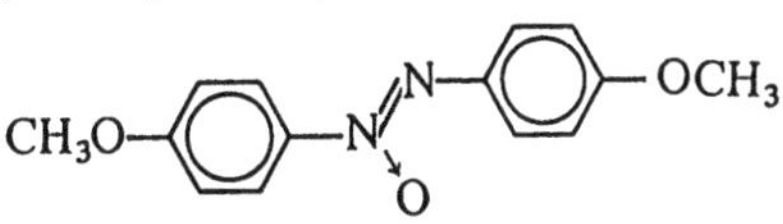

K 118 N 135.5 I

N-(*p*-methoxybenzylidene)-*p*′-butylaniline, MBBA

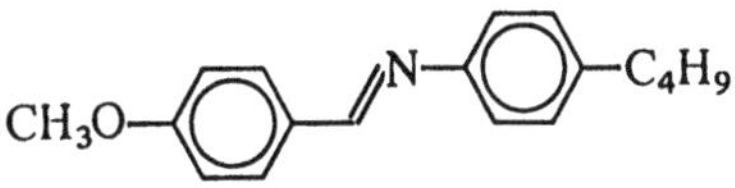

K 22 N 47 I

p-pentyl-*p*′-cyanobiphenyl, 5CB

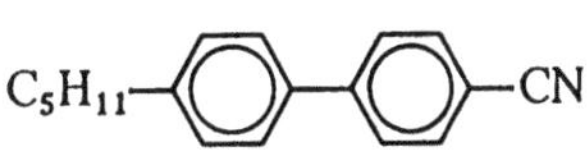

K 22.5 N 35 I

p-pentylphenyl-*trans*-*p*′-pentylcyclohexylcarboxylate

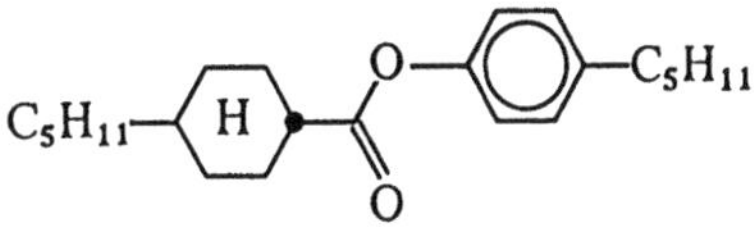

K 37 N 47 I

Its melting point and nematic range are typical of many of the long-known mesogenic compounds. The other examples are chosen from more recently synthesized materials that have their nematic range around room temperature. Many of the known mesogenic compounds have a structure somewhat similar to the examples of Table 1.1:

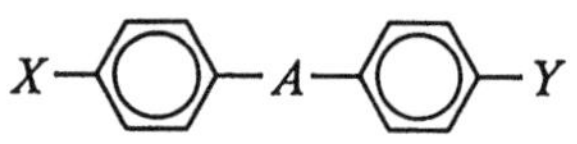

or

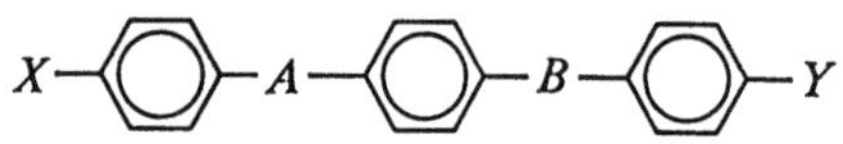

In these structures one can distinguish various parts:

i) Cyclic structures, often benzene rings, but also heterocyclic rings or cyclohexane rings are being used. Examples are given in Table 1.2.

Table 1.2. Various types of ring systems in mesomorphic compounds

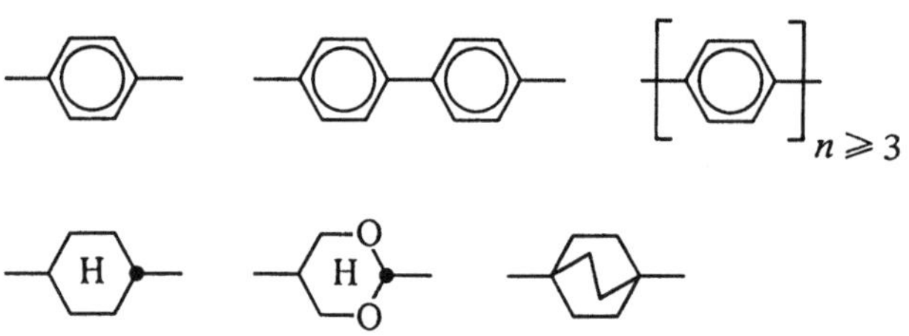

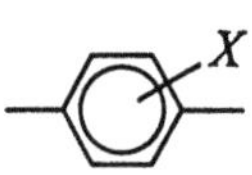

X: Cl, Br, I, OH, O*R*, OOC*R*, CH_3, CN, NO_2
R is C_nH_{2n+1}

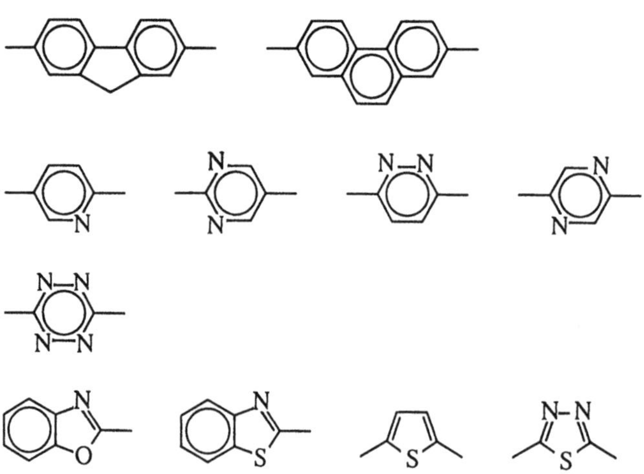

Table 1.3. Various types of bridging groups in mesomorphic compounds

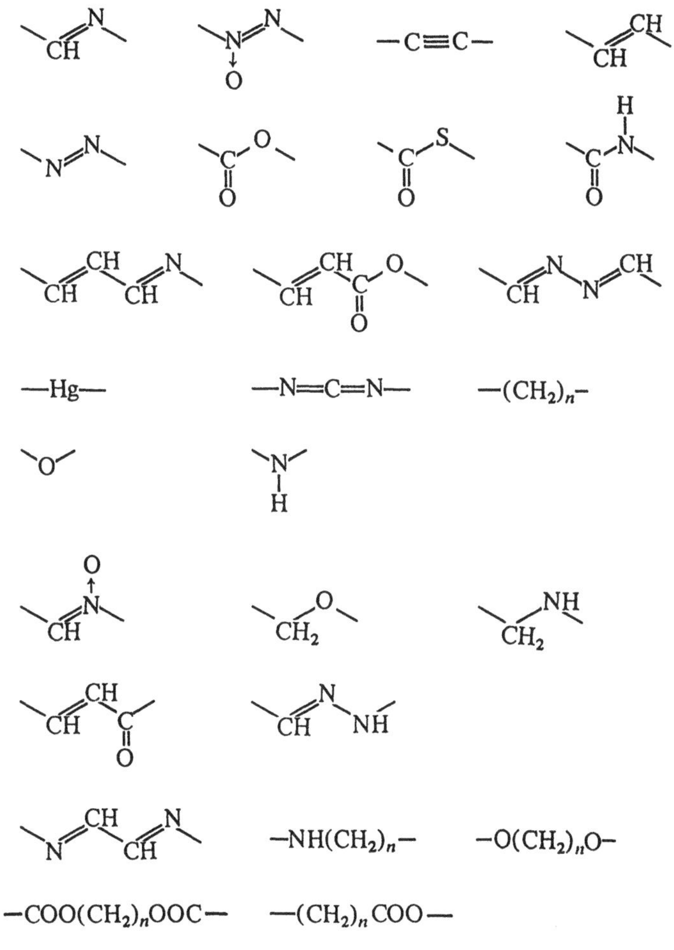

ii) Bridging groups A, B, connecting the rings. Some well-known examples are given in Table 1.3.

iii) End groups X, Y, that can be either aliphatic chains (alkyl, alkoxy) or a simple rigid group (−CN). Examples are given in Table 1.4.

The resulting molecular structures are generally not flat. Only for azobenzenes are the azo bridge group and the two benzene rings in one plane. For azoxybenzenes there is an angle of the order of 20° between the two benzene rings, while Schiff bases are strongly non-planar (Van der Veen and Grobben 1971, see Fig. 1.4). Biphenyls are also non-planar; here the angle between the two rings depends on the end substituents.

Table 1.4. Various types of end groups in mesomorphic compounds (*R*: *n*-alkyl group, *R*′: branched or unsaturated alkyl group)

$-OR$	$-R$	$-COOR$	$-COR$
$-OOCR$	$-OOCOR$	$-CH{=}CH-COOR$	
$-CN$	$-Cl$	$-NO_2$	
$-H$	$-F$	$-Br$	
$-J$	$-R'$	$-N{=}C{=}O$	
$-OH$	$-OR'$	$-OCOR'$	
$-NH_2$	$-COOR'$	$-CR{=}CR-COOR$	
$-SR$	$-NHR$	$-NR_2$	
	$-NHCPR$	$-N{=}C{=}S$	
	$-O(CH_2)_nOR$	$-OCF_3$	

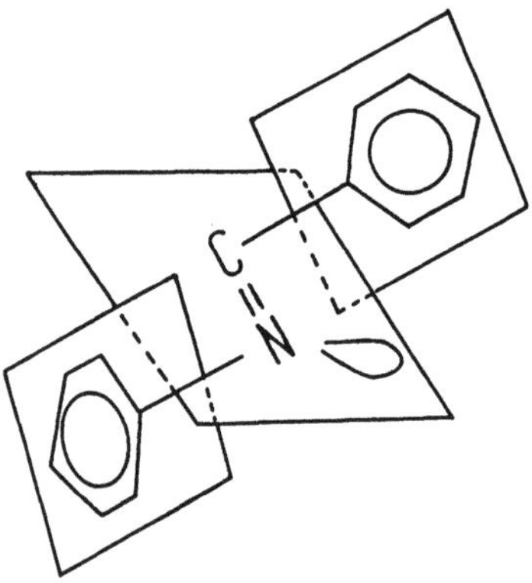

Fig. 1.4. Structure of benzylideneaniline

From the types of compound two requirements for mesomorphic behaviour can be extracted in an empirical way:

i) The compounds should have an anisotropic elongated form, as reflected in their length/width ratio L/W.
ii) The compounds should have a relatively rigid polarizable part.

The first rule can be illustrated by comparing normal-alkyl and branched-alkyl chains. In the latter case a strong reduction of T_c is observed. The same effect is observed if a molecule is broadened by substitution at *ortho* positions. The polarizable part is evidently present in the many cases where aromatic rings are involved. It seems that the larger L/W is [condition (i)], the less polarizability is required. In fact aliphatic ring systems can replace the aromatic rings while mesomorphic behaviour is still preserved. In a first approximation permanent dipoles are of less importance for mesomorphic behaviour. In fact there are well-known examples of non-polar nematic liquid crystals:

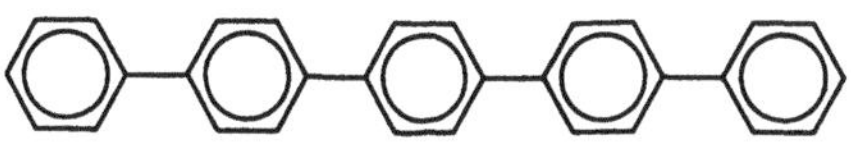

p-quinquephenyl K 380 N 431 I

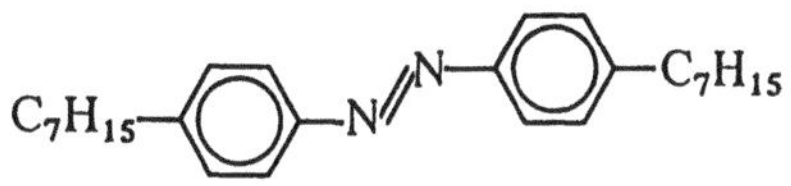

p, *p'*-diheptylazobenzene K 40 N 47 I

The other extreme situation is obtained when a terminal cyano group is present, in which case the molecule has a large dipole moment. This certainly has influence on the mesomorphic behaviour, which is thought to be related to the presence of antiparallel dipole-dipole correlations in this type of compounds.

The notion of mesophase stability refers to stability in comparison with the next phase at higher temperatures. Hence attention should be focussed on the clearing temperature T_C if the nematic stability of various compounds is compared. The range of existence of a mesophase depends, of course, also on the melting point. Unlike the clearing point the melting point often varies erratically in a homologous series of compounds. If the melting point is at a higher temperature than the clearing point it is in some cases still possible to observe a reversible clearing point in the supercooled state. This is called *monotropic* mesomorphic behaviour. This illustrates that, to obtain a useful mesomorphic range, a high clearing point is not sufficient. The melting point should be low enough as well.

Some examples of smectic liquid crystals are given in Table 1.5. In some cases up to five smectic phases can be observed (*polymorphism*); the

Table 1.5. Some examples of smectic liquid crystals

p,p'-dinonylazobenzene

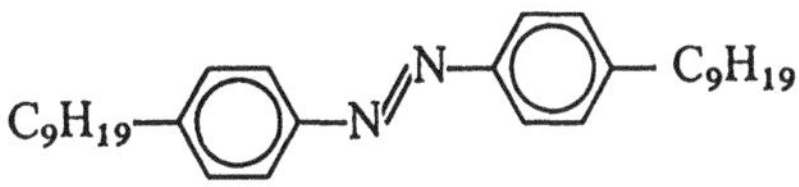

K 37 S_B 40 S_A 53 I

p,p'-diheptyloxyazoxybenzene, HOAB

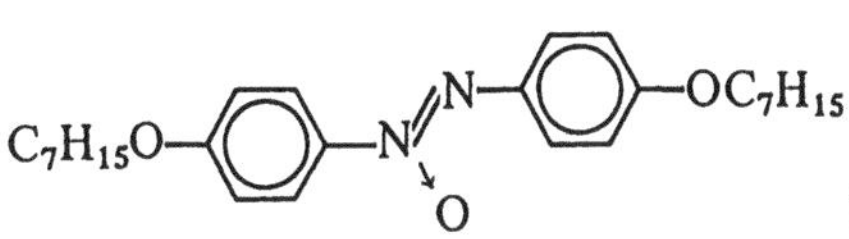

K74.5 S_C95.5 N 124 I

Terephthalylidene-bis-(*p*-butylaniline), TBBA

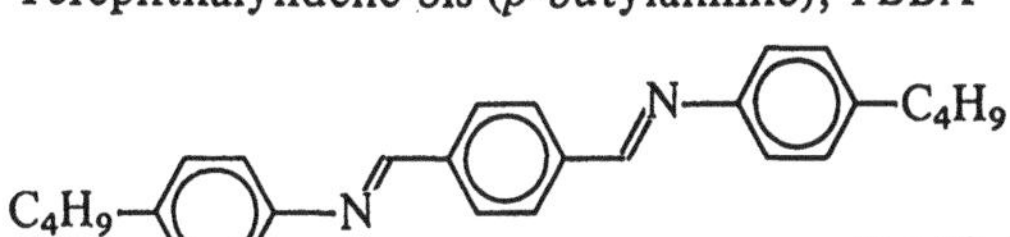

K 112.5 S_G 114 S_C 172.5 S_A 198.5 N 235.5 I (89.5 S_H 80 S_5)

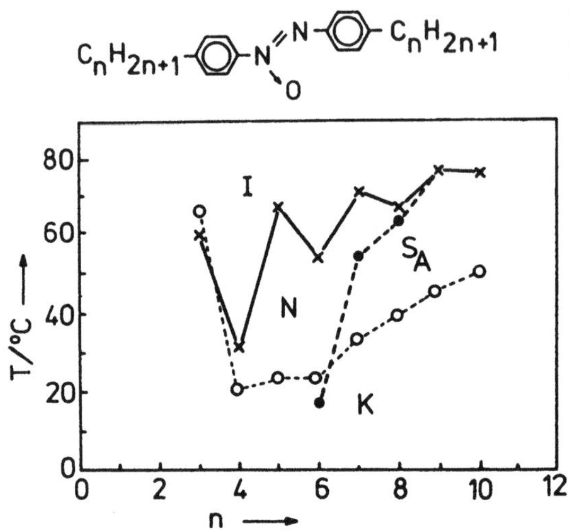

Fig. 1.5. Transition temperatures of the series of p,p'-dialkylazoxybenzenes (Van der Veen et al. 1973)

associated structures will be discussed in Chap. 3. The traditional way to discuss nematic *versus* smectic behaviour is to consider *homologous* series. A typical picture thus obtained is given in Fig. 1.5. For short alkyl chains one finds a nematic phase, at intermediate chain length a nematic and a smectic phase and at long chain length only smectic phases. This has led to models in which the ratio between the central (polarizable) aromatic core and the total molecular length in a homologous series (which depends on the length of the less polarizable aliphatic chains) is the determining factor. More recently it has become clear that this picture is not complete. In the case of an asymmetric bridging group (or otherwise asymmetric end chains) it is possible to study *isometric* series: the total length of the molecule is kept constant, while the length of one end chain is increased and the other decreased. Then one often finds smectic phases for one type of asymmetry only. Some extreme examples are:

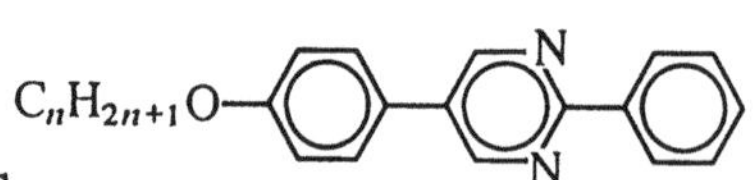

(I)

and

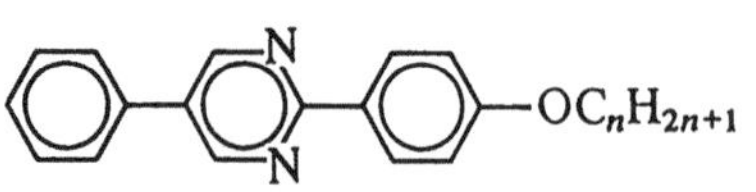

(II)

For series (I) all compounds studied (from $n = 2$ to $n = 10$) show only a smectic A phase. For series (II) all compounds studied (up to $n = 12$) show only a nematic phase. Evidently the asymmetry plays an important role. Only recently has this fact been taken into account in theoretical models.

Table 1.6. Some examples of chiral nematic liquid crystals

N-(*p*-ethoxybenzylidene)-*p*′-(*β*-methylbutyl)aniline

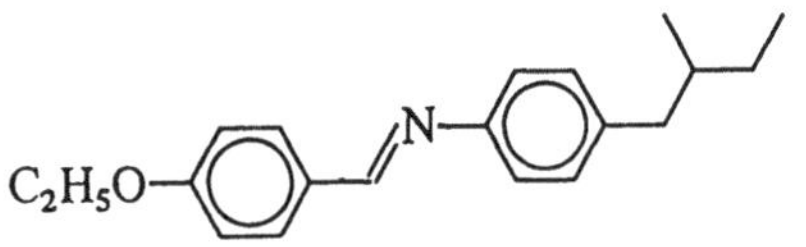

K 15 N* 60 I

Cholesteryl myristate

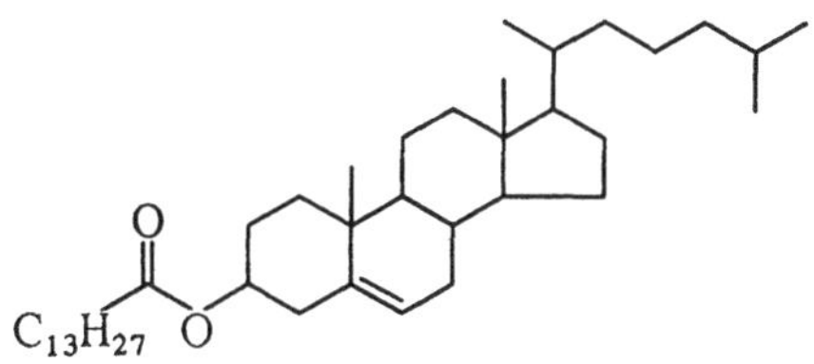

K 71 S_A 81 N* 86.5 I

Examples of chiral nematic liquid crystals are given in Table 1.6. An ordinary S_A phase can occur in a number of chiral nematics at temperatures below the N* region. Finally it is worthwhile to mention that Demus et al. (1983) have published a useful book with tables of mesomorphic compounds and their transition temperatures.

1.3 Classification of the Mesophases

The very existence of a liquid crystalline phase can obviously be established by visual inspection of the compound while it is heated. The mesophase is distinguished from the isotropic liquid by its turbid appearance, and from the crystalline solid by its flow properties. Smectic phases are often more viscous than a nematic phase. In order to determine the transition temperature a polarizing microscope usually is used, equipped with a heating stage in which a small sample between a glass plate and a cover slide is inserted.

Traditional methods to study phase transitions are the variation of the density (dilatometry) and of the specific heat (calorimetry). Classical dilatometric methods are rather time consuming and require fairly large samples for a good accuracy. More modern techniques, based on the oscillation time of a hollow V-shaped tube filled with the substance under consideration, make volumetric studies more attractive again. Provided a phase transition is first order, it can be detected by the associated discontinuity in the density.

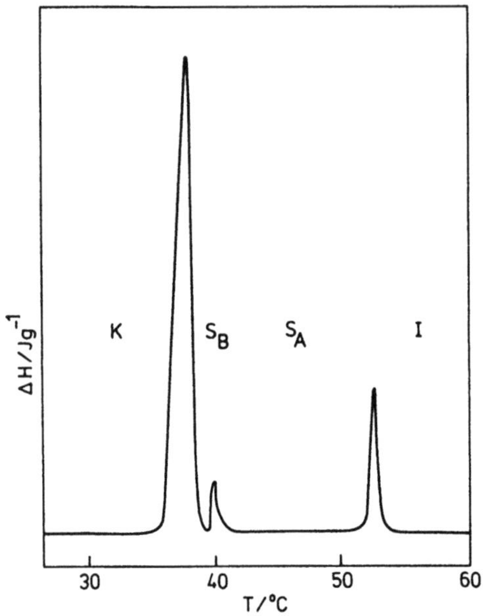

Fig. 1.6. DSC recording of p, p'-dinonylazobenzene (see Table 1.5)

Modern calorimetric methods such as differential scanning calorimetry (DSC) and differential thermal analysis (DTA) are by now well-established tools for studying mesomorphic systems. Figure 1.6 gives an example of a DSC curve. The NI transition is first order with heats of transition of the order of 500 J/mole. Transitions between different mesophases, for example the SN transitions, can be second order. However first order transitions with very small latent heats can also occur. A drawback of scanning calorimetric methods is that the latent heat and the pretransitional increase in the specific heat near a phase transition are lumped together into one peak. In order to establish the nature of a phase transition the true latent heat (or absence of a latent heat) must be measured. This can be done by adiabatic calorimetry, which is, however, much more time consuming.

The polarizing microscope is a classical and useful tool for the investigation of liquid crystals. Used as an *orthoscope* one looks at the object image of a thin mesomorphic layer between two glass plates. Depending on the boundary conditions and the type of phase, specific textures are observed, that provide a means of classifying the different phases. These textures will be discussed in some detail in the next chapters. Usually texture changes occur at the transitions between the various phases. This is best observed with decreasing temperature, for, when the sample is heated, the textures are sometimes reminiscent of the phases at lower temperatures. These *paramorphic* textures are, however, not stable over a longer period of time. A useful book with photographs of typical textures has been written by Demus and Richter (1978).

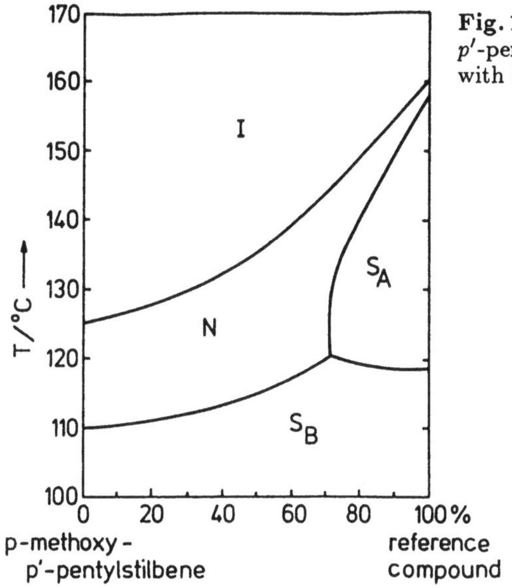

Fig. 1.7. Binary phase diagram of *p*-methoxy-*p*′-pentylstilbene and a reference compound with known smectic phases

Microscopy becomes a very powerful tool in combination with the investigation of the miscibility of binary mixtures. If two phases have a different symmetry they do not mix, and they are separated in a phase diagram by an equilibrium curve at which the transition takes place. This knowledge can be applied to binary mixtures of liquid crystals in order to establish whether two phases are different or not. In practice the method is especially used to classify the various smectic phases. An example is given in Fig. 1.7. The smectic phases of the reference compound are called A and B. By mixing *p*-methoxy-*p*′-pentylstilbene with this compound the smectic phase can be classified as smectic B, since it mixes with the corresponding phase of the reference compound over the whole concentration range. Strictly speaking this criterion of selective miscibility only applies if the two compounds have the same molecular symmetry. Hence, if two rather different compounds do *not* mix, this does not necessarily reflect a difference between the two phases. Conclusions can only be drawn from the miscibility of the two phases, and not from an observed lack of miscibility. The practical application of miscibility studies is assisted by some further considerations:

i) Some compounds show polymorphism with a large number of mesophases. This gives the certainty that these phases are really different.
ii) When the temperature is varied there is usually a fixed sequence in which the various phases are observed.

In practice a qualitative phase diagram can easily be obtained with the contact method. The two substances are brought in contact and the zone

Fig. 1.8. Conoscopic figure of a homeotropic liquid crystalline layer

of interdiffusion is observed through the microscope during cooling of the sample.

When investigating samples with a uniform director pattern it is often useful to use the polarizing microscope as a *conoscope:* a strongly convergent light beam is focussed in the plane of the object, and gives a directions image (as opposed to object image) or interference figure. In the conoscope the observer looks through rather than at the object at the stage of the microscope (Wahlstrom 1979). Figure 1.8 gives an example of a conoscopic figure of a sample with the director uniformly along the axis of the microscope. In a normal orthoscopic picture this would appear as pseudo-isotropic (dark) between crossed polarizers. In orthoscopy such a pseudo-isotropic sample can still be distinguished from a genuinely isotropic one by the observation of director fluctuations. These thermally excited fluctuations of the, on average uniform, director (see also Sect. 7.3) show up as intensity fluctuations very analogous to the noise on a television screen.

X-ray diffraction provides direct information about the residual positional order in liquid crystals. Samples with an overall random director pattern correspond to powdered samples in ordinary crystallography. Consequently reflections are observed at angles θ for which the Bragg condition is fulfilled

$$2d_{hkl} \sin\theta = m\lambda \quad , \quad m \text{ integer} \quad . \tag{1.7}$$

Here λ is the wavelength of the radiation and d_{hkl} a periodicity in the sample. The suffixes hkl refer to the Miller indices that indicate sets of planes in a certain direction. The width of the peaks depends obviously on the degree of long-range positional order. In liquids, where only short-range order is present, broad diffuse peaks will be observed. In the nematic and in the isotropic phase of mesogenic compounds the diffraction pattern consists of two diffuse rings. The inner ring at small Bragg angles is related to the length of the molecules, the outer ring to the average lateral distance between neighbouring molecules. In the various smectic phases the random distribution in the longitudinal direction of the molecules is replaced by a repeat distance corresponding to the thickness of the smectic layers. Consequently the diffuse inner ring (001 reflection) changes into a sharp reflection (see Fig. 1.9). The absence or weakness of higher orders (002, etc.) is an indication of the imperfection of the layer structure. From the 001 reflection the thickness of the smectic layers can be calculated via (1.7). For orthogonal smectics this value is usually slightly smaller than the length of a fully stretched molecule as calculated from models. The difference can be accounted for by the deviations from perfect orientational order.

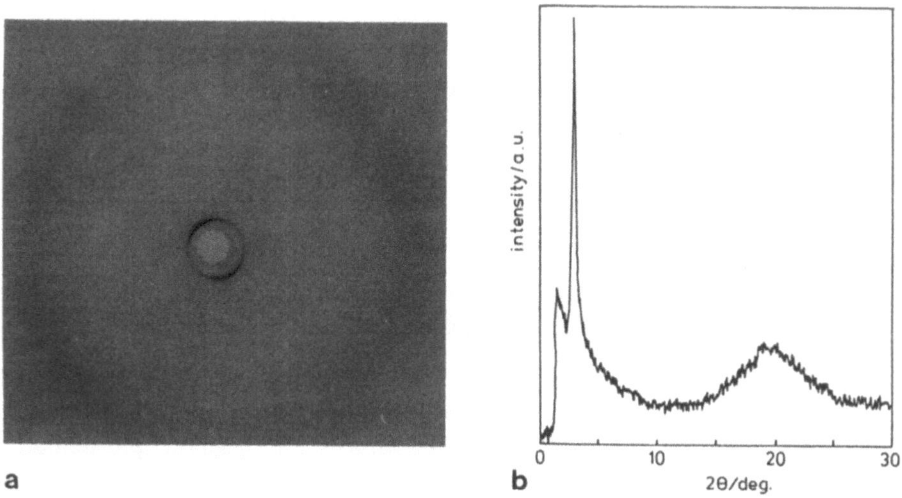

Fig. 1.9. Typical x-ray photograph **(a)** and associated radial intensity distribution **(b)** of a non-oriented smectic A phase. The variation in intensity of the inner ring indicates that the orientations were not completely random

In smectic A and C phases the peaks at higher Bragg angles ($\theta \approx 12°$) remain diffuse as in the nematic phase (liquid layers). For the other smectics various combinations of sharp peaks can be observed in this region; this will be discussed in some detail in Chap. 3. Obviously, more detailed information can be obtained from studies of either samples with a uniform director pattern (which are not necessarily single crystals from the point of view of positional order) or single crystals.

2. Nematic Liquid Crystals

In this chapter we describe in some detail the three-dimensional liquids among the thermotropic liquid crystals. First the features of nematics are summarized, and some thermodynamic properties of the nematic-isotropic phase transition are discussed. Then we consider the various types of textures that may occur. A nematic phase can be considered as a chiral nematic phase with infinite pitch. This similarity is stressed in the second section, which further summarizes the remarkable optical properties of chiral nematics. Finally, we discuss their textures and the factors that influence the pitch.

2.1 Nematics Proper

The essential features of the nematic phase of thermotropic liquid crystals consisting of elongated molecules can be summarized as follows:

a) The molecules are, on average, aligned with their long axes parallel to each other. Macroscopically this leads to a preferred direction, described by the director $\boldsymbol{n}$. For almost all thermotropic nematics known so far the nematic phase is uniaxial, i.e. there exists rotational symmetry around $\boldsymbol{n}$.

b) There is no long-range correlation between the centres of mass of the molecules, i.e. they have three-dimensional translational symmetry. Hence the nematic phase is a fluid. The viscosities are of the same order of magnitude as the viscosity of the isotropic liquid phase.

c) The axis of uniaxial symmetry has no polarity, even though the constituent molecules may be polar. This leads to the equivalence of $\boldsymbol{n}$ and $-\boldsymbol{n}$, and to the necessity of using a second-rank tensor as the order parameter.

Rather similar observations have been made for nematics consisting of disc-like molecules and for lyotropic nematics. These will be discussed in Sect. 4.2 and 4.4, respectively.

When a compound shows nematic behaviour, this phase usually occurs directly below the isotropic phase when the temperature is decreased. The two phases are separated by a first-order phase transition. Hence one can measure the associated latent heat ΔH and the density change $\Delta \varrho$. Reliable measurements of these quantities are still scarce. Classical results for ΔH obtained by adiabatic calorimetry have been given by Arnold (1964), amongst others for the series of p,p'-dialkoxyazoxybenzenes. As shown by Barrell et al. (1974) these results compare well with DSC results after proper correction for pretransitional heat effects. The transition entropy $\Delta \Sigma$ can be calculated directly from ΔH since the free energy difference between the nematic and the isotropic phase, $\Delta F = \Delta H - T\Delta\Sigma$, is zero at the phase transition, and consequently it follows that

$$\Delta\Sigma = \Delta H/T \quad . \tag{2.1}$$

Typical experimental results for two homologous series are shown in Fig. 2.1. It appears that T_c shows an odd-even effect in the sense that it alternates with increasing chain length m. For small values of m, $\Delta\Sigma$ alternates as well,

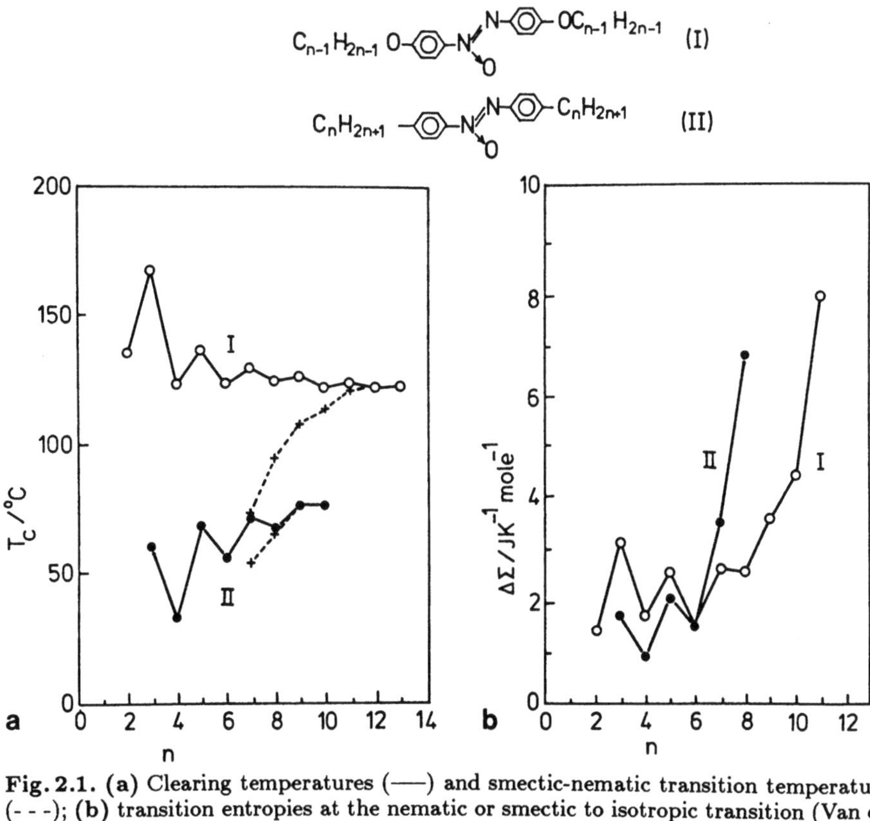

Fig. 2.1. (a) Clearing temperatures (——) and smectic-nematic transition temperatures (- - -); **(b)** transition entropies at the nematic or smectic to isotropic transition (Van der Veen et al. 1973), both for the series I and II shown above

along with T_c. As will be discussed in Sect. 12.2, simple theories predict

$$\Delta\Sigma \sim S_c^2 \quad , \tag{2.2}$$

where S_c is the value of the order parameter at T_c. Furthermore it should be noted that $\Delta\Sigma$ increases strongly for larger values of m. This can be ascribed to smectic-like short-range order in the nematic phase above the SN transition, that either disappears or is strongly reduced at the clearing point. A third small contribution to $\Delta\Sigma$ can be observed in homologous series where no smectic phase occurs (Leenhouts et al. 1979). This is attributed to a difference in flexibility of the alkyl chains of the molecules in the two phases.

The density changes at the NI transition are rather small: the relative density change $\Delta\varrho/\varrho_c$ is of the order of 0.3 % (Bahadur 1976). In homologous series there is again an odd-even alternation. As is to be expected, $\Delta\varrho/\varrho_c$ is larger for increasing values of S_c.

The NI transition is first order with a small volume change and a latent heat small compared to the latent heat at the melting point. As these values closely approximate the vanishing changes of a second order phase transition, many physical properties of the isotropic phase show critical behaviour (or pre-transitional behaviour) on cooling to temperatures close to T_c. These effects will be discussed in some detail in Sect. 12.3.

Physical measurements at constant volume are very useful, because their interpretation can rely on the fact that the intermolecular distances do not change. Nevertheless, pressure studies of liquid crystals are still relatively scarce. According to the Clausius-Clapeyron equation dT_c/dp depends on the thermodynamic quantities discussed so far in the following way

$$\frac{1}{T_c}\frac{dT_c}{dp} = \frac{\Delta V}{\Delta H} \quad . \tag{2.3}$$

Consequently T_c increases with increasing pressure. Limited information is available regarding the pressure dependence of S_c. Early reports, for PAA, indicated that, despite the increase of T_c, S_c is approximately constant. More recent results indicate a variation in S_c of about 10 % up to 5 kbar for MBBA and 5CB (Horn and Faber 1979). The difference in behaviour has been tentatively attributed to the presence of alkyl chains with some conformational freedom in the latter cases.

When cooling a mesomorphic compound from the isotropic phase, the nematic phase appears at the transition in the form of droplets. Without special measures the boundary conditions will vary over the substrate. With further cooling below T_{NI} this often leads to a *marbled* texture (Fig. 2.2), in

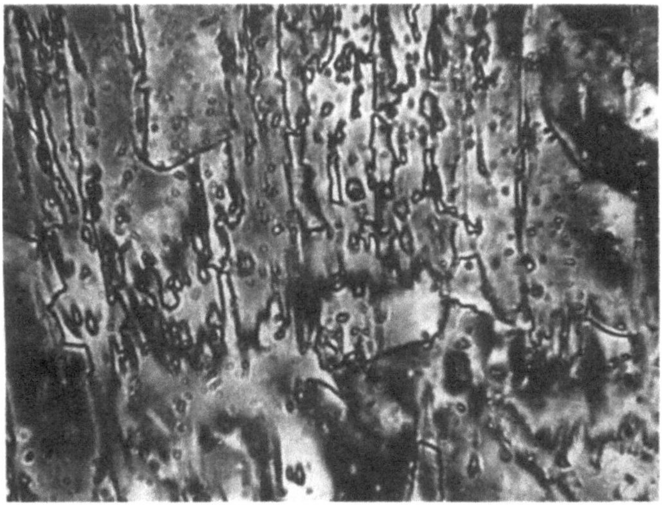

Fig. 2.2. Nematic marbled texture (crossed polarizers)

which many different uniform regions can be recognized. Between crossed polarizers the various uniform regions appear dark if the local preferred direction coincides with the direction of one of the polarizers. At the boundaries of these regions the director may vary in a discontinuous way. These discontinuities are called *disclinations,* in analogy with the dislocations in the solid state. Both line and point disclinations exist. In natural light, line disclinations are visible as threads floating in the liquid. The word nematic originates from this phenomenon. (The Greek word $\nu\eta\mu\alpha$ means thread.)

If the boundary conditions are somewhat relaxed (e.g. at temperatures close to $T_{\rm NI}$, or in general in thicker samples) a *schlieren* texture can often be found (Fig. 2.3). This texture is due to an overall non-uniform director pattern around a disclination. This can be a line disclination that is attached to one of the surfaces and is more or less perpendicular to it, or alternatively, a point disclination at the surface. Around this disclination the director rotates and between crossed polarizers one observes characteristic dark brushes that correspond to regions where the director is parallel to one of the polarizers. If one follows $\boldsymbol{n}$ over a closed circuit around one of the disclinations, the director rotates through an angle that must be a multiple of π. This is usually written as an angle $2\,\pi m$, where only textures according to the values $m = \pm\frac{1}{2}, \pm 1$ have been observed in nematics. The number of dark brushes that appear is $4|m|$, consequently 2 or 4 dark brushes are observed (compare Fig. 2.3). The director pattern around a disclination can be described in some detail within the framework of the continuum theory (Sect. 7.1).

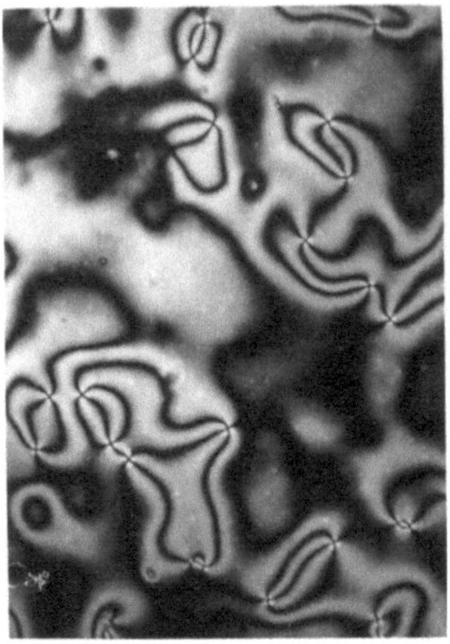

Fig. 2.3. Nematic schlieren texture (crossed polarizers)

In the above examples the textures were determined by arbitrarily occurring boundary conditions. By using surfactants or other physico-chemical surface treatments, very specific uniform boundary conditions can be obtained (see, for example, Cognard 1982). If the coupling is such that the resulting director pattern is perpendicular to the walls one speaks of a *homeotropic* texture. Observed through a conoscope this texture gives a symmetric cross, characteristic for a uniaxial crystal. In this way the homeotropic texture can easily be distinguished from the isotropic phase. In the *uniform planar* texture the director pattern is parallel to an easy direction created at the walls. An interesting variation is obtained if the upper glass plate is rotated with respect to the lower one. Provided that the boundary conditions remain fixed this leads to a *twisted planar* structure. Optically this texture behaves similarly to a long-pitch chiral nematic layer. The director pattern in these textures is schematically reproduced in Fig. 2.4.

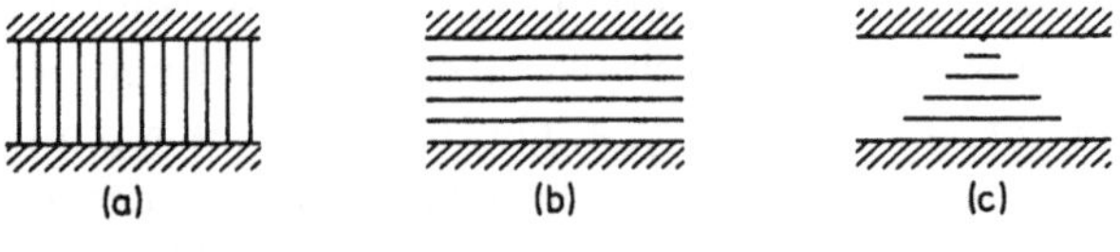

Fig. 2.4. Director field for a homeotropic (**a**), a uniform planar (**b**) and a twisted planar (**c**) nematic structure

2.2 Chiral Nematics

A chiral nematic phase may be found directly below the isotropic phase on the temperature scale. Confusion with a nematic phase can hardly arise, but care should be taken not to mix up a chiral nematic phase with a possible chiral smectic phase (see Sect. 3.2). Comparing the nematic and the chiral nematic phase, the following observations are important:

a) A particular compound never exhibits a phase transition from a nematic to a chiral nematic phase as a function of temperature.
b) Two different compounds, one with a nematic and the other with a chiral nematic phase, can possess the same type of smectic phase at lower temperatures.
c) Both phases are completely miscible. Addition of a small amount of an optically active compound (whether itself mesomorphic or not) changes a nematic phase into a long-pitch chiral nematic.
d) The x-ray diffraction patterns of nematic and chiral nematic phases are very similar. In both cases there is no long-range positional order that could lead to sharp reflections.
e) A mixture of two compounds of opposite chirality can give a nematic phase; at a certain composition the pitch becomes infinite (see Fig. 2.5). At this compensation point, there are no anomalies in any physical property that indicate a phase transition.

From these results it is clear that the chiral nematic phase is just a nematic phase of a somewhat different form arising from the chirality of the constituent molecules. Accordingly, the properties of the N^*I phase transition are very similar to those of a NI transition.

Chiral nematics have unique optical properties. These are best studied using a thin layer between two glass plates with tangential boundary conditions. Then the helix axis is perpendicular to the layer and one speaks

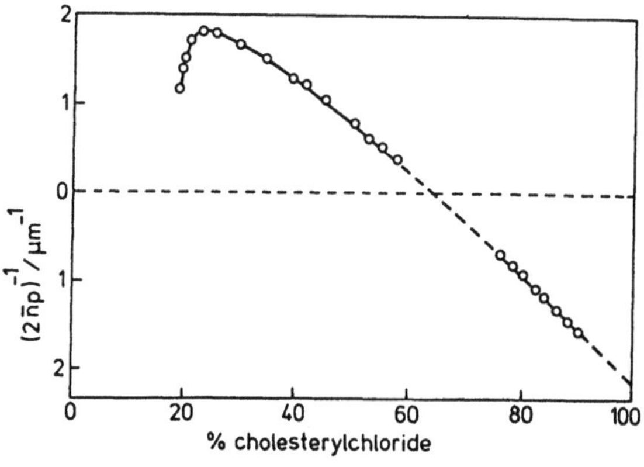

Fig. 2.5. Inverse pitch for a binary mixture of cholesterylchloride and cholesterylnonanoate

of a *planar* texture. In the direction normal to the layer the periodicity is $p/2$. The incident light is reflected according to Bragg's law $p \sin\theta = m\lambda'$, where m is an integer and λ' is the wavelength in the medium. For light incident parallel to the helix the electric vector vibrates perpendicular to the helix axis and $\lambda' = \lambda/\overline{n}$ with

$$\overline{n} = [\tfrac{1}{2}(n_{\|}^2 + n_{\perp}^2]^{1/2} \quad .$$

For oblique incidence a smaller value for the average refractive index applies. Normal incidence leads to a first-order reflection band around

$$\lambda_{\mathrm{max}} = \overline{n}p \quad . \tag{2.4}$$

Higher order reflections are forbidden in this situation. At oblique incidence the reflection shifts to shorter wavelengths, while in that case higher order reflections are also possible. When these reflections lie in the visible region, chiral nematics can show a wide variety of colours ("peacock" behaviour).

Because of the built-in screw sense of the planar chiral nematic structure, the refractive index is different for each of the two directions of rotation of circularly polarized light. The one for which the rotation of the electric vector exactly matches the helicoidal spiral, is completely reflected, while the other one of opposite rotational sense is transmitted. In the latter situation the electric vectors of incident and reflected light rotate in opposite directions (contrary to the situation of reflection of circularly polarized light against a mirror). As ordinary light can be decomposed into two circularly polarized components, the light reflected according to (2.4) will obviously be circularly polarized. In addition to the above properties the helical structure is also optically active. The optical rotatory power can be as high as 10^4 deg/cm, orders of magnitude higher than that of the constituent molecules. The optical properties of chiral nematics will be discussed in more detail in Sect. 7.4.

Much work has been done in relation to the intermolecular interactions that cause the helicoidal structure. An interesting case has been reported by Coates and Gray (1973) who synthesized the following substance

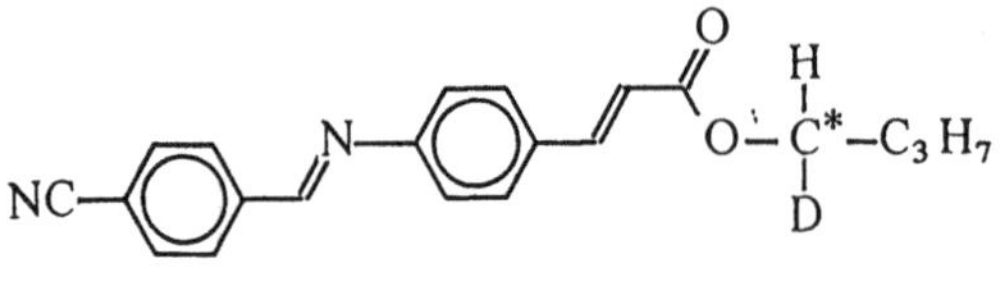

It shows a chiral nematic phase with $p = 155\,\mu\mathrm{m}$, while for D replaced by H just a nematic phase is observed. As the CH and CD bond lengths are very similar, this shows that steric repulsions are probably not essential for

the production of a helix. When a chiral substance is added to a nematic liquid at low concentrations c, the pitch p appears to vary linearly with c. This phenomenon is used to define the so-called helical twisting power:

$$h = (pc)^{-1} \quad . \tag{2.5}$$

In some cases addition of chiral molecules of a certain handedness to a nematic induces a chiral nematic with the opposite rotational sense, a quite remarkable phenomenon. Necessarily then at higher concentrations an inversion of the direction of rotation results (Hanson et al. 1975).

In general p decreases slightly with temperature. However, near a chiral nematic-smectic phase transition the pre-transitional formation of smectic clusters can make this temperature dependence very pronounced. This effect is advantageously used in thermography. By making a suitable mixture, a desired temperature range can be chosen such that the main reflection band is in the visible region. Small variations in temperature now show up as a change in colour.

As far as chiral nematic textures are concerned, the planar texture (helix perpendicular to the glass walls) has already been mentioned. An interesting variation can be obtained by taking a small-angle wedge (see Fig. 2.6a). At certain positions along the wedge the thickness equals an integer times $p/2$. Because of the fixed tangential boundary conditions the

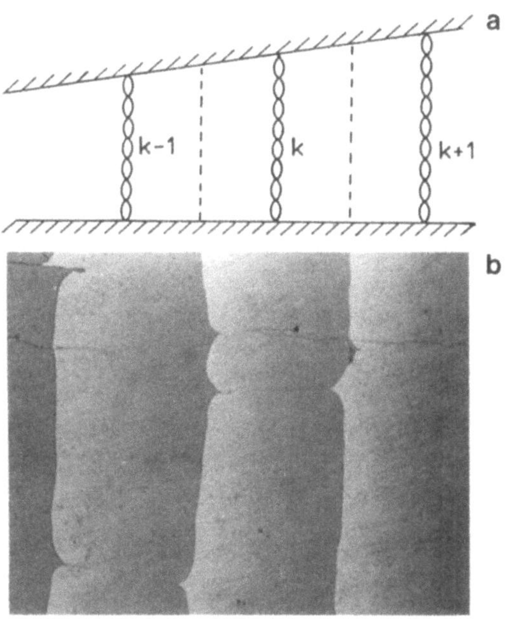

Fig. 2.6a,b. Grandjean structure; **(a)** schematic representation, **(b)** actual appearance (courtesy of C.J. Gerritsma)

pitch has to adjust in the intermediate regions, getting smaller or larger depending on whether the thickness decreases or increases, respectively. Consequently regions are observed where the pitch varies from $p_{\rm min}$ to $p_{\rm max}$. These regions are separated by a disclination line where $kp_{\rm max}/2$ equals $(k+1)p_{\rm min}/2$, k and $k+1$ being the number of half-pitches between the glass plates (see Fig. 2.6b). This arrangement is called a *Grandjean* wedge, and can be used to determine p.

In a layer without uniform boundary conditions or in the bulk, the helix axis is free to deform from a straight line into a curved one. The only restriction is that the periodicity is maintained. This leads to heavily deformed structures in which the helix axes of various regions make arbitrary angles with each other. The different regions are separated by disclination lines (Bouligand 1972). Macroscopically this texture, which is often called focal-conic, scatters light strongly.

Finally, it must be mentioned that for chiral nematics with a relatively short pitch, several different types of phase can often be observed between the isotropic and the N^* phase; these are the so-called "blue phases". The temperature region in which these blue phases are stable is often quite small, some tenths of a degree. Instead of one of the textures described above, an optically *isotropic* structure is observed showing colours in reflected light. Under a microscope a texture of uniformly coloured platelets is found in this region. Small transition enthalpies have been observed corresponding to at least three types of blue phases. From the high optical rotatory power one can conclude that a helicoidal structure is still present. More detailed investigations have been made (Stegemeier et al. 1986), and models have been proposed in which these properties are explained by some form of arrangement of the helices in a cubic superstructure, the lattice being defined in terms of singularities of the director field (Hornreich and Shtrikman 1980; Meiboom et al. 1981).

3. Smectic Liquid Crystals

In this chapter we consider the smectic mesophases, that are characterized by both orientational order of the long molecular axes and by a reduced positional order. The common feature of the various smectic mesophases is the existence of a one-dimensional (1D) density wave: the molecular centres are, on average, arranged in equidistant planes. Additional symmetry properties, that distinguish the various types of smectic phase as indicated by the subscripts A, B, ... I, are discussed in the first section. The second section is devoted to phases in which the director is orthogonal or tilted with respect to the liquid smectic layers. These phases are called smectic A and smectic C, respectively. Next, various types of order can be distinguished within the layers. These types of smectics are discussed in the remaining sections. Information is given about the structure and the resulting textures of the various smectic modifications.

3.1 Broken Symmetry and the Organization of Smectics

Common to all smectic phases is – in addition to the orientational order of the long molecular axes – the existence of a one-dimensional (1D) density wave. The molecular centres are, on average, arranged in equidistant planes, leading to what is loosely called a layer structure. In fact the word "layer" suggests far too much because the deviations from the equilibrium positions in the smectic planes can be quite large in these phases. Taking the density wave along the z axis, the distribution function now depends on the coordinate z as well. Because of the periodicity d in the z direction this function can be written as a Fourier series

$$\sum_{n=0}^{\infty} a_n \cos(2\pi nz/d) \quad ; \quad n \text{ integer} \quad , \qquad (3.1)$$

where the coefficients a_n depend on the orientation of a given molecule with respect to the z axis, and in the case that additional positional order exists within the layers also on the coordinates x and y. This means that smectics are described in a first approximation by an order parameter of the form

$$\varrho_1 = \langle \cos(2\pi z/d) \rangle \quad . \tag{3.2}$$

As already mentioned in Sect. 1.1 a sharp x-ray reflection is found in smectics at small scattering angles, corresponding to the repeat distance d ("layer thickness"). From a fundamental point of view it is quite interesting to note that for some of the smectic phases this peak is not a δ-function type Bragg peak, as would correspond to long-range order, but a weaker singularity (Litster 1980). While for true long-range order the positional correlation between two separate layers approaches a constant in the limit that the layers become infinitely apart, the observed singularity corresponds to an algebraic decay $z^{-\eta}$ of that correlation. Denoting in general the relevant correlation function by $G(\boldsymbol{r})$ the various situations that occur are summarized in Table 3.1. For liquids $G(\boldsymbol{r})$ decays rapidly to zero, with a correlation length of typically several molecular dimensions. Its Fourier transform, which gives the scattering cross section, is a broad Lorentzian. The interesting point to note here is that, in systems with less than three dimensions a situation of quasi long-range order can also exist, where $G(\boldsymbol{r})$ decays to zero rather slowly as $r^{-\eta}$ as $r \to \infty$ (no long-range order, η small positive). For a smectic A phase the corresponding singularity in the x-ray spectrum has been calculated by Caillé (1972), and is given in Table 3.1. The difference from true long-range order is subtle, and difficult to observe because the experimental x-ray reflection is always a convolution of the theoretical line shape and the resolution function of the spectrometer. Using special high-resolution techniques the result shown in Fig. 3.1 could be obtained. It follows from the expression for the distortion free energy of a smectic A phase that such a 1D density wave in a 3D medium is indeed unstable at finite temperatures (see Sect. 7.5). The mean-square fluctuations of displacements of the smectic layers from their equilibrium positions diverge logarithmically with the size of the sample. This means that fluctuations in the smectic layers are responsible for the absence of true long-range order; this is the so-called Landau-Peierls instability (Landau and Lifshitz 1980). Much of the recent theoretical and experimental interest in smec-

Table 3.1. Behaviour of the correlation function $G(\boldsymbol{r})$ and the related scattering cross section $S(\boldsymbol{q}-\boldsymbol{q}_0)$

	$G(\boldsymbol{r})$	$S(\boldsymbol{q}-\boldsymbol{q}_0)$
Long-range	constant	$\delta(\boldsymbol{q}-\boldsymbol{q}_0)$
Quasi long-range	$r^{-\eta}$	$(\boldsymbol{q}-\boldsymbol{q}_0)^{\eta-2}$
Short-range	$\exp(-r/\xi)$	$[\xi(q-q_0)^2+1]^{-1}$

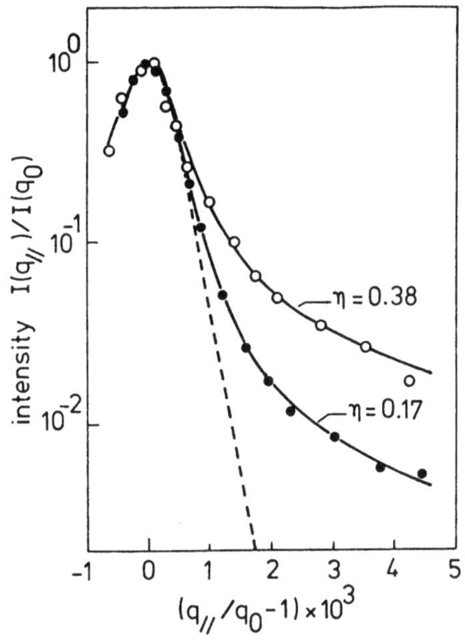

Fig. 3.1. Intensity profile of the 001 reflection of p,p'-octyloxycyanobiphenyl at two temperatures in the S_A phase. The dashed curve is the resolution function, which would be seen if there were a Bragg peak. Solid curves correspond to an algebraic decay (Als-Nielsen et al. 1980)

tics as model systems to study phase transitions is related to this type of instability.

Once the layer structure has been established, one can distinguish between orthogonal and tilted smectic phases. In the first case the director is perpendicular to the layers (or equivalently parallel to the direction of the density wave). Tilted phases are characterized by an angle between the director and the normal to the smectic planes. In general there will be a long-range order of the tilt directions within a layer. In this case the tilt directions of adjacent layers will be coupled as well, leading to an overall uniform tilt, and the resulting phase is a biaxial one. For chiral molecules this situation is modified; there is a small angle α between the tilt directions of adjacent layers. In this way a helical structure is formed, with a pitch $p = (2\pi/\alpha)d$. Macroscopically the normal to the layers is again an axis of uniaxial symmetry. These phases are called chiral smectics. Finally the possibility should be considered that the tilt directions within a layer vary with position. In that case one cannot expect long-range correlation of the tilt directions of adjacent layers. The resulting phase will be effectively uniaxial, and will only differ from the corresponding orthogonal phase by a somewhat smaller layer thickness d.

The various types of smectic phase can be classified according to their symmetry properties as indicated in Table 3.2. The two main columns arise from the occurrence of orthogonal and tilted phases. The various rows are due to differences in the organization of the molecules within the layers. The

Table 3.2. Characteristics of the various smectic phases

Structure within the layers	Orthogonal	Tilted $a>b^{*)}$	Tilted $a<b^{*)}$
Liquid	A	C	
Bond-orientational order	hexatic B	F†)	I†)
(Pseudo)-hexagonal packing	B	G	G'
Herringbone packing	E	H	H'

*) b is the unique axis of the monoclinic cell ($a \neq b \neq c$, $\alpha = \gamma = 90° \neq \beta$).
†) In this phase there is only locally a monoclinic lattice.

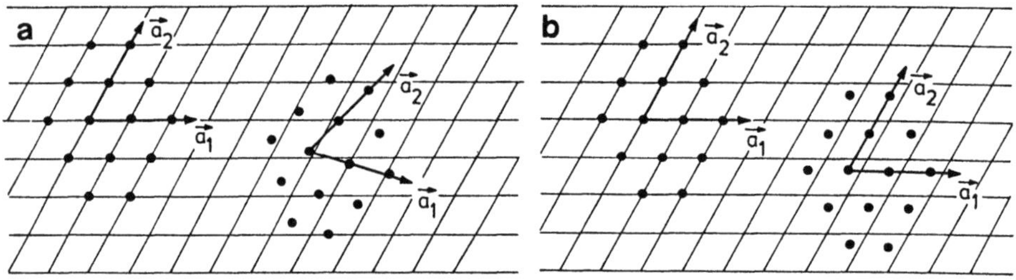

Fig. 3.2. Local 2D lattice for a liquid (**a**) and for an hexatic phase with bond-orientational order (**b**). In the latter case local lattice vectors are oriented in the same direction

first row (smectic A and C) corresponds to layers that can be considered as 2D liquids. There is only short-range correlation between the molecular positions within the layers over a distance ξ, typically corresponding to a few molecules. In Fig. 3.2a a picture is given of two of these short-range clusters of size ξ. These clusters possess a lattice-like structure. Because of the fluid character of the layers the distance between the centres of the clusters is not related to the local lattice spacing, neither are the intermolecular directions correlated. Now an additional type of ordering may exist within the liquid layers: a long-range order of the orientations of the clusters. This phenomenon is called bond-orientational order. This type of order, the so-called hexatic order mentioned in the second row of Table 3.2, does not violate the fluid character of the layer. The resulting phase is called a hexatic phase, and is illustrated in Fig. 3.2b. Finally, a solid phase results if the translational freedom also disappears and the centres of the clusters are separated by a multiple of the lattice spacing. Depending on whether the lath-like molecules rotate freely around their long axis or not, either a hexagonal close-packed or a herringbone-like 2D lattice is observed. These situations correspond to the last two rows of Table 3.2.

Once some form of order within the smectic layers has been established, one should inquire after the possible correlations between the layers. In practice there appear to be long-range correlations in the direction perpendicular to the layers, between the 2D bond-orientational order within the layers, so that the hexatic order is in fact 3D. The same is observed for the positional order within the layers, if present, which means that the last two rows of Table 3.2 refer simply to 3D crystals, though with highly anisotropic properties. Thus lamellar solids would be a better name than liquid crystals. One might wonder whether structures of stacked 2D-ordered layers without long-range coupling between the layers are in fact possible. For the smectic B phase there are some indications that this could be the case (Gane et al. 1981b). Finally the subdivision of the column of tilted phases in Table 3.2 refers to the tilt direction, to be discussed in more detail in Sect. 3.3.

3.2 Smectics with Liquid Layers

There are two types of smectic phase with layers of simple 2D liquids, smectic A and smectic C. Provided the smectic C phase is not a chiral one, they are distinguished by the fact that smectics A are optically uniaxial and smectics C biaxial. Both can occur as a single mesophase between the isotropic and the crystalline state, or in the combinations NS_A, NS_C, S_AS_C and NS_AS_C. With DSC a small heat peak is usually associated with the S_AS_C transition; this is due to a pre-transitional increase of the specific heat as no latent heat has been measured (Huang and Viner 1982). For the NS_A transition the pre-translational effect of the specific heat can vary strongly. If T_{AN} approches T_{NI}, the heat peak strongly increases until finally a latent heat also appears (first-order transition) (Thoen et al. 1984). For $T_{AN} \ll T_{NI}$ the effect is small and pre-transitional smectic order may be found over a large temperature range in the nematic phase.

Direct evidence for the layered structure is obtained from the x-ray diffraction patterns where a sharp reflection is observed corresponding to the layer thickness. Furthermore, sometimes terraced drops are formed on clean glass plates without a cover glass. This clearly indicates a layered structure, each step comprising many layers. A layered structure with fluid character within the layers leads to the so-called *focal-conic* textures. Due to the absence of positional correlations the layers can bend easily, the only restriction being that the layer thickness is maintained. As a result singularities in the director field may develop. The simplest example of such a structure is given in Fig. 3.3. Here the layers form closed cylinders around an axis, which is a line singularity. More intricate singularity patterns can be obtained by deforming the simple singular structure from a cylinder to a torus by closing the line singularity Γ_1. This leads to a second singular line Γ_2 perpendicular to the closed loop of the first one. The resulting singularities can often be actually observed as so-called focal lines, in the general

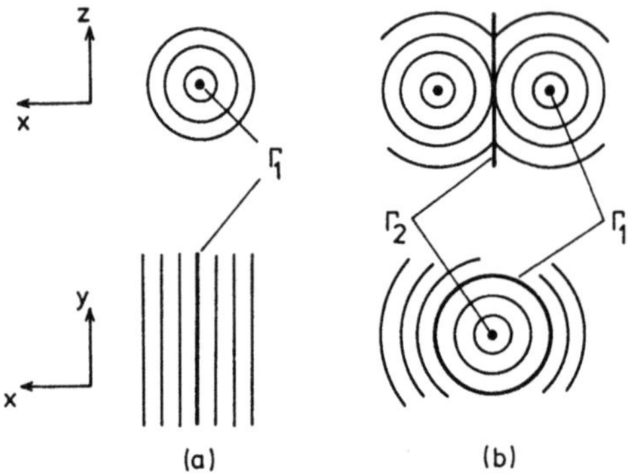

Fig. 3.3a,b. Construction of a focal-conic texture. **(a)** The smectic layers are bent and closed into a cylinder, giving a singular line Γ_1. **(b)** The cylinder is bent and closed into a torus; when space is filled this leads to a second singular line Γ_2 through the midpoint of the circle into which Γ_1 is deformed. In the most general situation Γ_1 is an ellipse and Γ_2 a conjugate hyperbola

case ellipses and conjugate hyperbolae. Figure 3.4a shows an example of a simple or *fan-shaped* focal-conic texture of smectic A. Sometimes the focal-conic units are arranged in a *polygon* type superstructure. In smectics C the focal-conic textures are usually much less perfect (Fig. 3.4b). On cooling through the SI transition the smectic phase appears in the form of small batons. They have axial symmetry but are seldom simply cylinders, and they are the nuclei of the focal-conic units. These batons can be used to

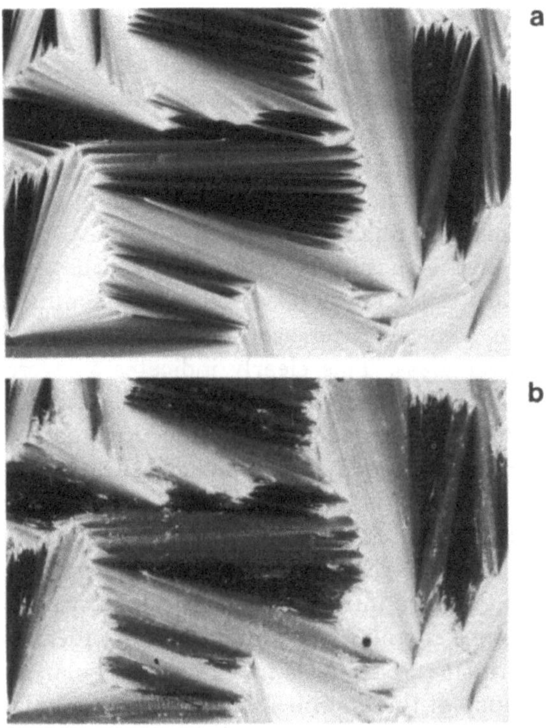

Fig. 3.4. (a) Fan-shaped focal-conic texture of smectic A; **(b)** Broken fan-shaped focal-conic texture of the same sample in the smectic C phase (courtesy of D. Demus)

distinguish an SI from an NI transition, as the nematic phase appears in the form of droplets.

3.2.1 Smectics A

A smectic A phase is characterized by at least one order parameter corresponding to the layer structure, for example ϱ_1 given by (3.2). The simplest model compatible with the macroscopic properties of the smectic A phase is as follows: The molecules are arranged in layers with a thickness about equal to the length of the molecule. The director coincides with the layer normal, while $\boldsymbol{n}$ and $-\boldsymbol{n}$ are still equivalent. This explains the uniaxiality of the optical properties. Consequently, apart from the *focal-conic* texture, smectics A also exist in the *homeotropic* texture. In that case it cannot be distinguished optically from a homeotropic nematic texture. In a thin layer with parallel tangential boundary conditions a uniform planar texture may be obtained. No twist can be induced with enantiomorphic molecules because this is incompatible with the symmetry of the smectic A phase.

Usually the layer thickness d, as obtained from the x-ray diffraction pattern, is somewhat less than the length l of the molecule. In many smectics A the difference amounts to about 5 % (Leadbetter 1979). This can be attributed to the imperfect orientational order: a value of $S<1$ will cause the projection of l on the director to be somewhat smaller than l itself. Conformational changes in the alkyl chains may also be important in this respect. In some cases the differences between d and l are found to be much larger (De Vries 1977). This phenomenon can be understood by assuming a tilted arrangement of the local director with respect to the layer normal. The tilt angle then must be of the order of 20°. To explain the macroscopic uniaxiality, long-range correlation of the second angle that specifies the direction of the tilt of the local director, must be absent. Probably this means that the tilt direction is not uniform within the layers. The alternative possibility of a uniform tilt within the layers while the tilt directions of the individual layers are uncorrelated, seems to be unlikely from a thermodynamic point of view.

So far, only symmetric molecules have been considered, or at least the assumption has been made that any asymmetry of the molecules is nullified at a local level. When the molecules, however, possess a dipole moment positioned at the end of the molecule, one can distinguish a head and a tail. Examples are *para*-CN or -NO_2 substituted compounds, for which a whole variety of smectic A phases has been observed (Hardouin et al. 1983). In the first place the molecules can minimize their dipole interaction via antiparallel dipole correlation, e.g.

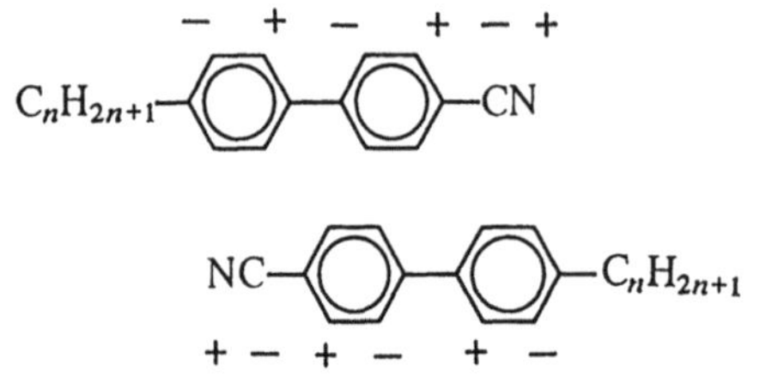

Consequently smectic A phases can be formed with a layer thickness, that is approximately given by $d \approx a + 2b$, where a and b are the length of the aromatic head and the aliphatic tail, respectively (Brownsey and Leadbetter 1980). Typically this leads to $d \approx 1.4\,l$ (with $l = a+b$). The dipole-dipole interaction has a profound effect on many physical properties (De Jeu 1983). For example, it may be the origin of the so-called reentrant nematic behaviour. This means that the following phases appear consecutively with decreasing temperature: nematic $\rightarrow$ smectic A (with $d \approx 1.4\,l$) $\rightarrow$ reentrant nematic.

With asymmetric molecules, phases with a broken up-down symmetry, i.e. long-range order of heads and tails, can also be expected. In principle such an effect can have two different causes: (i) dipolar effects and (ii) steric effects ("wedge"-shaped molecules). So far, only the dipolar case has been found to occur. A priori at least two additional types of smectic A phase with polarized layers may be expected (see Fig. 3.5), called the S_{Af} and the S_{A2} phase. The S_{Af} phase (f for ferroelectric) carries an overall electric polarization P. It has been discussed theoretically, but has not been observed. The S_{A2} phase is antiferroelectric and has a periodictiy of $d \approx 2\,l$ (Levelut et al. 1981). It does actually occur for compounds with a terminal CN or NO_2 group. In fact for such compounds several more variants of phases with broken up-down symmetry have been observed.

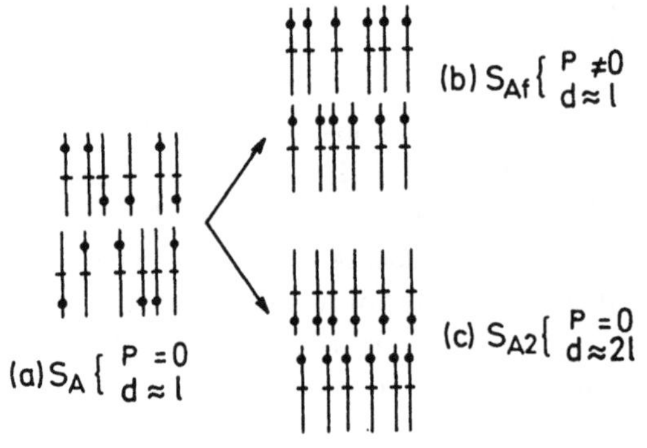

Fig. 3.5a–c. The two ways in which smectic layers with a broken up-down symmetry can be combined; S_{Af} is ferroelectric, S_{A2} is antiferroelectric

3.2.2 Smectics C

The characteristic property that distinguishes the smectic C phase from the A phase is its optical biaxiality, provided that the smectic C phase is not chiral. This optical biaxiality can be observed directly in a thin smectic C sample when the layers are parallel to the glass plates. In conoscopy the symmetric cross, typical of a homeotropic nematic or smectic A phase, changes into a split cross at an excentric position, typical of biaxial crystals. The generally accepted model for smectics C is sketched in Fig. 3.6. Within a layer the director $\boldsymbol{n}$ is uniformly tilted by an angle ω with respect to the layer normal. Secondly the tilt directions of the various layers are correlated in the same way. The tilt direction can be described by a C-director $\boldsymbol{c}$, which lies in the layers. Then $\boldsymbol{c}$ must be more or less uniform over macroscopic distances. Clearly the smectic C phase is still invariant with respect to replacing the director $\boldsymbol{n}$ by $-\boldsymbol{n}$. However, such a symmetry operation does not hold for the C-director. The states described by the directors $\boldsymbol{c}$ and $-\boldsymbol{c}$ are not equivalent.

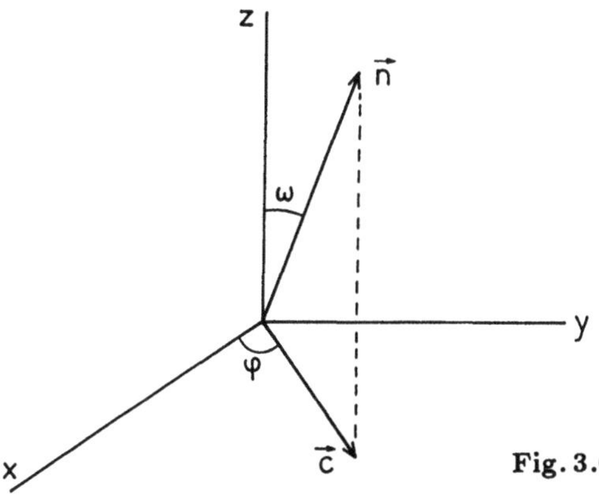

Fig. 3.6. Model of the smectic C phase

Homeotropic textures are not compatible with the biaxiality of the smectic C phase. This provides a useful criterion to distinguish between smectic C and A. Apart from the focal-conic texture already mentioned, additional textures are possible for smectic C due to its tilted structure described by the C-director:

(i) Non-uniformity of $\boldsymbol{c}$ gives rise to *schlieren* textures which are somewhat more disturbed than those of nematics (Fig. 3.7). The non-equivalence of $\boldsymbol{c}$ and $-\boldsymbol{c}$ implies directly that following a closed path around a singularity in the C-director field $\boldsymbol{c}$ must rotate over a multiple of 2π. Hence now only singularities with $m = \pm 1$ occur (four dark brushes).

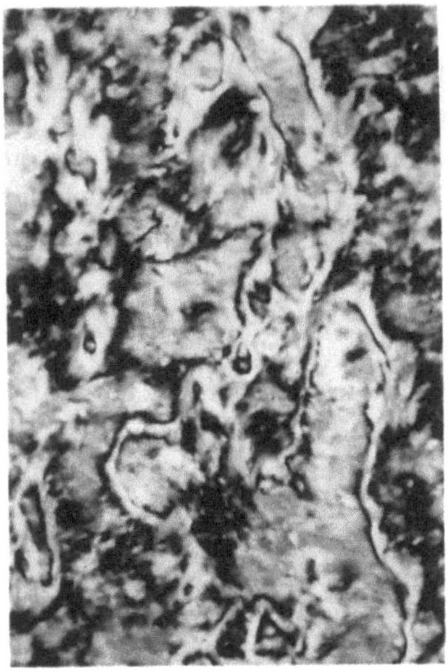

Fig. 3.7. Schlieren texture of a smectic C phase (courtesy of D. Demus)

(ii) In the case of enantiomorphic molecules a chiral smectic C phase occurs, denoted by C^*. The helix axis is parallel to the layer normal, while $\boldsymbol{c}$ in successive layers is systematically rotated over a small angle. The repetition length now equals p instead of $p/2$ as for chiral nematics; the pitch comprises many layers. For light propagating along the helix axis the optical properties and textures are very similar to those of chiral nematics. At oblique incidence there are differences however, since due to the tilt, the local optical properties of the C^* phase are triaxial rather than biaxial as in the N^* phase. The characteristic common property of the C^* and the C phase is the correlation between the tilt directions of adjacent layers (finite angle α or $\alpha = 0$). A chiral smectic C phase can also be obtained by adding some enantiomorphic molecules to an ordinary C phase. This provides another criterion to distinguish smectic C from smectic A.

The tilt angle provides a natural order parameter to distinguish smectic C from smectic A. Thus the order parameter is written as

$$\chi = \omega \exp(\mathrm{i}\phi) \quad , \tag{3.3}$$

where the azimuthal angle ϕ gives the direction of $\boldsymbol{c}$ in which the tilt takes place (Fig. 3.6). The tilt angle ω can show quite different values for various compounds. For HOAB (see Table 1.5) a tilt angle of the order of 40–50° has been observed, with little or no temperature dependence. This seems to

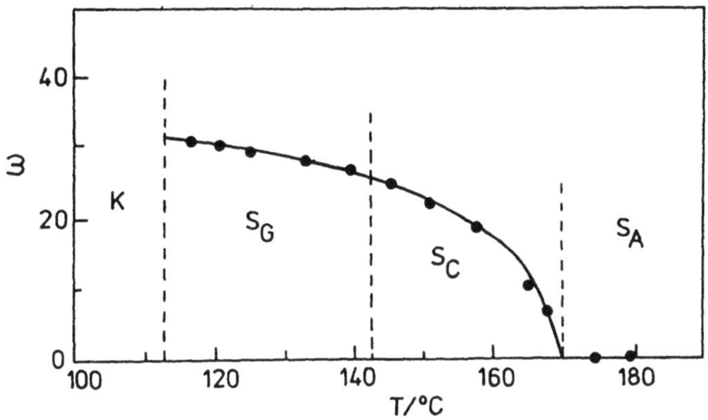

Fig. 3.8. Tilt angle in the smectic phases of TBBA as obtained from the 001 layer spacing in x-ray scattering (after Doucet et al. 1973)

be typical of many compounds with an S_C phase directly below an N phase. On the other hand in the case of an S_AS_C transition the tilt angle is found to increase from zero with decreasing temperature. A well-known example is the compound TBBA, for which this effect is illustrated in Fig. 3.8.

A molecular model for the smectic C phase has to describe among other properties the appearance of the tilt angle and its temperature dependent behaviour. In the case of non-polar compounds usually only orthogonal smectic phases are observed (De Jeu and Eidenschink 1983). From this it seems that permanent dipole moments are in general an important factor in producing the tilt, especially when they are situated outside the molecular centre and directed off the long molecular axis. Because of the absence of a long-range correlation between the rotations of the molecules around their long axis (at least for smectics C), the molecules may be taken to rotate freely around these axes in a first approximation. Consequently the interactions between permanent and induced dipoles must be the relevant ones, because those between permanent dipoles average out to zero. For the S_AS_C transition this can be illustrated with the following series of p, p' substituted azobenzenes (De Jeu 1977):

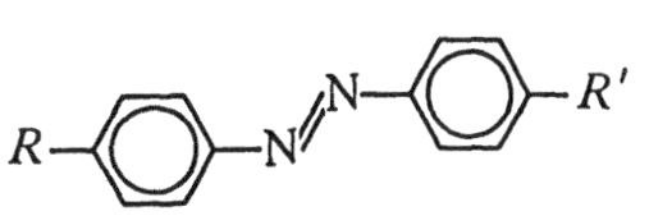

with

$$R = R' = OC_{n-1}H_{2n-1} \quad , \qquad (I)$$

$$R = OC_{n-1}H_{2n-1} \quad , \quad R' = C_nH_{2n+1} \quad , \qquad (II)$$

$$R = R' = C_nH_{2n+1} \quad . \qquad (III)$$

The differences in conformations between these series, and thus in steric interactions, are negligible. Series (I) with two dipole moments due to the oxygen atoms gives N and S_C phases, series (II) with only one dipole moment gives N, S_A and S_C phases, and finally in the non-polar series (III) only N and S_A phases are found. This information will be incorporated into the models for the S_C phase to be discussed in Sect. 14.2.

During recent years the chiral C^* phase has attracted much attention because of its potential application in display devices. As noted by Meyer (1977) the symmetry of this phase allows for ferroelectricity because the symmetry plane and the inversion centre of the ordinary smectic C phase have disappeared. The only symmetry left is a two-fold rotation axis parallel to the smectic layers and perpendicular to the director $\boldsymbol{n}$. If a transverse dipole moment is present, each smectic layer possesses an electric polarization parallel to this two-fold axis. Hence the director and polarization spiral together along the helix axis, leading to an overall zero bulk polarization. Howver, the polarization can be used advantageously in a thin cell (say 2 μm) in which the spiralling is suppressed (Clark and Lagerwall 1984). The surface conditions are made such that the director is confined to the plane of the surface, but without any specific direction. Then there are two possible stable orientations of $\boldsymbol{n}$ corresponding to the intersection of the cone of tilt angle ω with the plane of the surface (Fig. 3.9). These regions have director orientations making an angle 2ω with each other and can be distinguished optically. The net polarization associated with each region points up and down, respectively. Hence by altering the polarity of an applied field the director can be switched through the angle 2ω.

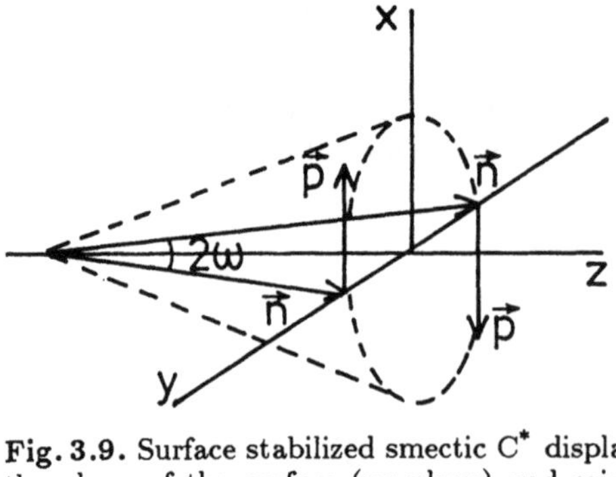

Fig. 3.9. Surface stabilized smectic C^* display configuration. The director is confined to the plane of the surface (yz plane) and originally in the smectic A phase along the z direction. Introducing a tilt angle ω then leads to two possible director orientations

3.3 Smectics with Bond-Orientational Order

As mentioned in the first section of this chapter, the concept of bond-orientational order provides a distinction between the hexatic-B and the crystalline-B phase (see Table 3.1). The relevant parameter for sixfold bond-orientational order is

$$\Psi = \exp[6\mathrm{i}\theta(\boldsymbol{r})] \quad , \tag{3.4}$$

where $\theta(\boldsymbol{r})$ is the angle between the local lattice vectors or "bonds" and some reference axis (see Fig. 3.2b). The associated long-range correlation function may decay algebraically to zero as $r^{-\eta}$ (2D hexatic phase) or exhibit true long-range order. Birgeneau and Litster (1978) have argued that the latter is the case in liquid crystals, because the interlayer coupling here actually makes the hexatic bond-orientational order three dimensional. Some years later these predictions were verified on a p, p'-substituted biphenyl (Pindak et al. 1981)

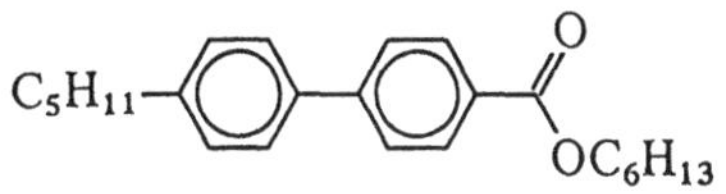

With decreasing temperature the following phase sequence was observed:

$$\mathrm{I}\,85\,\mathrm{S_A}\,68\,\mathrm{S_{Bhex}}\,60\,\mathrm{S_E} \quad .$$

From x-ray measurements on a free-standing film of about 100 layers with a uniform director pattern, one could conclude the existence of short-range positional correlations within the layers with a typical dimension $\xi \approx 100$ Å; this is to be compared with $\xi = 20$ Å in the S_A phase. The hexagonal pattern of six reflections is evidence for a long-range 3D sixfold bond-orientational order. The absence of interlayer positional correlation was confirmed by mechanical measurements, demonstrating that the hexatic-B phase (contrary to the crystalline-B phase) does not support a shear parallel to the planes (Pindak et al. 1982). Calorimetric measurements indicate that the transition $S_A S_{Bhex}$ can be second order (Viner et al. 1983).

Only relatively few compounds with a hexatic-B phase have been found. Goodby (1981) reported a compound with both a hexatic-B and a crystalline-B phase. No extensive information exists yet on the possible textures of the hexatic-B phase. Up to now it has always been found below a smectic A phase. The phase transition is characterized by rather subtle changes in the focal-conic texture at the phase transition. Of course, homeotropic textures can also occur.

Smectics F and I are the tilted analogues of the hexatic B-phase. Well studied are the homologues of TBBA, for example (Gane et al. 1981a):

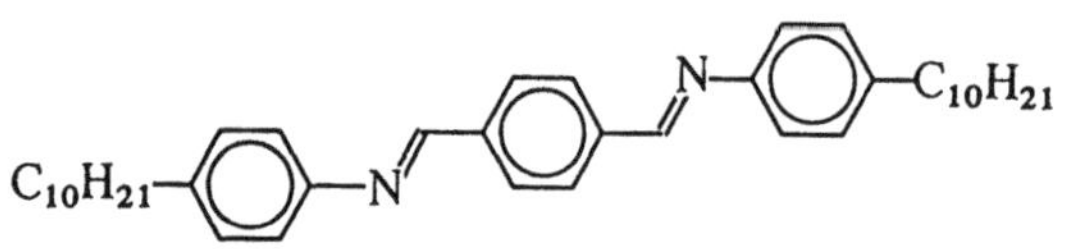

(TBDA)

K 73 S_G 120 S_F 150 S_I 156 S_C 191 S_A 192 I .

The powder x-ray diffraction patterns of S_F and S_I are intermediate between those of S_C and S_{Bcrys}. Apart from the 00l layer reflections only the lowest order $hk0$ reflections are observed (200, 110, etc.). These reflections coincide in a single ring at large Bragg angles. The peak is narrower than in S_C (see Fig. 3.10), but still relatively broad compared with the Bragg peak in an S_{Bcrys} phase, to be discussed in the next section. For S_I this peak is sharper than for S_F. Monodomain samples give six maxima, showing that both S_F and S_I have 3D long-range bond-orientational order. In fact S_F was the first example of this type of ordering, and is properly described as a stacked tilted hexatic phase. Using oriented samples no structure in the $hk0$ reflections is observed, showing that the layers are positionally uncorrelated. They can slide readily relative to each other (but not rotate), while preserving the direction of the crystal axes.

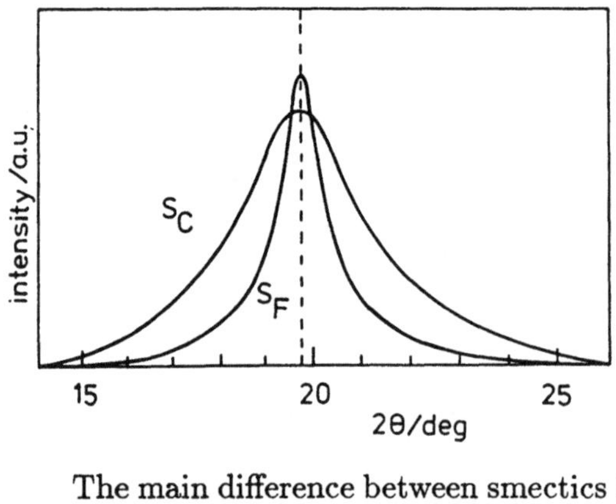

Fig. 3.10. Intensity profile of the high Bragg angle reflection in the smectic C and smectic F phase of TBPA, the pentyl homologue of TBBA (after Benatter et al. 1979)

The main difference between smectics F and I concerns the local lattice. For S_F the C-director points towards the middle of an edge of the local hexagon, for S_I towards a vertex (Fig. 3.11). Taking β as the unique angle of the local monoclinic cell, the difference in tilt direction leads to $a>b$ for S_F and to $b>a$ for S_I. The shape of the x-ray peak at large Bragg angles in the S_F phase can be well accounted for by an in-plane scattering function in the form of a Lorentzian. Hence the positional order shows an

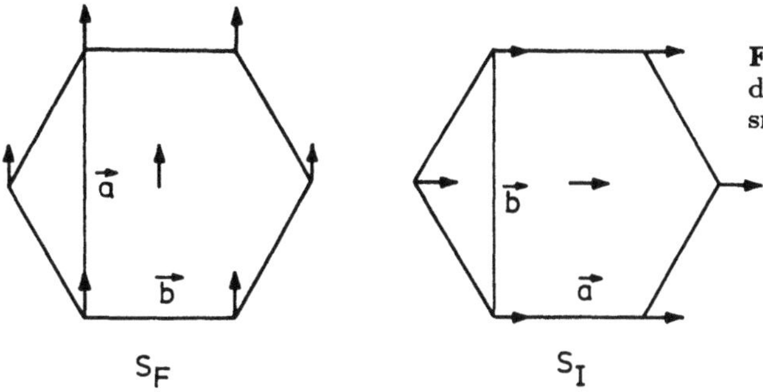

Fig. 3.11. Difference in tilt directions for smectic F and smectic I

exponential decay with a correlation length $\xi \approx 100\,\text{Å}$. In one particular case the corresponding S_I line was found to be narrower, in combination with pronounced wings (Gane et al. 1981a; Benatter et al. 1981). Attempts to describe the line-shape with a Lorentzian failed, whatever correlation length was taken. However, good fits were obtained employing a power-law (algebraic) decay, expected for a truly 2D crystal. In other cases, however, no such a difference between S_F and S_I shows up.

The results just mentioned pose the general question of whether it is possible to have 2D positionally ordered layers without the lattice becoming three dimensional. In principle this could occur if the 2D lattice refers to the aromatic cores of the molecules, while decoupling of the layers is accomplished due to melting of the alkyl chains. Evidence for this is available from electron paramagnetic resonance with spin probes, carried out in p,p'-octadecyloxyazoxybenzene:

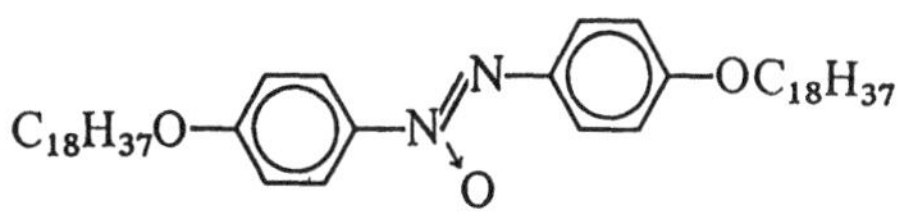

K 94 S_I 99 S_C 155 I

using as spin probe a similar molecule with an additional radical group situated half way along the alkyl chain. The result for the order parameter at this position indicates that the alkyl chain in the S_I phase is as disordered as in the S_C phase, in contrast to the situation in the solid (Dvolaitsky et al. 1973).

Microscopically the textures of the S_F and S_I phase are related to those of the S_C phase. On cooling, the broken fan-shaped (focal-conic) texture of the S_C phase usually develops additional defects (Biering et al. 1980) at the S_CS_F transition, leading to what has been called a striated or checkerboard fan texture. Alternatively an S_C schlieren texture persists into the S_F phase. For the S_I phase there is less information. The differences in texture from the S_F phase are often very subtle and hardly observable (Richter et al.

1981). Transition enthalpies are relatively large between S_C and S_F or S_I. Very small latent heats are found for the phase transitions S_IS_F, S_FS_G and S_IS_G (Wiegeleben et al. 1980; Pelzl et al. 1981).

3.4 Smectics with Ordered Layers

The last two rows of Table 3.2 include the smectics with 2D long-range positional order within the layers. Here again the question arises about the possible long-range positional correlation between the layers, in which case the order becomes 3D. For the S_B phase this has been an open question for some time, and still is for some substances. In many cases, however, the 3D nature of the positional ordering has been established by x-ray diffraction (Leadbetter et al. 1979a). This has been confirmed by mechanical measurements showing that the S_B phase can support a shear parallel to the layers (Cagnon and Durand 1980; Pindak et al. 1982). For the remaining phases S_E, S_G and S_H the 3D nature of positional order is directly evident from the types of reflection observed in x-ray diffraction. These phases are distinguished from simple crystals by a large amount of disorder and from the ability of the molecules to reorientate themselves. This latter property is evident from, for example, dielectric relaxation effects (Kresse 1982). Thus a better name would be *lamellar plastic crystals.*

The basic structure of these phases involves, at the local level, a herringbone-type of packing of the molecules. Its origin has been considered by Doucet (1979). The fundamental assumption is that the main contribution to the packing within the layers comes from steric forces. Consider a molecular cross section parallel to the smectic layers. Then the central aromatic core of a typical smectogenic molecule can be approximated by an envelope of the shape given in Fig. 3.12a. Now three possible packings can be imagined that are the most compact (see Figs. 3.12b–d). These packings can be distinguished only in that the herringbone-packing of Fig. 3.12d allows a nearly hexagonal organization of the alkyl chains that are perpendicular to the plane of the drawing. Taking the freedom of rotation of these chains into account, this construction leaves an optimal freedom to pack and is thus the most likely one. In the S_E and S_H phases the herringbone packing is long range, whereas it is only local in the S_B and S_G phases. In the latter case there is a dynamic equilibrium between the various local herringbone structures. The time-scale involved is 10^{-11} s (Dianoux and Volino 1979), leading to an effectively hexagonal symmetry for S_B and S_G.

The various phases with ordered layers can be distinguished by their x-ray patterns and their textures. The basic differences are summarized here. As an example PBAPC, *p*-phenylbenzylidene-*p*′-aminopentylcinnamate, is

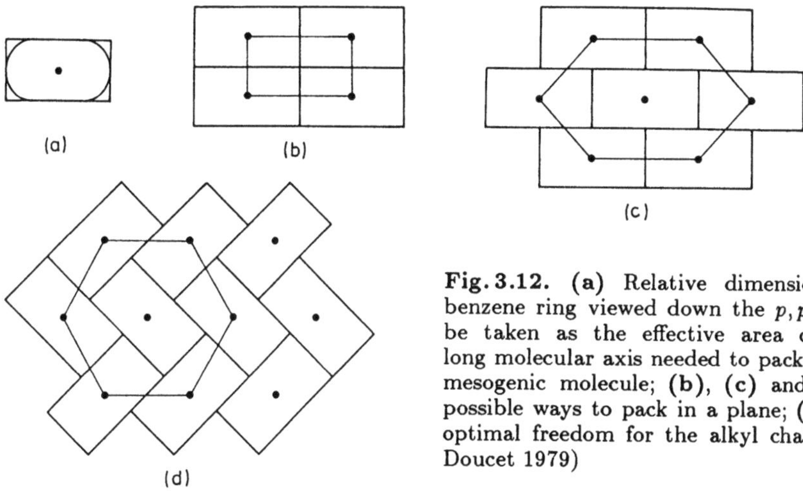

Fig. 3.12. **(a)** Relative dimensions of a benzene ring viewed down the p,p'-axis, to be taken as the effective area down the long molecular axis needed to pack a typical mesogenic molecule; **(b)**, **(c)** and **(d)** are possible ways to pack in a plane; **(d)** leaves optimal freedom for the alkyl chains (after Doucet 1979)

considered

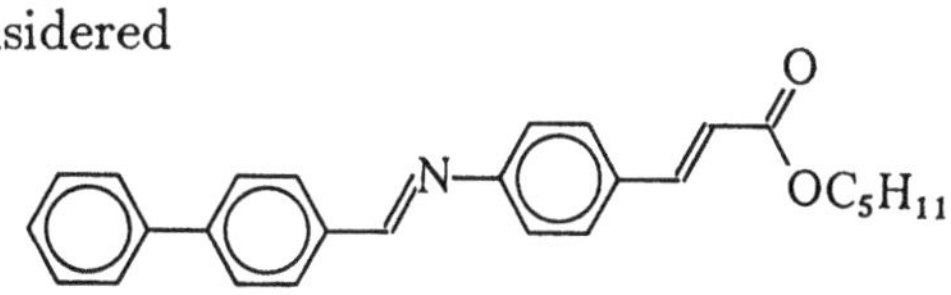

$$K\,92\,S_E\,102\,S_B\,168\,S_A\,204\,I \quad .$$

Apart from the peak at small Bragg angles corresponding to the layer spacing, the differences show up at higher angles as is illustrated in Fig. 3.13.

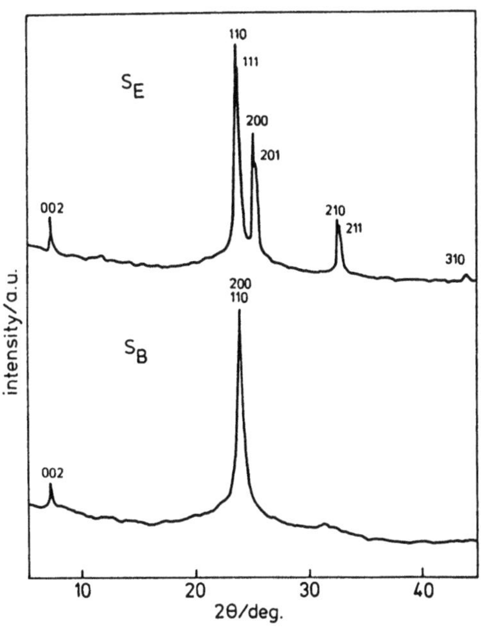

Fig. 3.13. Powder x-ray pattern of the smectic B and smectic E phase of PBAPC (after Doucet et al. 1975)

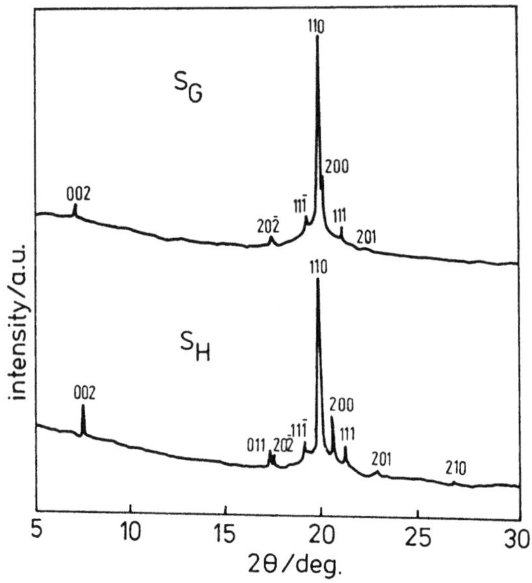

Fig. 3.14. Powder x-ray pattern of the smectic G and smectic H phase of TBBA (after Doucet et al. 1974)

In the S_B phase the 200 and 110 reflections are superimposed due to the hexagonal symmetry, leading to one diffraction ring in the powder pattern. In the S_E phase the hexagon is distorted, leading to a separation of the peaks. In fact the symmetry is orthorhombic. The presence of two glide mirrors imposes some extinction conditions on the reflections (Doucet et al. 1975).

TBBA (see Table 1.5) has, besides the S_A and S_C phases already discussed, an S_G and an S_H phase at lower temperatures. The powder diffraction patterns of these two phases are rather similar (see Fig. 3.14). Both have monoclinic symmetry. The S_G phase is C-face-centred, which leads to specific extinction conditions: reflections hkl with $h+k=2n+1$ (n integer) are absent. This distinguishes the pattern from that observed for the S_H phase. Here these reflections are in fact observed (011 and 210 in Fig. 3.14), but not the reflection $h0l$ with $h=2n+1$. 2D positionally ordered layers without interlayer correlation would give only reflections $hk0$ and $00l$. From the occurrence of reflections hkl the appearance of 3D positional order can be concluded immediately.

Positionally ordered layers no longer bend easily, and no focal-conic textures can be observed. Instead, mosaic textures are found (see Fig. 3.15 for the case of S_B) consisting of various regions with a uniform director pattern. In addition, homeotropic textures may be observed for the uniaxial phases and schlieren textures for the biaxial phases. The differences in mosaics of the various phases are often rather subtle. Clearly defined textures as in Fig. 3.15 are only observed if the phase develops from a nematic

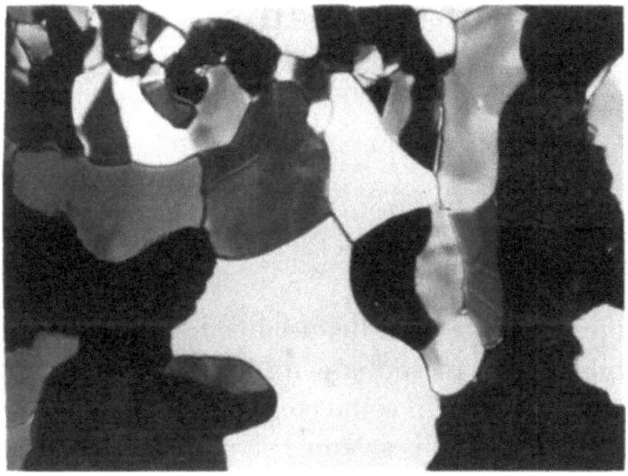

Fig. 3.15. Mosaic texture of a smectic B phase

or isotropic phase. Otherwise the mosaics will retain to a large extent the texture of the preceding high temperature smectic phase.

4. Other Liquid Crystalline Systems

In this chapter some rather different topics will be discussed. To begin with we will show that the possibilities for mesophase formation are not exhausted with nematic and smectic phases. In some compounds, liquid crystals with a cubic structure may be found. These cubic structures are rather rare in comparison to the other mesophases to be discussed here. Then in Sect. 4.2 mesophases consisting of disc-like molecules are discussed. This field of research is relatively young. Apart from a nematic phase, disc-like molecules may show several types of columnar phase (2D solid, 1D liquid). Polymers can often give liquid crystals, and form a field of increasing interest. In the third section of this chapter we shall summarize some basic information for the non-specialist, emphasizing similarities with mesophases of molecules of low molecular weight. Finally, we shall give some basic information about lyotropic liquid crystals. This state of matter consists of rod-like or disc-like molecules or aggregates of molecules in solution. Again we have to restrict ourselves strongly, and we refer to various other books for more detailed information.

4.1 Cubic Thermotropic Mesophases

Several examples exist of mesogenic compounds that form an optically isotropic liquid crystalline phase. One case has already been mentioned: the cholesteric blue phases. Such a behaviour can also exist for compounds without optical activity, in which case no special arrangement of helices can apply. Examples are the substituted alkoxybiphenyl carboxylic acids:

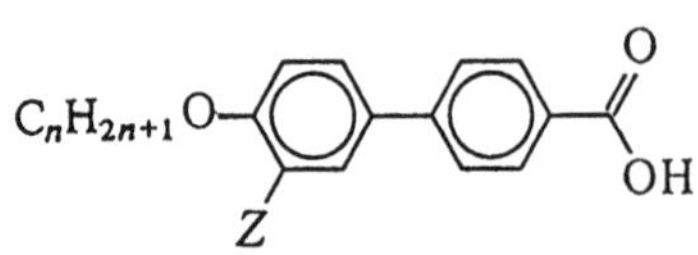

Z is CN, NO_2 .

For Z is NO_2 the higher homologues form both S_A and S_C phases. For $n = 16$ a further phase is observed between S_A and S_C. That phase has been called smectic D (Demus et al. 1980). For $n = 18$ the phase sequence $S_C S_D I$ is found with increasing temperature. The S_D phase is in fact not smectic, because no evidence for a layer structure exists. X-ray investigations, as well as the optical isotropy, indicate a cubic structure. Carboxylic acids are well known to give rise to liquid crystalline phases due to dimerization of the hydroxylic groups (Kolbe and Demus 1968). The compounds described above possess in addition a strong dipole moment due to the CN- or NO_2-group. Both these features are probably involved in the appearance of the cubic lattice structure.

The most extensive x-ray study has been made by Etherington et al. (1986) for Z is CN and $n = 18$. It leads to a cubic unit cell with a lattice parameter of 86 Å and containing approximately 700 molecules. An interpretation has been given in terms of micelles sited at special positions compatible with the space group P23. In the micellar model based on P23 the distance between micellar centres is 60 Å for a unit cell of 86 Å. This value seems very reasonable for the dimers involved, as the individual molecules have a maximum (fully extended) length of about 35 Å.

More recently optically isotropic (cubic) mesophases were observed in the series (Demus et al. 1981)

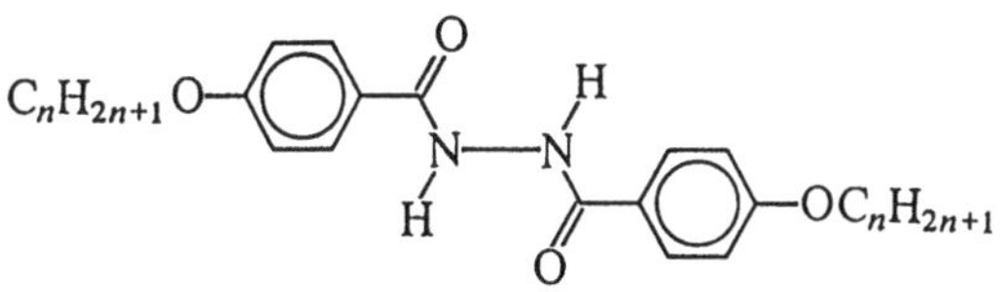

$(n = 8, 9, 10)$.

This time they are observed at temperatures *below* an S_C phase. Again groups are present that can be supposed to form intermolecular hydrogen bonds CO...HN. For $n = 8$ the cubic structure has a lattice constant $a = 45.7$ Å and comprises about 115 molecules. These phases seem to be different from the smectic D phase mentioned above.

A common feature of these cubic mesophases is that, on cooling, the nucleation is slow. Unlike transitions between other liquid crystalline phases, considerable supercooling can occur. Nevertheless the associated latent heats appear to be relatively small.

4.2 Mesophases of Disc-Like Molecules

Up to now thermotropic liquid crystals consisting of compounds with a rod-like or lath-like shape have been considered. It has long been realized that mesophases consisting of disc-like molecules must be possible too. In fact such phases have been observed during the pyrolysis of organic materials. During this process at approximately 300°C spherulites grow which show an optical anisotropy. The idea that these are liquid crystalline phases has been substantiated by various observations: optical anisotropy, nematic-like texture, low viscosity and alignment by a magnetic field (Zimmer and White 1982). Analysis shows that mixtures of many different compounds with a disc-like structure are involved. The occurrence of the mesophase is important for the production process of carbon fibers (Otani 1981).

The first single-component system of disc-like molecules showing nematic behaviour (N_D) has been reported by Tinh et al. (1979). The molecules have the following structure (hexa-alkyloxybenzoates of triphenylene):

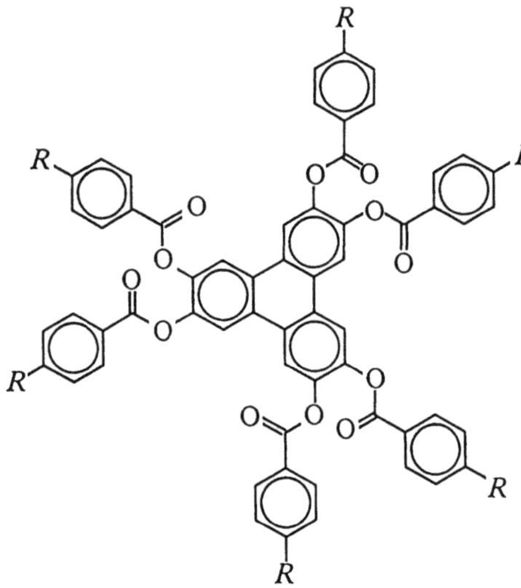

R is C_8H_{17} : K 183 N_D 192 I
R is $C_7H_{15}O$: K 168 N_D 253 I

The original classification as nematic was based on the following observations: (i) high fluidity, (ii) schlieren textures, (iii) diffuse rings in the x-ray pattern, (iv) orienting effect in a magnetic field. As expected, such an N_D phase is not miscible with an ordinary N phase consisting of rod-like molecules. In the mean time some information about the physical properties of this type of mesophase has become available (Mourey et al. 1982). For the heptyloxy-substituted compound a low birefringence is found which is negative ($\Delta n \approx -0.08$), and moreover a positive dielectric anisotropy ($\varepsilon_{\|} \approx 4$, $\varepsilon_{\perp} \approx 3.3$) and a positive conduction anisotropy ($\sigma_{\|}/\sigma_{\perp} \approx 1.5$). The elastic constant K_1 is of the same order of magnitude as in nematics of rod-like molecules (see Sect. 6.4), while the viscosities are two orders of magnitude higher. Compounds of disc-like molecules with a nematic phase at a considerably lower temperature (truxene derivatives) also exist (Tinh et al. 1981):

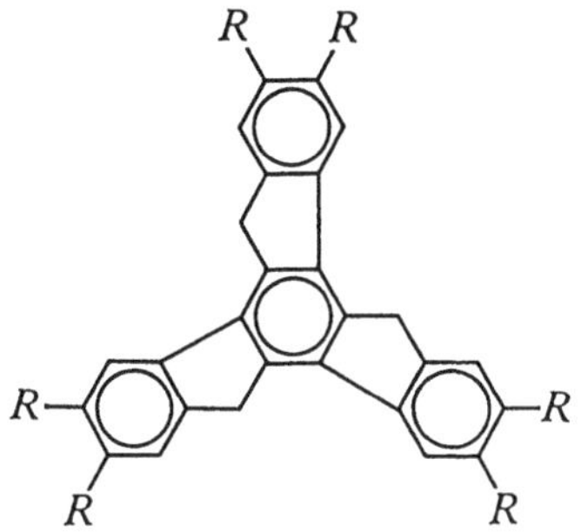

R is $OOCC_{10}H_{12}$: D 64 N_D 89 D .

In this case both at higher and lower temperatures a so-called columnar mesophase is found. Reviews summarizing the various types of molecules investigated so far have been given by Dubois and Billard (1982) and Destrade et al. (1983).

Apart from the N_D phase, disc-like molecules can give rise to mesophases in which the molecules are arranged in columns: these are the *canonic* mesophases. An example of such a phase is illustrated in Fig. 4.1. The discs in the columns form a 1D liquid (disordered columns), while the columns in turn form a 2D lattice. Various types of canonic mesophases have been found. The main distinctions are up to now:

– Orthogonal or tilted phases, referring to the angle between the director and the axis of the columns.
– The arrangement of the columns: h: hexagonal, r: rectangular.
– The organization of the centres of mass of the molecules in the columns: d: disordered (liquid columns), o: ordered.

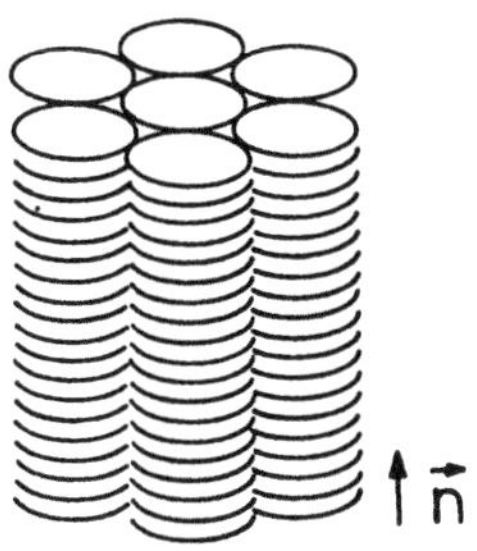

Fig. 4.1. Schematic picture of a columnar mesophase

If in the latter case a long-range correlation between the positions in the different columns exists the phases are in fact (anisotropic) crystalline solids. The resulting phases are summarized in Table 4.1, where the appropriate entry indicates that this type of phase has actually been observed. The appearance of disordered columns is probably related to the disorder of the side chains of the molecules (Rutar et al. 1982).

Table 4.1. Various types of organization of canonic mesophases

Lattice of columns	Structure within columns	Orthogonal	Tilted
Hexagonal	disordered	D_{hd}	–
	ordered	D_{ho}	–
Rectangular	disordered	D_{rd}	D_t
	ordered	–	–

In fact the columnar phases were discovered prior to the observation of a nematic phase for disc-like molecules. The first example of a mesophase of disc-like molecules is (Chandrasekhar et al. 1979):

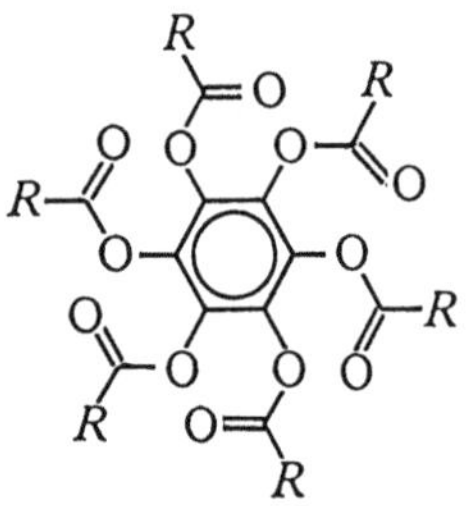

R is C_6H_{13} : $K\,81\,D_{rd}\,87\,I$
R is C_7H_{15} : $K\,80\,D_{rd}\,83\,I$.

The x-ray pattern shows a diffuse outer ring, corresponding to the average distance between the discs in the columns, and three sharp peaks corresponding to the 2D lattice of the columns. If R is C_6H_{13} these can be accounted for on the basis of a rectangular lattice with $a = 28.5$ Å and $b = 17.7$ Å. This implies a deviation of about 15 % from hexagonal symmetry for which $a/b = \sqrt{3}$. As the molecules under consideration are centrosymmetric this means that the director is at least locally tilted with respect to the cylinder axes. X-ray results for canonic mesophases have been reviewed by Levelut (1983).

The field of mesophases of disc-like molecules is in rapid development. Similar to the discussion for rod-like molecules in Sect. 3.2, phases with a broken up-down symmetry can be imagined. For disc-like molecules it is difficult to introduce dipoles perpendicular to the ring. However, these types of molecules can be modified such that sterically up and down can be distinguished. Such phases have indeed recently been observed (Malthête et al. 1985; Levelut et al. 1986). Other interesting topics are the appearance of a chiral N_D phase (Malthête et al. 1981), and of reentrant behaviour, which means that a certain phase appears twice on the temperature scale. An example of such a phase sequence is $K\,D_r\,N_D\,D_r\,N_D\,I$ which is found for some of the truxene derivatives mentioned earlier. The textures of the different mesophases have also been studied (Frank and Chandrasekhar 1980). Descriptions of the phase transitions with disc-like molecules have been given by Kats (1978) and Feldkamp et al. (1981) in terms of mean field theory.

4.3 Liquid Crystalline Polymers

For compounds of low molecular weight, the liquid crystalline phase is separated by first-order phase transitions from the crystalline phase at lower temperatures and from the isotropic liquid phase at higher temperatures. The latter transition can also be found in the case of polymers, but no crystallization is observed. On cooling, the liquid crystalline polymer is transformed into a polymer glass, the transition being indicated by a bend in the VT-curve. At the transition into the polymer glass the orientational order is frozen in without any changes.

In order to synthesize macromolecules with liquid crystalline properties suitable mesogenic building blocks (monomers) must be used. These mesogenic building blocks should be provided with appropriate functional end groups A and B, in order to obtain a polymer. In fact two rather different situations occur:

i) If A and B are capable of performing a condensation reaction, the mesogenic groups themselves build up the polymer main chain:

ii) If A is capable of performing an addition polymerisation reaction, the polymer main chain is built up by A, and the mesogenic groups are fixed as side chains to the polymer main chain:

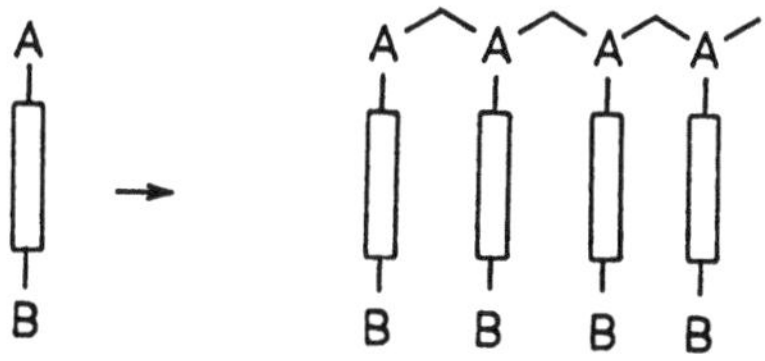

4.3.1 Liquid Crystalline Main-Chain Polymers

A schematic view of the uniaxial distribution of semiflexible polymer chains in a nematic and in a smectic phase, respectively, is presented in Fig. 4.2. The overall flexibility of the polymer main chain can be influenced by introducing flexible spacers (e.g. alkyl chains) between the functional groups A, B and the rigid mesogenic groups. If A and B are directly attached to

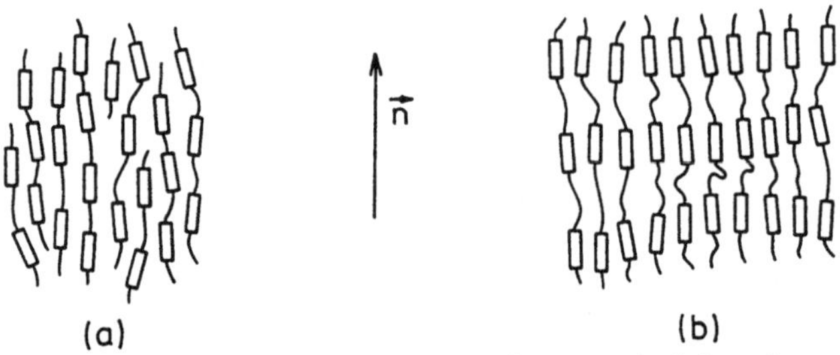

Fig. 4.2. Schematic representation of a nematic **(a)** and a smectic phase **(b)** composed of main-chain polymers

these groups a rigid rod-like structure is produced for the polymer chain. The interesting physical properties are obtained for the semi-flexible polymers. Well-known examples are the copolyesters prepared by the acidolysis of polyethylene terephtalate (PET) with n-hydroxybenzoic acid (PHB):

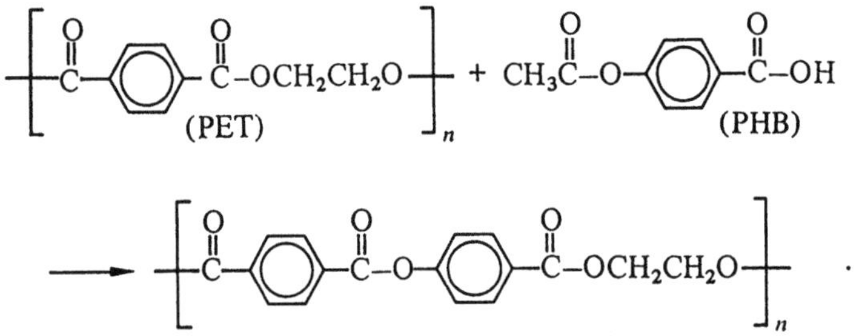

As shown, rigid polarizable segments are introduced reminiscent of the kind of mesogenic cores found in low molecular weight liquid crystals. These segments are separated by flexible $-OCH_2CH_2O-$ groups, whose number depends on the ratio PHB and PET. Injection molded copolyesters of this type have highly anisotropic mechanical properties owing to the orientation of the polymer chains during the molding process. The tensile strength of PET, as a function of the amount of PHB, is given in Fig. 4.3. The frozen-in orientational order when mesomorphic solutions are extruded causes the mechanical behaviour to be very different from that of conventional materials, and makes these materials of increasing technological interest. Very high tensile strengths of the order of 140 GPa (compare 200 GPa for steel) have been reached with polyamide fibres (trade names: X-500, Kevlar, Twaron). In fact many of these aromatic polyamides are only liquid crystalline in solution, and could be called lyotropic.

Another series of liquid crystalline polymers are the synthetic polypeptides that spontaneously form mesophases in a variety of solvents. The rod-

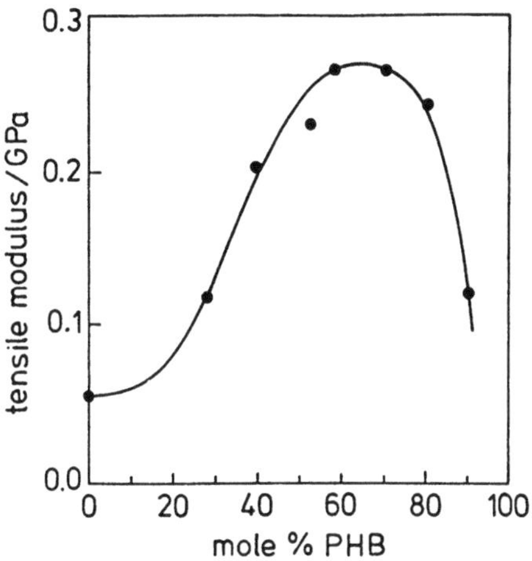

Fig. 4.3. Tensile strength of a copolyester of PET and PHB (after Jackson and Kuhfuss 1976)

like polypeptide conformation is that of an α helix. Their behaviour is in many ways identical to that of smaller mesogens, except that all dynamic effects in these viscous solutions are several orders of magnitude slower. The helical polymers are chiral, and their normal textures in the mesophase correspond to those of chiral nematics (see also Sect. 4.4).

4.3.2 Liquid Crystalline Side-Chain Polymers

A description of the liquid crystalline behaviour of these polymers involves the consideration of two aspects. First of all no rigid segments are present in the polymer main chain. The freedom of the chain segments to rotate around the various bonds gives a tendency towards a statistically distributed chain conformation. The second aspect concerns the tendency of the rigid mesogenic side groups towards orientational order, thus restricting the conformational freedom of the main chain. Clearly it depends on the detailed molecular structure which tendency dominates. The crucial point is the linkage between the polymer main chain and the side groups. If the mesogenic groups are directly coupled to the polymer main chain, the tendency towards a statistically distributed chain conformation hinders any anisotropic orientation of the side group. Moreover, this effect is enhanced by the steric hindrance of the relatively voluminous structures which further suppresses any orientational order. The importance of these two effects can be diminished by introducing flexible spacers between the polymer main chain and the mesogenic side groups. In this way the motions of the polymer main chain are decoupled to a certain extent from the possible orientational order of the side groups. In addition more possibilities are available to avoid steric hindrance, and liquid crystalline phases may be observed.

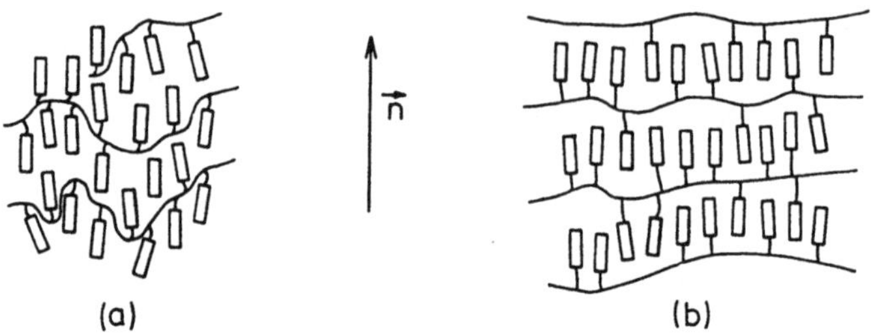

Fig. 4.4. Schematic representation of a nematic **(a)** and a smectic phase **(b)** composed of side-chain polymers

These concepts have proved to be very successful for the synthesis of liquid crystalline polymers. Nematic (both chiral and non-chiral) and smectic phases can be obtained, schematically shown in Fig. 4.4. If the spacers are sufficiently long, the type of phase is related to that of the mesogenic monomer: it changes from nematic to smectic with increasing length of the end substituent (Table 4.2). Actually, a tendency to higher order always exists if the monomers are converted to a polymer. If nematic polymers are obtained, the corresponding monomers often exhibit none or only a monotropic nematic phase. Starting with nematic monomers mainly smectic polymer phases are obtained. For the polymers the NI phase transition is always found at higher temperatures than for the corresponding monomers. Because of the decrease in the translational and rotational freedom of the mesogenic groups, this tendency is to some extent not surprising. Interestingly, the main chain itself evidently has a less important effect on the orientational order.

Table 4.2. Some nematic and smectic liquid crystalline side chain polymers (Finkelman 1980); G indicates glass phase

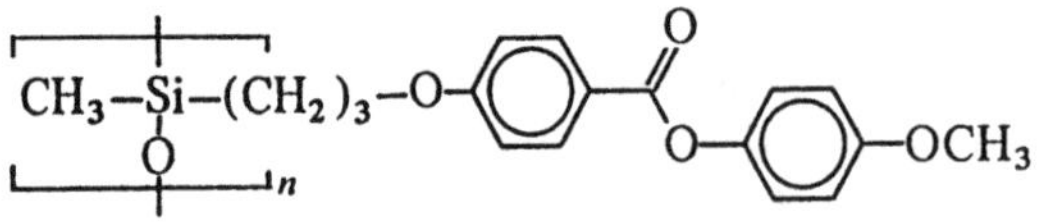	G 36 N 101 I
$-OC_6H_{13}$	G 30 S 101 I
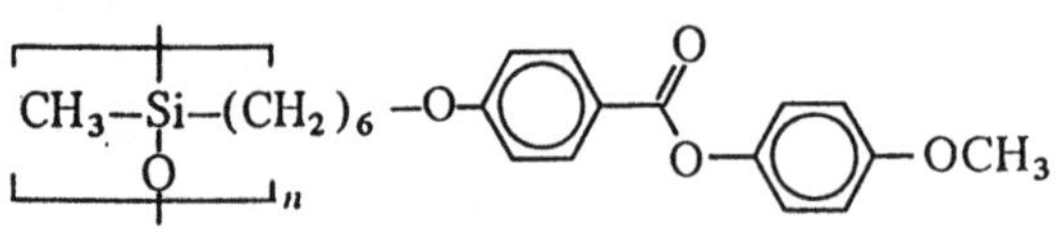	G 5 S 46 N 118 I

The physical properties of liquid crystal polymers have hardly been studied yet. Liquid crystal side chain polymers can be oriented in electric and magnetic fields. The possibility of freezing in the obtained structure by going below the glass temperature, may enable various applications to be realized. Some general references to polymer liquid crystals are Blumstein (1978), Finkelman (1980, 1983) and Ciferri et al. (1982).

4.4 Lyotropic Liquid Crystals

As mentioned in Sect. 1.1, liquid crystals are called lyotropic if the concentration of the constituent units in a solvent is an important variable. Lyotropic liquid crystals can be formed by two rather different types of building blocks. The first possibility is large rod-like macromolecules in solution. As was mentioned in Sect. 4.3 in connection with aromatic polyamides, these polymers give liquid crystalline phases in suitable solvents. In these cases the rods are semiflexible. More rigid rods that give liquid crystalline phases are found in some systems of biological interest. A well-known example is tobacco mosaic virus (TMV), which has the shape of a cylinder with a length of 300 nm and a diameter of 18 nm. It is built from a large number of protein sub-units arranged around an RNA helix (Stubbs et al. 1977; Kreibig and Wetter 1980). Other systems that have been well studied are polypeptides, in particular poly-γ-benzyl-L-glutamate (PBLG):

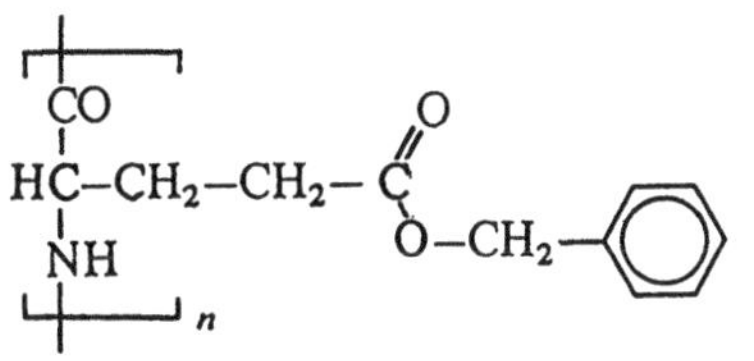

In a large number of organic solvents it takes on a rodlike α-helical confirmation, which is stabilized by intermolecular hydrogen bonds. At volume fractions above 0.1, PBLG orders spontaneously with the rod axes more or less parallel. In fact it is a cholesteric phase with a pitch of the order of 0.01 to 1 mm (Samulski 1978).

Secondly the various liquid crystalline phases formed by amphiphilic molecules in solution should be discussed briefly. Molecules are called amphiphilic because of the existence of two distinct parts with rather different properties. The hydrophilic (polar) head attracts water, while the lipophilic tail (often a hydrocarbon chain) avoids water. Some examples that are of importance for liquid crystals are given in Table 4.3. The amphiphilic character of these compounds is shown clearly by their solubility

Table 4.3. Some examples of amphiphilic molecules

Soap-like:	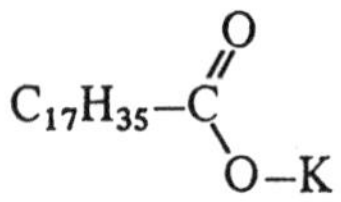	Potassium stearate
General soap:	$[CH_3–(CH_2)_n–CO–O–]_m^- X^{m+}$	X is m-valent cation
Phospholipids:	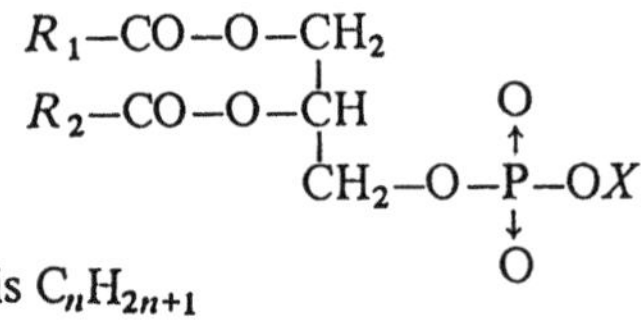	
	R_1, R_2 is C_nH_{2n+1} X is H : phosphatidic acid X is $CH_2–CH_2–N^+–(CH_3)_3$: lecithin	

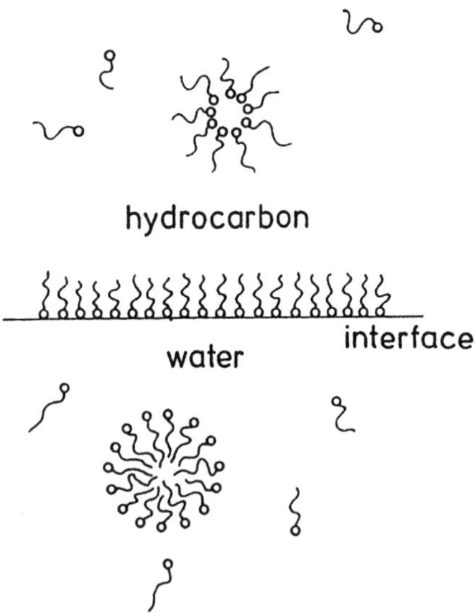

Fig. 4.5. Organization of an amphiphilic compound with polar head and apolar alkyl chain in micelles and at a water-hydrocarbon interface

properties in polar solvents (water) and non-polar solvents (hydrocarbons) respectively. This effect is schematically indicated in Fig. 4.5. At an interface water/hydrocarbon the molecules form a smectic-like layer. Away from the interface microdomains may be formed, called micelles. Often the sequence with increasing concentration is as follows: at low concentrations one finds isolated amphiphilic molecules; above the critical micellar concentration (CMC-point) micelles are formed, first spherical in shape, and at higher concentrations rod-like or disc-like. Figure 4.5 shows a cross-section of such a micelle that may comprise 20–100 molecules. Depending on the environment one distinguishes normal micelles (in water, heads at the outside) and inverse micelles (in hydrocarbons, heads at the inner side). These two possibilities should be kept in mind throughout this section, where usually only the solutions in water will be mentioned. A review of surfactant-water liquid crystal systems has been given by Tiddy (1980).

Amphiphilic compounds are important for several reasons. They have a strong influence on the properties of the medium in which they are dispersed, mainly due to the possibility of layer-formation at interfaces. Organic compounds that cannot be dissolved in water may do so in the presence of a small quantity of an amphiphilic compound. The solute goes inside the micelles which then swell and can grow to hundreds of nm in diameter. At some stage one could speak of small drops of a hydrocarbon phase encapsulated by surfactant layers within a water-like phase, or *vice versa*. As long as the system does not need to be described by a two-phase system one refers to microemulsions (Mittal 1977). Well-known of course, is the case of soap in water which dissolves grease.

If the percentage of water is reduced, the micellar solutions give way to a succession of liquid crystalline phases, of which the lamellar and hexagonal phases are the best known. As an example Fig. 4.6 shows schematically

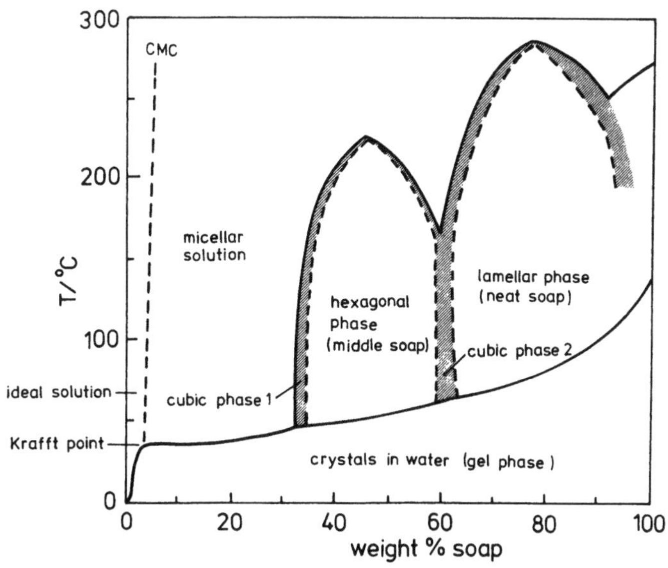

Fig. 4.6. Simplified phase diagram of a typical soap-water system

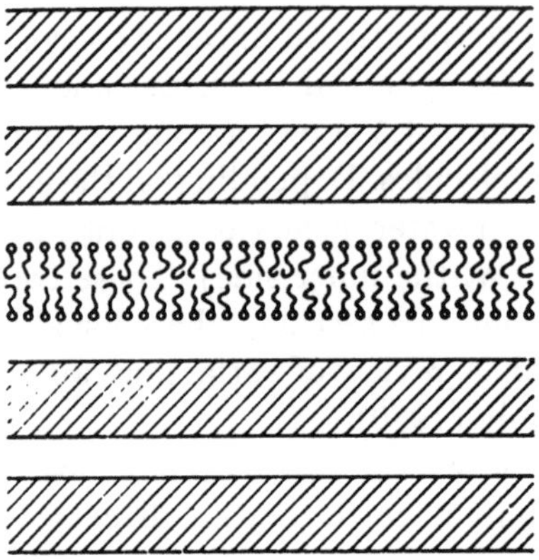

Fig. 4.7. Schematic view of the lamellar mesophase (neat soap) (after Skoulios 1978)

Fig. 4.8. Schematic view of the hexagonal mesophase (middle soap) (after Skoulios 1978)

the various phases observed in a typical soap/water system. In a certain temperature range a lamellar phase is usually formed at small water concentrations. This phase is pictured in Fig. 4.7 and is also called "neat soap". At intermediate water concentrations the lamellae will not be stable any more. One finds a liquid crystal phase with very long cylindrical micelles that are arranged in a hexagonal array ("middle soap"). This is pictured in Fig. 4.8. Finally at higher water concentrations one arrives at an isotropic micellar solution. Intermediate between these phases different types of isotropic liquid crystal phases may be formed. From x-ray diffraction they appear to have some form of a cubic arrangement of the anisotropic micelles (Fontell 1981).

The structure of the liquid crystal phases requires that the alkyl chains have a relatively large freedom to rotate around the various C-C bonds, so that different conformations can be adopted ("molten" alkyl chains). Consequently, these phases occur only above a certain temperature (lower full line in Fig. 4.6). Below this temperature one finds an emulsion of crystals in water (gel-phase). In fact both the polar heads and the alkyl chains are positionally disordered in the liquid crystal phases described so far. Intermediate situations with ordered polar heads and disordered chains can exist for the pure soap and at very low water concentrations (Skoulios 1978).

While the hexagonal phase can be considered as the limiting case when rod-like micelles have become infinitely long, the lamellar phase corresponds to the limit of infinite disc-like micelles. In the intermediate case of finite rods or discs nematic phases can exist. This occurs especially in multicomponent systems, for example, when a salt is added to a ternary system of a charged amphiphile, water, and a long-chain alcohol. Various types of nematic phase have actually been observed (Yu and Saupe 1980a):

i) The N_L-phase, in which disc-like micelles are the constituent entities that are orientationally ordered. The diamagnetic anisotropy is negative. The visco-elastic properties have been studied by Haven et al. (1981).
ii) The N_C-phase, in which rod-like micelles are the ordered units. It has a positive diamagnetic anisotropy.

The transitions between each of these phases and the related lamellar and hexagonal phase appear to be continuous. Yu and Saupe (1980b) discovered a third nematic phase between the N_C and the N_L phase, which is biaxial. It may consist either of lath-like micelles that are formed at the transition between rods and discs, or of a mixture of rod- and disc-like micelles.

Attempts have been made to rationalize the structures of the different phases in terms of intermolecular forces. Various types of forces have to be considered:

a) The interaction between hydrocarbon and water, which leads to a segregation of the polar and apolar parts. This tends to decrease the area per chain at the interface.
b) Electrostatic interactions, related to the presence of charges.
c) Entropy contributions from the disordered conformations of the chains.
d) Steric factors related to possible differences in the dimensions of the polar and apolar parts.

The last point has been studied especially by Skoulios (1978) with x-ray diffraction, determining the area available to a molecule and to its hydrophilic or lipophilic part.

To investigate the molecular behaviour in relation to the structure of the phase, magnetic resonance is a suitable method to use (Charvolin and Tardieu 1978). For example, the linewidth in proton or deuterium magnetic resonance (PMR, DMR) gives information about the mobility of the molecules within the micellar units. A typical result is given in Fig. 4.9. Interestingly, the cubic phase, which is highly viscous in contrast to the fluid micellar solution, appears as a liquid through PMR investigations. This behaviour can be attributed to motional averaging of the dipolar interactions in all three directions. Evidently the finite rods are arranged here such that rapid diffusion in this structure is equivalent to an isotropic motion. This is in contrast to the hexagonal and lamellar structure where an equivalent averaging in three dimensions cannot occur. As a second example the results for the order parameter of the chains in the lamellar phase are given in Fig. 4.10 as measured by DMR. A striking feature is that, with the exception of the first and the last groups, all the other methylenes have nearly the same order parameter. This behaviour differs from that of an isolated chain, fixed at one end, where each segment is expected to have a greater orientational

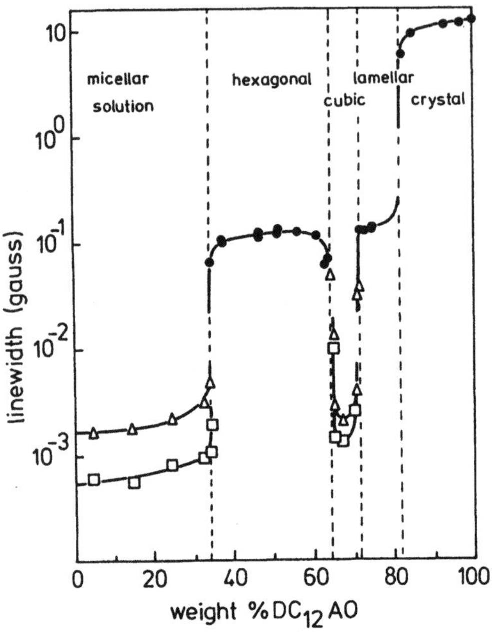

Fig. 4.9. Linewidth in the various phases of dimethyldodecylamine oxide in D_2O at 30°C as observed in proton magnetic resonance (after Lawson and Flautt 1968)

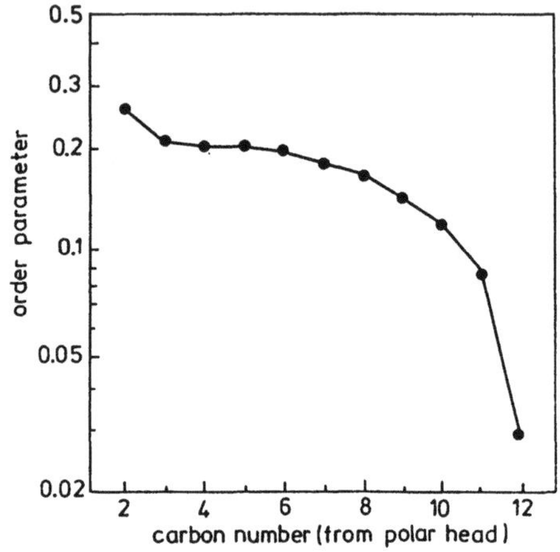

Fig. 4.10. Order parameter of the CD bond in an oriented lamellar sample of deuterated potassium laurate-water (after Charvolin and Tardieu 1978)

freedom than the preceding one, and the order decreases continuously. This difference can be attributed to the steric repulsion between neighbouring chains in the layers, which prevents the orientational order from decreasing along the chain.

Finally the importance of lyotropic liquid crystals for biological systems should be mentioned here. For reviews of this subject the reader is referred to Bouligand (1978) and Chapman (1978).

Part II

Continuum Theory

5. Static Continuum Theory

The purpose of this chapter is to describe liquid crystals in terms of a macroscopic theory of elasticity. For reasons of simplicity we will mainly pay attention to the "simplest" liquid crystals, i.e. nematics. This class of liquid crystals is described macroscopically by an orientational tensor field. The organization of this chapter is as follows. First we give a short introduction to tensor analysis for the reader's convenience. Then we show that the orientational order of a nematic must be described by a tensor of the second rank, the so-called macroscopic tensor order parameter. Next we derive the theory of curvature elasticity, i.e. the distortion free energy density of a nematic, and we show that three elastic constants are required in order to describe the curvature elastic properties of nematics. These elastic constants are connected with splay, twist and bend deformations, respectively. The extension of the theory of elasticity or continuum theory to the case of smectics is only briefly mentioned. Finally we derive the explicit expressions for the splay, twist and bend elastic constants starting from the general expression for the free energy density of a nematic and we give results for some specific free energy densities.

5.1 Cartesian Tensors

The physical quantities that appear in the theory of liquid crystals are tensors of a certain rank, as all physical quantities should be. Tensors of rank zero and rank one have specific names. They are called scalars and vectors, respectively. The state of a physical system is described by specifying the values of the components of the relevant tensors in a particular coordinate system. The tensors that appear in the theory of liquid crystals are particularly simple: they are Cartesian tensors. The purpose of this section is to briefly review the properties of Cartesian tensors. A detailed treatise of tensor calculus can be found in Levi-Civita's superb book "The absolute differential calculus" (1977).

In tensor calculus it is quite advantageous to make use of the Einstein summation convention. This means that the appearance of repeated Greek indices implies a summation over all possible values of these indices. Consider e.g. the index α, which may take the values, 1, 2, and 3; then, according to the summation convention,

$$A_\alpha B_\alpha = A_1 B_1 + A_2 B_2 + A_3 B_3 \quad .$$

The starting point of the present introduction to Cartesian tensors is the ordinary three-dimensional Euclidean space with a basis consisting of the unit vectors $\boldsymbol{e}_\alpha$ with $\alpha = 1, 2, 3$, where

$$\boldsymbol{e}_\alpha \cdot \boldsymbol{e}_\beta = \delta_{\alpha\beta} \quad , \quad \alpha, \beta = 1, 2, 3$$

and $\boldsymbol{e}_1 \times \boldsymbol{e}_2 = \boldsymbol{e}_3$, $\boldsymbol{e}_2 \times \boldsymbol{e}_3 = \boldsymbol{e}_1$, $\boldsymbol{e}_3 \times \boldsymbol{e}_1 = \boldsymbol{e}_2$. The symbol $\delta_{\alpha\beta}$ is called the Kronecker delta, which is defined in all coordinate systems as

$$\delta_{\alpha\beta} = \begin{cases} 0, & \text{if} \quad \alpha \neq \beta \\ 1, & \text{if} \quad \alpha = \beta \end{cases} \quad .$$

An arbitrary vector $\boldsymbol{r}$ can be written as

$$\boldsymbol{r} = r_\alpha \boldsymbol{e}_\alpha \quad ,$$

where the numbers r_α correspond to the components of $\boldsymbol{r}$ in the cartesian coordinate system spanned by the basis vectors $\boldsymbol{e}_\alpha$.

Next the original cartesian coordinate system is rotated. Consequently the basis vectors are rotated. The resulting new basis vectors $\overline{\boldsymbol{e}}_\alpha$ can be expressed in terms of the original ones by

$$\overline{\boldsymbol{e}}_\alpha = R_{\alpha\beta} \boldsymbol{e}_\beta \quad \text{with}$$

$$R_{\alpha\beta} = \overline{\boldsymbol{e}}_\alpha \cdot \boldsymbol{e}_\beta \quad .$$

The matrix R, which is composed of the nine elements $R_{\alpha\beta}$, is called the transformation matrix. Clearly the element $R_{\alpha\beta}$ is the βth component of $\overline{\boldsymbol{e}}_\alpha$ in the original coordiate system.

The elements of the inverse matrix R^{-1} are determined by considering the inverse transformation

$$\boldsymbol{e}_\alpha = R^{-1}_{\alpha\beta} \overline{\boldsymbol{e}}_\beta \quad \text{with}$$

$$R^{-1}_{\alpha\beta} = \boldsymbol{e}_\alpha \cdot \overline{\boldsymbol{e}}_\beta = \overline{\boldsymbol{e}}_\beta \cdot \boldsymbol{e}_\alpha = R_{\beta\alpha} \quad .$$

This relation immediately shows that the determinant of the transformation matrix R equals one. So-called improper rotations, i.e. det $R = -1$, are not considered here.

It is clear that a rotation affects the components of a vector $\boldsymbol{r}$ as well. The components $\bar{r}_\alpha$ of a vector $\boldsymbol{r}$ in the rotated coordinate frame are expressed, in terms of the original components r_α, using the identity

$$\boldsymbol{r} = \bar{r}_\alpha \bar{\boldsymbol{e}}_\alpha = r_\beta \boldsymbol{e}_\beta = r_\beta R^{-1}_{\beta\alpha} \bar{\boldsymbol{e}}_\alpha \quad ,$$

or in other words

$$\bar{r}_\alpha = R^{-1}_{\beta\alpha} r_\beta = R_{\alpha\beta} r_\beta \quad .$$

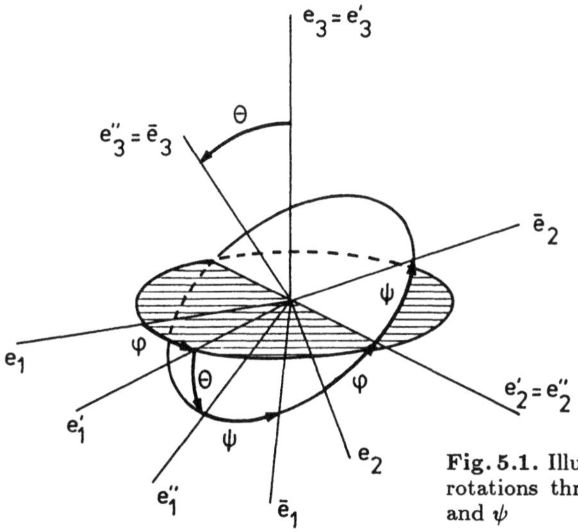

Fig. 5.1. Illustration of the three successive rotations through the Eulerian angles ϕ, θ and ψ

In order to obtain the general expression for the transformation matrix R an arbitrary rotation of the original coordinate system must be carried out. Such a rotation involves three successive rotations around coordinate axes and consequently three Eulerian angles, ϕ, θ and ψ as illustrated in Fig. 5.1, are required. Usually a rotation over an angle ϕ around $\boldsymbol{e}_3$ is carried out first giving rise to a new basis set

$$\boldsymbol{e}'_\alpha = R_{\alpha\beta}(\phi)\boldsymbol{e}_\beta \quad \text{with}$$

$$R(\phi) = \begin{pmatrix} \cos\phi & \sin\phi & 0 \\ -\sin\phi & \cos\phi & 0 \\ 0 & 0 & 1 \end{pmatrix}$$

i.e. $\boldsymbol{e}'_3 = \boldsymbol{e}_3$. Next a rotation over an angle θ around $\boldsymbol{e}'_2$ is carried out giving rise to a new basis set

$$e''_\alpha = R'_{\alpha\beta}(\theta)e'_\beta \quad \text{with}$$

$$R'(\theta) = \begin{pmatrix} \cos\theta & 0 & -\sin\theta \\ 0 & 1 & 0 \\ \sin\theta & 0 & \cos\theta \end{pmatrix}$$

i.e. $e''_2 = e'_2$. Finally a rotation over an angle ψ around e''_3 is carried out giving rise to the new basis set

$$\overline{e}_\alpha = R''_{\alpha\beta}(\psi)e''_\beta \quad ,$$

where $R''_{\alpha\beta}(\psi) = R_{\alpha\beta}(\psi)$ and $\overline{e}_3 = e''_3$. Consequently the general expression is given by

$$\overline{e}_\alpha = R_{\alpha\beta}(\psi,\theta,\phi)e_\beta \quad \text{with} \tag{5.1}$$

$$R_{\alpha\beta}(\psi,\theta,\phi) = R_{\alpha\mu}(\psi)R'_{\mu\nu}(\theta)R_{\nu\beta}(\phi) \tag{5.2}$$

or

$$R(\psi,\theta,\phi) = \begin{pmatrix} \cos\theta\cos\phi\cos\psi - \sin\phi\sin\psi & \cos\theta\sin\phi\cos\psi + \cos\phi\sin\psi & -\sin\theta\cos\psi \\ -\cos\theta\cos\phi\sin\psi - \sin\phi\cos\psi & -\cos\theta\sin\phi\sin\psi + \cos\phi\cos\psi & \sin\theta\sin\psi \\ \sin\theta\cos\phi & \sin\theta\sin\phi & \cos\theta \end{pmatrix} . \tag{5.3}$$

Note that the transformation matrix may also be obtained by carrying out the rotations in the initial coordinate system provided that the order of the rotations is inverted.

A vector $\boldsymbol{r}$ may now be defined as an object, which has three components, that transform according to

$$\overline{r}_\alpha = R_{\alpha\beta}r_\beta \quad ,$$

when the coordinate system is rotated.

A cartesian tensor $\tilde{\boldsymbol{T}}$ of the second rank is defined as an object with $3^2 = 9$ components $T_{\alpha\beta}$ with $\alpha, \beta = 1,2,3$, which transform according to

$$\overline{T}_{\alpha\beta} = R_{\alpha\gamma}R_{\beta\delta}T_{\gamma\delta} \quad ,$$

when the coordinate system is rotated.

A cartesian tensor $\tilde{\boldsymbol{T}}$ of the nth rank is defined to be an object, which has 3^n components $T_{\alpha_1\alpha_2\ldots\alpha_n}$ with $\alpha_1, \alpha_2, \ldots, \alpha_n = 1,2,3$, which transform according to

$$\overline{T}_{\alpha_1\alpha_2\ldots\alpha_n} = R_{\alpha_1\beta_1}R_{\alpha_2\beta_2}\cdots R_{\alpha_n\beta_n}T_{\beta_1\beta_2\ldots\beta_n} \tag{5.4}$$

when the coordinate system is rotated. It follows directly from this definition that a scalar is a tensor of rank zero, whereas a vector is a tensor of rank one.

The sum of two tensors $\tilde{\boldsymbol{S}}$ and $\tilde{\boldsymbol{T}}$ of the same rank n is again a tensor $\tilde{\boldsymbol{U}}$ of the rank n with components

$$U_{\alpha_1\alpha_2\ldots\alpha_n} = S_{\alpha_1\alpha_2\ldots\alpha_n} + T_{\alpha_1\alpha_2\ldots\alpha_n} \quad .$$

The product of a tensor $\tilde{\boldsymbol{S}}$ of rank m with a tensor $\tilde{\boldsymbol{T}}$ of rank n gives a tensor $\tilde{\boldsymbol{U}}$ of the rank $m+n$ with elements

$$U_{\alpha_1\alpha_2\ldots\alpha_m\beta_1\beta_2\ldots\beta_n} = S_{\alpha_1\alpha_2\ldots\alpha_m}T_{\beta_1\beta_2\ldots\beta_n} \quad .$$

A tensor of rank n yields a tensor of rank $n-2$ by means of the operation of contraction, which consists of equating two indices of the tensor and next applying the summation convention. Clearly such an operation can be repeated several times provided that the rank n of the original tensor is sufficiently large. A combination of the operations multiplication and contraction gives rise to an operation called the composition or inner multiplication of two tensors. Consider the tensors $\tilde{\boldsymbol{S}}$ and $\tilde{\boldsymbol{T}}$ of rank 2. A third tensor $\tilde{\boldsymbol{U}}$ of rank 2 can be constructed from these tensors by means of the following composition

$$U_{\alpha\beta} = S_{\alpha\gamma}T_{\gamma\beta} \quad .$$

Finally it should be mentioned that the rank of a tensor can be enlarged from n to $n+1$ by means of the process of differentiation, i.e. the operation of differentiation turns the tensor $S_{\beta\gamma}$ of rank 2 into the tensor $\partial_\alpha S_{\beta\gamma}$ of rank 3.

An example of a tensor of the second rank is the Kronecker delta. This is a tensor with the property

$$\overline{\delta}_{\alpha\beta} = R_{\alpha\gamma}R_{\beta\delta}\delta_{\gamma\delta} = \delta_{\alpha\beta} \quad .$$

An example of a tensor of the third rank is the Levi-Civita tensor, which is defined in all coordinate systems as

$$\varepsilon_{\alpha\beta\gamma} = \begin{cases} 1, & \text{if} \quad \alpha\beta\gamma = 123, 231, 312 \\ -1, & \text{if} \quad \alpha\beta\gamma = 321, 213, 132 \\ 0, & \text{otherwise} \end{cases} \quad .$$

The tensor character directly follows from

$$\bar{\varepsilon}_{\alpha\beta\gamma} = R_{\alpha\mu} R_{\beta\nu} R_{\gamma\varrho} \varepsilon_{\mu\nu\varrho} = \varepsilon_{\alpha\beta\gamma} \det R = \varepsilon_{\alpha\beta\gamma}$$

as $\det R = 1$ for all proper rotations.

Using the Levi-Civita tensor the αth component of the vector product $\boldsymbol{C} = \boldsymbol{A} \times \boldsymbol{B}$ is given by

$$C_\alpha = \varepsilon_{\alpha\beta\gamma} A_\beta B_\gamma$$

or in components

$$C_1 = A_2 B_3 - A_3 B_2, \quad C_2 = A_3 B_1 - A_1 B_3, \quad C_3 = A_1 B_2 - A_2 B_1 \quad .$$

Finally the product of two Levi-Civita tensors is considered here. These products obey

$$\begin{aligned} \varepsilon_{\alpha\beta\gamma}\varepsilon_{\mu\nu\varrho} = \; & \delta_{\alpha\mu}\delta_{\beta\nu}\delta_{\gamma\varrho} + \delta_{\alpha\nu}\delta_{\beta\varrho}\delta_{\gamma\mu} + \delta_{\alpha\varrho}\delta_{\beta\mu}\delta_{\gamma\nu} - \delta_{\alpha\varrho}\delta_{\beta\nu}\delta_{\gamma\mu} \\ & - \delta_{\alpha\nu}\delta_{\beta\mu}\delta_{\gamma\varrho} - \delta_{\alpha\mu}\delta_{\beta\varrho}\delta_{\gamma\nu} \quad , \end{aligned} \tag{5.5a}$$

$$\varepsilon_{\alpha\beta\gamma}\varepsilon_{\alpha\mu\nu} = \delta_{\beta\mu}\delta_{\gamma\nu} - \delta_{\beta\nu}\delta_{\gamma\mu} \quad , \tag{5.5b}$$

$$\varepsilon_{\alpha\beta\gamma}\varepsilon_{\alpha\beta\mu} = 2\delta_{\gamma\mu} \quad , \tag{5.5c}$$

$$\varepsilon_{\alpha\beta\gamma}\varepsilon_{\alpha\beta\gamma} = 6 \quad . \tag{5.5d}$$

It is worthwhile to end this section with a citation from Weyl's classic "Raum-Zeit-Materie" (1918) which reads "Das Eindringen in den Tensorkalkül hat – abgesehen von der Angst vor Indizes, die überwunden werden muss – gewiss seine begrifflichen Schwierigkeiten. Formal ist aber die Rechenmethodik von der äussersten Einfachheit, viel einfacher z.B. als der Apparat der elementaren Vektorrechnung". (See Reference for translation.)

5.2 Macroscopic Order Parameter

All liquid crystalline states are characterized by an orientational order of the molecules. In fact this is the only aspect in which the nematic and isotropic phases differ. As already mentioned in Sect. 1.1, the orientational order must be described by a tensor of the second rank. In this section a macroscopic approach will be presented, i.e. the order parameter is constructed independent of any assumption regarding the interactions of the constituent molecules. A tensor order parameter can be obtained in the following way. Application of some field $\boldsymbol{X}$ results in a response $\boldsymbol{Y}$ of the system given by

$$Y_\alpha = T_{\alpha\beta} X_\beta \quad , \tag{5.6}$$

where $T_{\alpha\beta}$ is a symmetric tensor, i.e. $T_{\alpha\beta} = T_{\beta\alpha}$, and X_α and Y_α denote the components of $\boldsymbol{X}$ and $\boldsymbol{Y}$, respectively, in a given coordinate system. As an example $\boldsymbol{X}$ and $\tilde{\boldsymbol{T}}$ may represent respectively the external magnetic field $\boldsymbol{B}$ and the susceptibility tensor $\tilde{\chi}$; then $\mu_0^{-1}\boldsymbol{Y}$ is the magnetization $\boldsymbol{M}$, where μ_0 denotes the permeability of the vacuum. Clearly $\tilde{\boldsymbol{T}}$ is diagonal in a properly chosen coordinate system, i.e.

$$\tilde{\boldsymbol{T}} = \begin{pmatrix} T_1 & 0 & 0 \\ 0 & T_2 & 0 \\ 0 & 0 & T_3 \end{pmatrix} \quad . \tag{5.7}$$

The elements T_1, T_2 and T_3 are temperature dependent. The tensor order parameter is obtained by extracting the anisotropic part of $\tilde{\boldsymbol{T}}$. This can be accomplished by putting $\sum_{i=1}^{3} T_i = T$. Then the elements T_1, T_2, and T_3 can be expressed as

$$\begin{aligned} T_1 &= \tfrac{1}{3}T(1 - Q_1 + Q_2) \quad , \\ T_2 &= \tfrac{1}{3}T(1 - Q_1 - Q_2) \quad , \\ T_3 &= \tfrac{1}{3}T(1 + 2Q_1) \quad . \end{aligned} \tag{5.8}$$

Accordingly the diagonalized version of the tensor $\tilde{\boldsymbol{T}}$ is

$$T_{\alpha\beta} = T(\tfrac{1}{3}\delta_{\alpha\beta} + Q_{\alpha\beta}) \quad , \tag{5.9}$$

where the tensor $\tilde{\boldsymbol{Q}}$ with elements $Q_{\alpha\beta}$ is called the tensor order parameter. The diagonal representation of the tensor order parameter is given by

$$\tilde{\boldsymbol{Q}} = \begin{pmatrix} -\frac{1}{3}(Q_1 - Q_2) & 0 & 0 \\ 0 & -\frac{1}{3}(Q_1 + Q_2) & 0 \\ 0 & 0 & \frac{2}{3}Q_1 \end{pmatrix} \quad . \tag{5.10}$$

The tensor $\tilde{\boldsymbol{Q}}$ is called the tensor order parameter because of the appearance of the two order parameters Q_1 and Q_2. The isotropic liquid is described by $Q_1 = Q_2 = 0$. The anisotropic liquid with uniaxial symmetry is described by only one order parameter. In this case the unique axis is conventionally chosen along the basis vector $\boldsymbol{e}_3$ and the medium is described by $Q_1 \neq 0$ and $Q_2 = 0$. Changing the uniaxial symmetry into a biaxial one requires the introduction of the second independent order parameter Q_2. Clearly both order parameters depend on the temperature.

As an example consider the suceptibility tensor $\tilde{\boldsymbol{\chi}}$ in a uniaxial medium with its symmetry axis along the basis vector $\boldsymbol{e}_3$, i.e.

$$\tilde{\boldsymbol{\chi}} = \begin{pmatrix} \chi_\perp & 0 & 0 \\ 0 & \chi_\perp & 0 \\ 0 & 0 & \chi_\parallel \end{pmatrix} \quad . \tag{5.11}$$

Rewriting gives

$$\chi_{\alpha\beta} = (\chi_\parallel + 2\chi_\perp)(\tfrac{1}{3}\delta_{\alpha\beta} + Q_{\alpha\beta}) \tag{5.12}$$

with $Q_1 = \Delta\chi/(\chi_\parallel + 2\chi_\perp)$ and $Q_2 = 0$, where $\Delta\chi = \chi_\parallel - \chi_\perp$ denotes the anisotropy in the susceptibility.

The general expression for the tensor order parameter is obtained by an arbitrary rotation of the coordinate system. The starting point is a cartesian coordinate system with basis vectors $\overline{\boldsymbol{e}}_\alpha$ giving rise to a diagonal representation of the tensor order parameter. The elements of this representation are $\overline{Q}_{\alpha\beta}$. Next an arbitrary rotation is carried out leading to a coordinate system with basis vectors $\boldsymbol{e}_\alpha$. Now the elements $Q_{\alpha\beta}$ of the tensor $\tilde{\boldsymbol{Q}}$, with respect to the new coordinate system, are given by

$$Q_{\alpha\beta} = \overline{R}_{\alpha\gamma}\overline{R}_{\beta\delta}\overline{Q}_{\gamma\delta} \quad . \tag{5.13}$$

Using

$$\overline{R}_{\alpha\beta} = \boldsymbol{e}_\alpha \cdot \overline{\boldsymbol{e}}_\beta$$

it follows directly that

$$\begin{aligned} Q_{\alpha\beta} = & -\tfrac{1}{3}(Q_1 - Q_2)(\boldsymbol{e}_\alpha \cdot \overline{\boldsymbol{e}}_1)(\boldsymbol{e}_\beta \cdot \overline{\boldsymbol{e}}_1) - \tfrac{1}{3}(Q_1 + Q_2)(\boldsymbol{e}_\alpha \cdot \overline{\boldsymbol{e}}_2)(\boldsymbol{e}_\beta \cdot \overline{\boldsymbol{e}}_2) \\ & + \tfrac{2}{3}Q_1(\boldsymbol{e}_\alpha \cdot \overline{\boldsymbol{e}}_3)(\boldsymbol{e}_\beta \cdot \overline{\boldsymbol{e}}_3) \quad . \end{aligned} \tag{5.14}$$

This is the general expression of the tensor order parameter for a biaxial medium. A considerable simplification of this expression is obtained if the restriction is made to nematics with uniaxial symmetry. In this case the direction of the unique axis, which is given by the director $\boldsymbol{n}$, coincides with one of the basis vectors belonging to the cartesian coordinate system in which $\tilde{\boldsymbol{Q}}$ is diagonal. Here it is assumed that $\boldsymbol{n} = \overline{\boldsymbol{e}}_3$ or $\boldsymbol{n} = -\overline{\boldsymbol{e}}_3$, i.e.

$Q_2 = 0$. Thus (5.14) reduces to

$$\begin{aligned} Q_{\alpha\beta} &= -\tfrac{1}{3}Q_1(\boldsymbol{e}_\alpha \cdot \overline{\boldsymbol{e}}_\gamma)(\boldsymbol{e}_\beta \cdot \overline{\boldsymbol{e}}_\gamma) + Q_1(\boldsymbol{e}_\alpha \cdot \boldsymbol{n})(\boldsymbol{e}_\beta \cdot \boldsymbol{n}) \\ &= -\tfrac{1}{3}Q_1\delta_{\alpha\beta} + Q_1(\boldsymbol{e}_\alpha \cdot \boldsymbol{n})(\boldsymbol{e}_\beta \cdot \boldsymbol{n}) \quad , \end{aligned} \tag{5.15}$$

where the following relation has been used

$$(\boldsymbol{e}_\alpha \cdot \overline{\boldsymbol{e}}_\gamma)(\boldsymbol{e}_\beta \cdot \overline{\boldsymbol{e}}_\gamma) = \overline{R}_{\alpha\gamma}\overline{R}^{-1}_{\gamma\beta} = \delta_{\alpha\beta} \quad .$$

Next the tensor $\tilde{\boldsymbol{N}}$ is introduced, which has the elements

$$N_{\alpha\beta} = n_\alpha n_\beta \quad ,$$

where $n_\alpha = \boldsymbol{e}_\alpha \cdot \boldsymbol{n}$, i.e. n_α is the projection of $\boldsymbol{n}$ on the αth basis vector of an arbitrary cartesian coordinate system. The general expression for the tensor order parameter $\tilde{\boldsymbol{Q}}$ for a uniaxial medium is now given by

$$Q_{\alpha\beta} = Q_1(N_{\alpha\beta} - \tfrac{1}{3}\delta_{\alpha\beta}) \quad . \tag{5.16}$$

The order parameter Q_1, that appears in expression (5.16), depends on the temperature. The tensor order parameter components $Q_{\alpha\beta}$ have been written here in terms of the tensor components $N_{\alpha\beta}$ intentionally, instead of the more familiar $n_\alpha n_\beta$. It is thus stressed that the relevant fluctuating quantities refer to the tensor components $N_{\alpha\beta}$, instead of the vector components n_α. The description of fluctuations in terms of $\boldsymbol{n}(\boldsymbol{r})$ necessarily gives rise to contradictions due to the invariance of the nematic state to the replacement of $\boldsymbol{n}(\boldsymbol{r})$ by $-\boldsymbol{n}(\boldsymbol{r})$ (see Sect. 7.2).

It is also possible to define an order parameter with values between 0 (isotropic phase) and 1 (perfectly aligned nematic phase), provided that the values of the tensor $\tilde{\boldsymbol{T}}$ are known in the perfectly aligned phase. For example consider the susceptibility tensor (5.11). This tensor can also be written as

$$\chi_{\alpha\beta} = \tfrac{1}{3}(\chi_\parallel + 2\chi_\perp)\delta_{\alpha\beta} + \Delta\chi_{\alpha\beta} \quad \text{with} \tag{5.17}$$

$$\Delta\tilde{\boldsymbol{\chi}} = \Delta\chi_{\max} S \begin{pmatrix} -\frac{1}{3} & 0 & 0 \\ 0 & -\frac{1}{3} & 0 \\ 0 & 0 & \frac{2}{3} \end{pmatrix} \quad , \tag{5.18}$$

where $\Delta\chi_{\max}$ denotes the anisotropy in the susceptibility in the perfectly aligned phase. Now the order parameter $S = \Delta\chi / \Delta\chi_{\max}$ is defined between 0 (isotropic phase) and 1 (perfectly aligned phase). The connection with the original tensor order parameter is

$$Q_{\alpha\beta} = \Delta\chi_{\alpha\beta}/(\chi_{\|} + 2\chi_{\perp}) \quad . \tag{5.19}$$

The relation between the order parameter S as defined here and the orientational distribution as mentioned in Sect. 1.1 will be discussed in some detail in Chap. 9.

5.3 The Frank Free Energy

At this stage it will be clear that the orientation of a uniaxial liquid crystal must be described in terms of the tensor $\tilde{\boldsymbol{N}}$, whereas the degree of ordering is expressed by the order parameter Q_1. In many practical circumstances constraints are imposed upon the nematic liquid, for example by boundary conditions at the walls of the container or by external fields. Generally these constraints give rise to a liquid crystalline material where the orientation varies over the sample. These distortions take place over macroscopic distances (typically a few microns), and can often be observed optically. On the other hand the degree of ordering $Q_1(T)$ will be hardly affected by these slow variations of the orientation. Hence it is useful to consider a liquid crystalline material with a slowly varying orientation, such that the medium remains locally uniaxial while $Q_1(T)$ does not depend on the position $\boldsymbol{r}$. Consequently the liquid crystal can be described by a tensor field $\tilde{\boldsymbol{N}}(\boldsymbol{r})$ of the second rank. This consideration is the basis of the continuum theory of liquid crystals developed by Oseen (1933), Zocher (1933) and Frank (1958).

The continuum theory is a macroscopic phenomenological theory of liquid crystals dealing with a slowly varying tensor field $\tilde{\boldsymbol{N}}(\boldsymbol{r})$. Note that the usual term, i.e. director field, is misleading, since the relevant field concerns a tensor of the second rank instead of a vector. To some extent the theory is the analogue of the classical elasticity theory of solids. In the latter theory the solid undergoes strains under the action of applied forces. The appearing deformation in turn brings about restoring forces, that oppose the change in distance between neighbouring points in the material. In a nematic liquid the deformation concerns the change in orientation between neighbouring points. The relevant generalized forces, that are brought about by the deformation, oppose the curvature. Frank (1958) refers to these generalized forces as torque stresses and assumes a linear relationship between these stresses and the curvature strains, provided the latter are sufficiently small. Clearly his assumption is an equivalent of Hooke's law.

The starting point of the continuum theory is the free energy density $f(\boldsymbol{r})$ belonging to the tensor field $\tilde{N}(\boldsymbol{r})$. This expression is then expanded around the state of uniform parallel alignment. Because of the smallness of the spatial derivatives of the field $\tilde{N}(\boldsymbol{r})$ it suffices to retain only the first and second order terms, $\partial_\alpha N_{\beta\gamma}(\boldsymbol{r})$ and $\partial_\alpha \partial_\beta N_{\gamma\delta}(\boldsymbol{r})$ respectively, where $\partial_\alpha = \partial/\partial\alpha$ and the Greek indices denote x, y or z. This approximation is equivalent to assuming a linear relation between stresses and strains. In consequence the continuum theory postulates the following expression for the free energy density of a locally uniaxial liquid crystal

$$\begin{aligned} f(\boldsymbol{r}) = f_0 &+ \overline{L}_{\alpha\beta\gamma}(\boldsymbol{r})\partial_\alpha N_{\beta\gamma}(\boldsymbol{r}) + \overline{L}_{\alpha\beta\gamma\mu\nu\varrho}(\boldsymbol{r})[\partial_\alpha N_{\beta\gamma}(\boldsymbol{r})][\partial_\mu N_{\nu\varrho}(\boldsymbol{r})] \\ &+ \overline{L}_{\alpha\beta\gamma\delta}(\boldsymbol{r})\partial_\alpha \partial_\beta N_{\gamma\delta}(\boldsymbol{r}) \quad , \end{aligned} \tag{5.20}$$

where f_0 is the free energy density relating to the state of uniform alignment. The physics of the problem dictates that the tensors $\overline{L}_{\alpha\beta\gamma}(\boldsymbol{r})$, $\overline{L}_{\alpha\beta\gamma\delta}(\boldsymbol{r})$ and $\overline{L}_{\alpha\beta\gamma\mu\nu\varrho}(\boldsymbol{r})$ must be composed of all possible combinations of $N_{\alpha\beta}(\boldsymbol{r})$, $\delta_{\alpha\beta}$ and $\varepsilon_{\alpha\beta\gamma}$. Clearly the state of uniform alignment is described by

$$\partial_\alpha N_{\beta\gamma}(\boldsymbol{r}) = \partial_\alpha \partial_\beta N_{\gamma\delta}(\boldsymbol{r}) = 0.$$

It is customary to rewrite the free energy density (5.20) in terms of the components $n_\alpha(\boldsymbol{r})$ of the local director $\boldsymbol{n}(\boldsymbol{r})$. The contribution to the free energy density due to the distortion of the director field is

$$\begin{aligned} f_{\mathrm{d}}(\boldsymbol{r}) = L_{\alpha\beta}(\boldsymbol{r})\partial_\alpha n_\beta(\boldsymbol{r}) &+ L_{\alpha\beta\gamma\delta}(\boldsymbol{r})[\partial_\alpha n_\beta(\boldsymbol{r})][\partial_\gamma n_\delta(\boldsymbol{r})] \\ &+ L_{\alpha\beta\gamma}(\boldsymbol{r})\partial_\alpha \partial_\beta n_\gamma(\boldsymbol{r}) \quad . \end{aligned} \tag{5.21}$$

This expression is known as the Frank free energy density and is obtained by using $N_{\alpha\beta}(\boldsymbol{r}) = n_\alpha(\boldsymbol{r}) n_\beta(\boldsymbol{r})$. The tensors $L_{\alpha\beta}(\boldsymbol{r})$, $L_{\alpha\beta\gamma}(\boldsymbol{r})$ and $L_{\alpha\beta\gamma\delta}(\boldsymbol{r})$ depend on the order parameter Q_1. The possible forms of these tensors are determined by the scalar nature of the free energy density as well as its invariance on replacing $\boldsymbol{n}(\boldsymbol{r})$ by $-\boldsymbol{n}(\boldsymbol{r})$, i.e. the absence of head-tail effects. The following tensors give an essentially different and non-zero contribution to the Frank free energy density.

(i) Because of the head-tail symmetry the tensor $L_{\alpha\beta}(\boldsymbol{r})$ must take the form

$$L_{\alpha\beta}(\boldsymbol{r}) = k\varepsilon_{\alpha\beta\gamma} n_\gamma(\boldsymbol{r}) \quad ,$$

where the coefficient k is some function of the order parameter Q_1. The contribution to the Frank free energy density is

$$k\varepsilon_{\alpha\beta\gamma} n_\gamma(\boldsymbol{r}) \partial_\alpha n_\beta(\boldsymbol{r}) \quad . \tag{5.22}$$

Tensors like $n_\alpha(\boldsymbol{r})n_\beta(\boldsymbol{r})$ and $\delta_{\alpha\beta}$ do not need to be considered. Their corresponding terms do not contribute to the free energy density because (using n_α as shorthand for $n_\alpha(\boldsymbol{r})$)

$$n_\alpha n_\beta \partial_\alpha n_\beta = \tfrac{1}{2} n_\alpha \partial_\alpha (n_\beta n_\beta) = 0 \quad , \quad \text{since} \quad n_\beta n_\beta = 1 \quad ;$$

$\delta_{\alpha\beta}\partial_\alpha n_\beta = \partial_\alpha n_\alpha$ does not appear because of head-tail symmetry.

(ii) The tensor $L_{\alpha\beta\gamma\delta}(\boldsymbol{r})$ takes the forms

$$l_1 n_\alpha n_\gamma \delta_{\beta\delta} \quad ; \quad l_2 \delta_{\alpha\beta}\delta_{\gamma\delta} \quad ; \quad l_3 \delta_{\alpha\gamma}\delta_{\beta\delta} \quad ; \quad l_4 \delta_{\alpha\delta}\delta_{\beta\gamma} \quad .$$

The coefficients l_1, l_2, l_3 and l_4 are functions of the order parameter Q_1. The relevant combination with $(\partial_\alpha n_\beta)(\partial_\gamma n_\delta)$ gives the following contributions to the Frank free energy density:

$$\begin{aligned} &l_1 n_\alpha n_\beta (\partial_\alpha n_\gamma)(\partial_\beta n_\gamma) \quad ; \quad && l_2 (\partial_\alpha n_\alpha)(\partial_\beta n_\beta) \quad ; \\ &l_3 (\partial_\alpha n_\beta)(\partial_\alpha n_\beta) \quad ; && l_4 (\partial_\alpha n_\beta)(\partial_\beta n_\alpha) \quad . \end{aligned}$$

It can be verified that all other possible combinations for constructing $L_{\alpha\beta\gamma\delta}(\boldsymbol{r})$ either give rise to the same invariants or produce zero. For example, the tensor $n_\alpha n_\beta \delta_{\gamma\delta}$ does not contribute, as follows from

$$n_\alpha n_\beta \delta_{\gamma\delta}(\partial_\alpha n_\beta)(\partial_\gamma n_\delta) = \tfrac{1}{2} n_\alpha [\partial_\alpha (n_\beta n_\beta)](\partial_\gamma n_\gamma) = 0 \quad , \quad \text{for} \quad n_\beta n_\beta = 1 \quad .$$

Furthermore, terms of the form $\partial_\alpha(n_\alpha \partial_\beta n_\beta)$ and $\partial_\alpha(n_\beta \partial_\beta n_\alpha)$ may be always added to or substracted from the free energy density, because they only contribute to the surface energy according to Gauss's theorem,

$$\int \partial_\alpha (n_\alpha \partial_\beta n_\beta) d\boldsymbol{r} = \int (n_\alpha \partial_\beta n_\beta) dS_\alpha \quad ,$$

where $\int \boldsymbol{A} \cdot d\boldsymbol{S}$ denotes a surface integral with $d\boldsymbol{S}$ being normal to the surface at each point. This means that only three instead of four distinct invariants appear, as the invariants associated with l_2 and l_4 are related to each other via such a surface term

$$(\partial_\alpha n_\alpha)(\partial_\beta n_\beta) - (\partial_\alpha n_\beta)(\partial_\beta n_\alpha) = \partial_\alpha (n_\alpha \partial_\beta n_\beta) - \partial_\alpha (n_\beta \partial_\beta n_\alpha) \quad .$$

(iii) The tensor $L_{\alpha\beta\gamma}(\boldsymbol{r})$ must transform according to

$$l_5 n_\alpha n_\beta n_\gamma \quad ; \quad l_6 \delta_{\alpha\beta} n_\gamma \quad ; \quad l_7 \delta_{\alpha\gamma} n_\beta \quad ; \quad l_8 \delta_{\beta\gamma} n_\alpha \quad ;$$

where the coefficients l_5, l_6, l_7 and l_8 are functions of the order parameter

Q_1. Combining these tensors with $\partial_\alpha \partial_\beta n_\gamma$ and using Gauss's theorem, the same three distinct invariants are obtained as under (ii).

Summarizing, four basic invariants appear in the expression for the Frank free energy density, namely:

$$\varepsilon_{\alpha\beta\gamma} n_\gamma \partial_\alpha n_\beta \quad ; \quad n_\alpha n_\beta (\partial_\alpha n_\gamma)(\partial_\beta n_\gamma) \quad ; \quad (\partial_\alpha n_\alpha)(\partial_\beta n_\beta) \quad ;$$
$$(\partial_\alpha n_\beta)(\partial_\alpha n_\beta) \quad .$$

This means that the general expression for the Frank free energy density is given by

$$\begin{aligned} f_\mathrm{d}(\boldsymbol{r}) = & \, k \varepsilon_{\alpha\beta\gamma} n_\gamma \partial_\alpha n_\beta + k_1 (\partial_\alpha n_\alpha)(\partial_\beta n_\beta) + k_2 (\partial_\alpha n_\beta)(\partial_\alpha n_\beta) \\ & + k_3 n_\alpha n_\beta (\partial_\alpha n_\gamma)(\partial_\beta n_\gamma) \quad , \end{aligned} \tag{5.23}$$

where the coefficients k, k_1, k_2 and k_3 are certain functions of the order parameter Q_1.

Now it is customary to express the Frank free energy density in terms of a partly different set of invariants, namely $\operatorname{div} \boldsymbol{n}(\boldsymbol{r})$, $\boldsymbol{n}(\boldsymbol{r}) \cdot \operatorname{curl} \boldsymbol{n}(\boldsymbol{r})$ and $[\boldsymbol{n}(\boldsymbol{r}) \times \operatorname{curl} \boldsymbol{n}(\boldsymbol{r})]^2 = [\operatorname{curl} \boldsymbol{n}(\boldsymbol{r})]^2 - [\boldsymbol{n}(\boldsymbol{r}) \cdot \operatorname{curl} \boldsymbol{n}(\boldsymbol{r})]^2$. Neglecting surface terms it follows that

$$\varepsilon_{\alpha\beta\gamma} n_\gamma \partial_\alpha n_\beta = \boldsymbol{n} \cdot \operatorname{curl} \boldsymbol{n} \quad ,$$

$$\partial_\alpha n_\alpha = \operatorname{div} \boldsymbol{n} \quad ,$$

$$n_\alpha n_\beta (\partial_\alpha n_\gamma)(\partial_\beta n_\gamma) = (\operatorname{curl} \boldsymbol{n})^2 - (\boldsymbol{n} \cdot \operatorname{curl} \boldsymbol{n})^2 = (\boldsymbol{n} \times \operatorname{curl} \boldsymbol{n})^2 \quad ,$$

$$\begin{aligned} (\partial_\alpha n_\beta)(\partial_\alpha n_\beta) & = (\operatorname{div} \boldsymbol{n})^2 + (\operatorname{curl} \boldsymbol{n})^2 \\ & = (\operatorname{div} \boldsymbol{n})^2 + (\boldsymbol{n} \cdot \operatorname{curl} \boldsymbol{n})^2 + (\boldsymbol{n} \times \operatorname{curl} \boldsymbol{n})^2 \quad . \end{aligned}$$

Concluding, the Frank free energy density becomes

$$\begin{aligned} f_\mathrm{d}(\boldsymbol{r}) = & \, k \boldsymbol{n}(\boldsymbol{r}) \cdot \operatorname{curl} \boldsymbol{n}(\boldsymbol{r}) + \tfrac{1}{2} K_1 [\operatorname{div} \boldsymbol{n}(\boldsymbol{r})]^2 + \tfrac{1}{2} K_2 [\boldsymbol{n}(\boldsymbol{r}) \cdot \operatorname{curl} \boldsymbol{n}(\boldsymbol{r})]^2 \\ & + \tfrac{1}{2} K_3 [\boldsymbol{n}(\boldsymbol{r}) \times \operatorname{curl} \boldsymbol{n}(\boldsymbol{r})]^2 \quad , \end{aligned} \tag{5.24}$$

where $K_1 = 2(k_1 + k_2)$, $K_2 = 2k_2$ and $K_3 = 2(k_3 + k_2)$. The constants K_1, K_2 and K_3 are called the elastic constants for splay, twist and bend respectively.

The three basic deformations associated with K_1, K_2 and K_3 are shown in Fig. 5.2. The experimentally observed stability of a uniform director pat-

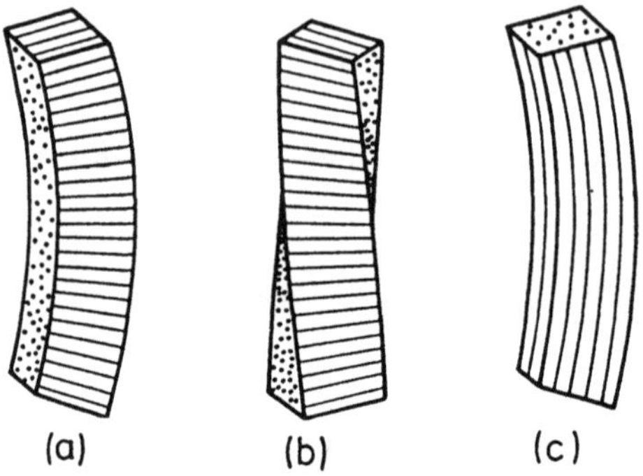

Fig. 5.2a–c. The three basic deformations of a uniaxial nematic: **(a)** splay, **(b)** twist, **(c)** bend

tern requires that K_1, K_2 and K_3 are all positive. The elastic constants have values of the order of 10^{-11} newton. The most reliable values for the elastic constants are derived from experiments involving the competing effects of field alignment and wall alignment on the sample. These so-called Frederiks-transitions will be considered in Chap. 6, where some typical values of the elastic constants for different compounds will also be discussed.

Some attention should be paid to the term $k\boldsymbol{n}(\boldsymbol{r})\cdot\operatorname{curl}\boldsymbol{n}(\boldsymbol{r})$ in the expression for the Frank free energy density. This term only contributes in the case of chiral nematics. The usual nematics do not contain this term, i.e. $k=0$, because their undistorted state is just the state of uniform alignment. The chiral nematic behaviour originates from the chirality of the molecules meaning that the intermolecular interaction contains an additional twist-producing term (see Sect. 13.5). Combining the term in question with the twist term gives:

$$\begin{aligned} &kn(\boldsymbol{r})\cdot\operatorname{curl}\boldsymbol{n}(\boldsymbol{r})+\tfrac{1}{2}K_2[\boldsymbol{n}(\boldsymbol{r})\cdot\operatorname{curl}\boldsymbol{n}(\boldsymbol{r})]^2 \\ &\quad=\tfrac{1}{2}K_2[\boldsymbol{n}(\boldsymbol{r})\cdot\operatorname{curl}\boldsymbol{n}(\boldsymbol{r})+t_0]^2-\tfrac{1}{2}K_2t_0^2 \end{aligned} \tag{5.25}$$

with $t_0=k/K_2$. Consequently a position-dependent director field now has a lower free energy than the situation of uniform parallel alignment. This position-dependent director field appears to be a helix. Calling the direction of the helix wave vector $\boldsymbol{q}_0$ the z axis, the director field is given by

$$n_x=\cos(q_0z)\quad;\quad n_y=\sin(q_0z)\quad;\quad n_z=0\quad; \tag{5.26}$$

where $q_0=|\boldsymbol{q}_0|$ denotes the magnitude of the helix wave vector. Such a director field gives rise to $[\operatorname{div}\boldsymbol{n}(\boldsymbol{r})]^2=[\boldsymbol{n}(\boldsymbol{r})\times\operatorname{curl}\boldsymbol{n}(\boldsymbol{r})]^2=0$ and $\boldsymbol{n}(\boldsymbol{r})\cdot\operatorname{curl}\boldsymbol{n}(\boldsymbol{r})=-q_0$. The Frank free energy density belonging to this director field is

$$f_{\rm d}(\boldsymbol{r})=\tfrac{1}{2}K_2(q_0-t_0)^2-\tfrac{1}{2}K_2t_0^2\quad. \tag{5.27}$$

This expression is minimal for $q_0 = t_0$, i.e. a chiral nematic with a pitch $p_0 = 2\pi/t_0$ is the equilibrium state. In this description a positive pitch gives rise to a right-handed helix. In agreement with the experimental description given in Sect. 2.2 the continuum theory describes the chiral nematic state as a twisted nematic phase.

In order to describe the smectic phase in terms of a continuum theory, an additional term must be added to the Frank distortion free energy density (5.24). The appearance of this term is due to the partly solid-like character of the smectic state and arises from positional deformations. For reasons of simplicity only the continuum theory of the smectic A phase will be discussed here. Before dealing with smectic A, however, the elastic energy of a solid with the corresponding symmetry will be given.

According to the theory of elasticity, see e.g. Landau and Lifshitz (1959a), the general form of the free energy density of a deformed crystal is

$$f_{\mathrm{el}}(\boldsymbol{r}) = \tfrac{1}{2}\lambda_{\alpha\beta\gamma\delta}u_{\alpha\beta}u_{\gamma\delta} \quad , \tag{5.28}$$

where $\lambda_{\alpha\beta\gamma\delta}$ is a tensor of rank four, called the elastic modulus tensor, satisfying $\lambda_{\alpha\beta\gamma\delta} = \lambda_{\gamma\delta\alpha\beta}$ and $u_{\alpha\beta}$ is the strain tensor. For small deformations the strain tensor is given by

$$u_{\alpha\beta} = \tfrac{1}{2}(\partial_\alpha u_\beta + \partial_\beta u_\alpha) \quad , \tag{5.29}$$

where the vector field $\boldsymbol{u}(\boldsymbol{r})$ describes the displacement of a point having a position vector $\boldsymbol{r}$ before the deformation. Since the strain tensor is symmetric, the number of different components of the elastic modulus tensor is in general 21 as follows directly from

$$\lambda_{\alpha\beta\gamma\delta} = \lambda_{\beta\alpha\gamma\delta} = \lambda_{\alpha\beta\delta\gamma} = \lambda_{\beta\alpha\delta\gamma} \quad . \tag{5.30}$$

If the solid possesses a uniaxial symmetry axis and a mirror plane perpendicular to this axis the number of independent components is reduced to 5. Calling the unit vector in the direction of the uniaxial axis $\boldsymbol{a}$ and two unit vectors perpendicular to the uniaxial axis $\boldsymbol{b}$ and $\boldsymbol{c}$ respectively, where $b_\alpha c_\alpha = 0$ and $a_\alpha = \varepsilon_{\alpha\beta\gamma}b_\beta c_\gamma$, the elastic free energy density can be written as

$$\begin{aligned} f_{\mathrm{el}}(\boldsymbol{r}) = [&\tfrac{1}{2}\lambda_1 A_{\alpha\beta}A_{\gamma\delta} + \tfrac{1}{2}\lambda_2 B_{\alpha\beta}B_{\gamma\delta} + \tfrac{1}{4}\lambda_3(A_{\alpha\beta}B_{\gamma\delta} + B_{\alpha\beta}A_{\gamma\delta}) \\ &+ \tfrac{1}{4}\lambda_4(B_{\alpha\gamma}B_{\beta\delta} + B_{\alpha\delta}B_{\beta\gamma}) \\ &+ \tfrac{1}{8}\lambda_5(A_{\alpha\gamma}B_{\beta\delta} + A_{\alpha\delta}B_{\beta\gamma} + B_{\alpha\gamma}A_{\beta\delta} + B_{\alpha\delta}A_{\beta\gamma})]u_{\alpha\beta}u_{\gamma\delta} \quad , \end{aligned} \tag{5.31}$$

where the tensor components $A_{\alpha\beta}$ and $B_{\alpha\beta}$ are given by

$$A_{\alpha\beta} = a_\alpha a_\beta \quad ; \quad B_{\alpha\beta} = \tfrac{1}{2}(b_\alpha b_\beta + c_\alpha c_\beta) \quad .$$

Thus the general expression for the elastic free energy density in a solid with the symmetry mentioned contains five moduli of elasticity. Now the elastic free energy density of a smectic A follows directly by incorporating into (5.31) the fluid character in planes perpendicular to the uniaxial axis. This means that the material does not oppose any elastic deformation in directions perpendicular to this axis, i.e. the corresponding elastic constants must be zero, $\lambda_2 = \lambda_3 = \lambda_4 = \lambda_5 = 0$. Thus the elastic free energy density of a smectic A is

$$f_{\text{el}}(\boldsymbol{r}) = \tfrac{1}{2}\lambda_1 A_{\alpha\beta} A_{\gamma\delta} u_{\alpha\beta} u_{\gamma\delta} \quad . \tag{5.32}$$

The uniaxial symmetry of the smectic A is due to the presence of a second rank orientational tensor field $\tilde{\boldsymbol{N}}$ with components $N_{\alpha\beta} = n_\alpha n_\beta$, where the director $\boldsymbol{n}$ is parallel or antiparallel to $\boldsymbol{a}$. The orientational deformations are coupled to the positional deformations by correlating the displacement vector $\boldsymbol{u}(\boldsymbol{r})$ with the local director $\boldsymbol{n}(\boldsymbol{r})$. This coupling is brought about mathematically by the constraint that the director must remain perpendicular to the distorted liquid layers. In order to derive the relationship between $\boldsymbol{n}(\boldsymbol{r})$ and $\boldsymbol{u}(\boldsymbol{r})$ three neighbouring points are considered, whose position vectors before the deformation are given by respectively $\boldsymbol{r}$, $\boldsymbol{r} + s\boldsymbol{b}$, and $\boldsymbol{r} + t\boldsymbol{c}$, where s and t are small quantities. Clearly these points span a small surface element with area st whose normal is parallel to $\boldsymbol{n}$. After the deformation the points in question are situated at $\boldsymbol{r}+\boldsymbol{u}(\boldsymbol{r})$, $\boldsymbol{r}+s\boldsymbol{b}+\boldsymbol{u}(\boldsymbol{r}+s\boldsymbol{b})$ and $\boldsymbol{r} + t\boldsymbol{c} + \boldsymbol{u}(\boldsymbol{r} + t\boldsymbol{c})$ respectively. Because of the smallness of s and t the corresponding surface element is spanned by the vectors

$$\begin{aligned} r_\alpha + sb_\alpha + u_\alpha(\boldsymbol{r} + s\boldsymbol{b}) - [r_\alpha + u_\alpha(\boldsymbol{r})] &= sb_\alpha + sb_\beta \partial_\beta u_\alpha \quad , \\ r_\alpha + tc_\alpha + u_\alpha(\boldsymbol{r} + t\boldsymbol{c}) - [r_\alpha + u_\alpha(\boldsymbol{r})] &= tc_\alpha + tc_\beta \partial_\beta u_\alpha \quad . \end{aligned}$$

The surface element $d\boldsymbol{S}$, which is given by the outer product of these two vectors, up to linear terms in the first-order derivatives of $\boldsymbol{u}$ is

$$\begin{aligned} dS_\alpha &= \varepsilon_{\alpha\beta\gamma}(sb_\beta + sb_\mu \partial_\mu u_\beta)(tc_\gamma + tc_\nu \partial_\nu u_\gamma) \\ &= st(a_\alpha + \varepsilon_{\alpha\beta\gamma} b_\beta c_\mu \partial_\mu u_\gamma + \varepsilon_{\alpha\beta\gamma} c_\gamma b_\mu \partial_\mu u_\beta) \quad . \end{aligned}$$

Consequently the local director is given by

$$\begin{aligned} n_\alpha(\boldsymbol{r}) = a_\alpha &+ \varepsilon_{\alpha\beta\gamma}(b_\beta c_\mu \partial_\mu u_\gamma + c_\gamma b_\mu \partial_\mu u_\beta) \\ &- a_\alpha c_\beta c_\gamma \partial_\gamma u_\beta - a_\alpha b_\beta b_\gamma \partial_\gamma u_\beta \quad . \end{aligned} \tag{5.33}$$

In order to facilitate the calculations, a particular coordinate system is chosen with the x, y and z directions coinciding with the $\boldsymbol{b}$, $\boldsymbol{c}$ and $\boldsymbol{a}$ axes respectively. In this representation the local director becomes according to (5.33):

$$n_x = -\partial_x u_z \quad ; \quad n_y = -\partial_y u_z \quad ; \quad n_z = 1 \quad .$$

This means that, up to quadratic terms in the displacement vector $\boldsymbol{u}$, the Frank free energy density is given by

$$f_{\mathrm{d}}(\boldsymbol{r}) = \tfrac{1}{2}K_1(\partial_x^2 u_z + \partial_y^2 u_z)^2 + \tfrac{1}{2}K_3[(\partial_x\partial_z u_z)^2 + (\partial_y\partial_z u_z)^2] \quad . \tag{5.34}$$

To this term the elastic free energy density (5.32) must be added. In the coordinate system chosen, this is expressed by

$$f_{\mathrm{el}}(\boldsymbol{r}) = \tfrac{1}{2}\lambda_1(\partial_z u_z)^2 \quad . \tag{5.35}$$

It is worthwhile to remark here that the expression (5.34) can be obtained by imposing the constraint that $\boldsymbol{n}\cdot\mathrm{curl}\,\boldsymbol{n}$ must remain zero in the distorted state. This can be equivalently expressed by stating that the twist constant K_2 must diverge in the smectic A phase. Neglecting the terms $\partial_x\partial_z u_z$ and $\partial_y\partial_z u_z$ with respect to $\partial_z u_z$, i.e. assuming K_3 to diverge as well, the total distortion free energy density of a smectic A is approximately given by

$$f(\boldsymbol{r}) = f_{\mathrm{el}}(\boldsymbol{r}) + f_{\mathrm{d}}(\boldsymbol{r}) = \tfrac{1}{2}\lambda_1(\partial_z u_z)^2 + \tfrac{1}{2}K_1(\partial_x^2 u_z + \partial_y^2 u_z)^2 \quad . \tag{5.36}$$

Hence the elastic properties of a smectic A are essentially determined by the two constants λ_1 and K_1, where $(K_1/\lambda_1)^{1/2}$ has the dimension of a length, which is comparable to the molecular length in a number of cases. Note that expression (5.36) does not contain derivatives of u_x and u_y; this is due to the symmetry of the smectic A phase and the liquid character of the smectic layers.

According to Oseen (1933) the only distortions, that must be taken into account in a first-order approximation for a description of the smectic A state, are specific undulations of the smectic layers. These undulations are required to be such that the interlayer distance is kept constant and the director remains normal to the layer. In the present formulation the Oseen assumption boils down to also requiring $\partial_z u_z = 0$. Consequently the Oseen description of the smectic A state imposes the following constraint on the allowed distortions

$$\mathrm{curl}\,\boldsymbol{n}(\boldsymbol{r}) = 0$$

as follows directly from (5.34). An equivalent way of expressing this constraint is to state that the elastic constants K_2 and K_3 of a smectic A are large enough to prevent twist and bend distortions. From the above discussion it is clear that this is rigorously true for K_2 and to a very good approximation for K_3.

5.4 General Properties of the Director Field

The experimentally determined sets of elastic constants appear to differ from nematic to nematic. Consequently every liquid crystal has its own descriptive director field, whose free energy density originates from the intermolecular interactions. The derivation of this director field from first principles is a prohibitively difficult problem. However, it is quite possible to determine the elastic constants of a given director field. Such a calculation fits naturally into the macroscopic theory of liquid crystals.

Each director field $\tilde{\boldsymbol{N}}(\boldsymbol{r})$ is described by its corresponding free energy density. Here only director fields with free energy densities based upon an interaction between the components of the tensors at the positions $\boldsymbol{r}$ and $\boldsymbol{r}+\boldsymbol{\varrho}$ are considered. As will be discussed later on, such a restriction is related to the assumption that the relevant intermolecular interactions in nematics are two-body interactions. The general form of the free energy density is obtained as follows. Consider the director field at the positions $\boldsymbol{r}$ and $\boldsymbol{r}+\boldsymbol{\varrho}$, where in general $\tilde{\boldsymbol{N}}(\boldsymbol{r}) \neq \tilde{\boldsymbol{N}}(\boldsymbol{r}+\boldsymbol{\varrho})$. Next, define the unit vector $\boldsymbol{u}=\boldsymbol{\varrho}/\varrho$ with $\varrho=|\boldsymbol{\varrho}|$. The free energy density can now be constructed with the aid of the three tensors $\tilde{\boldsymbol{N}}(\boldsymbol{r})$, $\tilde{\boldsymbol{N}}(\boldsymbol{r}+\boldsymbol{\varrho})$ and $\boldsymbol{u}$. It immediately follows that these three tensors give rise to the following four different rotational invariants

$$
\begin{array}{ll}
N_{\alpha\beta}(\boldsymbol{r})u_\alpha u_\beta \quad ; & N_{\alpha\beta}(\boldsymbol{r}+\boldsymbol{\varrho})u_\alpha u_\beta \quad ; \\
N_{\alpha\beta}(\boldsymbol{r})N_{\alpha\beta}(\boldsymbol{r}+\boldsymbol{\varrho}) \quad ; & N_{\alpha\beta}(\boldsymbol{r})N_{\beta\gamma}(\boldsymbol{r}+\boldsymbol{\varrho})u_\alpha u_\gamma \quad .
\end{array}
$$

Clearly the free energy density can be expanded in terms of these invariants. Using $N_{\alpha\beta}=n_\alpha n_\beta$, however, only the following three simple invariants appear to be relevant for the construction of the general term of the free energy density, namely

$$
\boldsymbol{n}(\boldsymbol{r})\cdot\boldsymbol{u} \quad , \quad \boldsymbol{n}(\boldsymbol{r}+\boldsymbol{\varrho})\cdot\boldsymbol{u} \quad , \quad \boldsymbol{n}(\boldsymbol{r})\cdot\boldsymbol{n}(\boldsymbol{r}+\boldsymbol{\varrho}) \quad .
$$

On the basis of these three invariants the general term of the polynomial expansion of the interaction energy between the local tensors $\tilde{\boldsymbol{N}}(\boldsymbol{r})$ and $\tilde{\boldsymbol{N}}(\boldsymbol{r}+\boldsymbol{\varrho})$ is

$$
-J^{abc}(\varrho)[\boldsymbol{n}(\boldsymbol{r})\cdot\boldsymbol{u}]^a[\boldsymbol{n}(\boldsymbol{r}+\boldsymbol{\varrho})\cdot\boldsymbol{u}]^b[\boldsymbol{n}(\boldsymbol{r})\cdot\boldsymbol{n}(\boldsymbol{r}+\boldsymbol{\varrho})]^c \quad ,
$$

where a, b and c are either zero or a natural number, while $a+c$ and $b+c$ must be even because of the head-tail symmetry. Furthermore, $J^{abc}(\varrho)$ is a coupling constant depending on the distance ϱ between the positions considered. Consequently the general expression for the free energy density is given by

$$f(\boldsymbol{r}) = -\frac{1}{2}\sum_{a,b,c}\int d\boldsymbol{\varrho}\, J^{abc}(\varrho)[\boldsymbol{n}(\boldsymbol{r})\cdot\boldsymbol{u}]^a[\boldsymbol{n}(\boldsymbol{r}+\boldsymbol{\varrho})\cdot\boldsymbol{u}]^b \times [\boldsymbol{n}(\boldsymbol{r})\cdot\boldsymbol{n}(\boldsymbol{r}+\boldsymbol{\varrho})]^c \quad , \tag{5.37}$$

where $\int d\boldsymbol{\varrho}$ denotes the integral over the volume of the system. The free energy density of the state with uniform parallel alignment, f_0, is obtained by simply putting $\boldsymbol{n}(\boldsymbol{r}) = \boldsymbol{n}(\boldsymbol{r}+\boldsymbol{\varrho}) = \boldsymbol{n}$. Changing to polar coordinates, i.e. $d\boldsymbol{\varrho} = \varrho^2 d\varrho\, d^2u$, where d^2u denotes the solid angle, the free energy density of the unperturbed state can be written as

$$f_0 = -\frac{1}{2}\sum_{a,b,c}\int_0^\infty d\varrho\,\varrho^2 J^{abc}(\varrho)\int d^2u(\boldsymbol{n}\cdot\boldsymbol{u})^{a+b} \quad . \tag{5.38}$$

The integration over the solid angle can be performed without any difficulty. Choosing the z axis along $\boldsymbol{n}$ it follows that

$$\int d^2u(\boldsymbol{n}\cdot\boldsymbol{u})^{a+b} = 2\pi\int_0^\pi d\theta\,\sin\theta(\cos\theta)^{a+b} = 4\pi/(a+b+1)$$

or

$$f_0 = -2\pi\sum_{a,b,c}(a+b+1)^{-1}\int_0^\infty d\varrho\,\varrho^2 J^{abc}(\varrho) \quad . \tag{5.39}$$

The elastic constants are obtained by expanding the free energy density (5.37) around the state of uniform parallel alignment. Using

$$n_\alpha(\boldsymbol{r}+\boldsymbol{\varrho}) = n_\alpha(\boldsymbol{r}) + \varrho_\beta\partial_\beta n_\alpha(\boldsymbol{r}) + \tfrac{1}{2}\varrho_\beta\varrho_\gamma\partial_\beta\partial_\gamma n_\alpha(\boldsymbol{r})$$

the Frank free energy density is given by

$$\begin{aligned} f_{\mathrm{d}}(\boldsymbol{r}) = &-\frac{1}{2}\sum_{a,b,c}J(a,b,c)\frac{1}{4\pi}\int d^2u[\boldsymbol{n}(\boldsymbol{r})\cdot\boldsymbol{u}]^a \\ &\times\left\{\frac{1}{2}b[\boldsymbol{n}(\boldsymbol{r})\cdot\boldsymbol{u}]^{b-1}u_\alpha u_\beta u_\gamma\partial_\beta\partial_\gamma n_\alpha(\boldsymbol{r})\right. \\ &+\frac{1}{2}b(b-1)[\boldsymbol{n}(\boldsymbol{r})\cdot\boldsymbol{u}]^{b-2}u_\alpha u_\beta u_\gamma u_\delta[\partial_\alpha n_\beta(\boldsymbol{r})][\partial_\gamma n_\delta(\boldsymbol{r})] \\ &\left.+\frac{1}{2}c[\boldsymbol{n}(\boldsymbol{r})\cdot\boldsymbol{u}]^b n_\alpha(\boldsymbol{r})u_\beta u_\gamma\partial_\beta\partial_\gamma n_\alpha(\boldsymbol{r})\right\} \end{aligned} \tag{5.40}$$

with

$$J(a,b,c) = 4\pi \int_0^\infty d\varrho\, \varrho^4 J^{abc}(\varrho) \quad .$$

The easiest way to extract the elastic constants from expression (5.40) is to choose the director $\boldsymbol{n}(\boldsymbol{r})$ along the z axis, i.e. $n_z^2(\boldsymbol{r}) = 1$ and $\partial_\alpha n_z(\boldsymbol{r}) = 0$. Next the resulting expression is compared with the corresponding expression of the Frank free energy density in terms of the elastic constants (5.24). For details of this calculation the reader is referred to the appendix at the end of this section. The elastic constants for splay, twist and bend respectively are given by

$$K_1 = \frac{1}{2} \sum_{a,b,c} \frac{J(a,b,c)}{(a+b+1)(a+b+3)} \left(\frac{3ab}{a+b-1} + c \right) \quad , \tag{5.41a}$$

$$K_2 = \frac{1}{2} \sum_{a,b,c} \frac{J(a,b,c)}{(a+b+1)(a+b+3)} \left(\frac{ab}{a+b-1} + c \right) \quad , \tag{5.41b}$$

$$K_3 = \frac{1}{2} \sum_{a,b,c} \frac{J(a,b,c)}{(a+b+3)} \left(\frac{ab}{a+b+1} + c \right) \quad . \tag{5.41c}$$

This means that director fields whose free energy density (5.37) does not contain any terms of the type $[\boldsymbol{n}(\boldsymbol{r}) \cdot \boldsymbol{n}(\boldsymbol{r}+\boldsymbol{\varrho})]^c$, always have a fixed ratio between the elastic constants of splay and twist, namely $K_1/K_2 = 3$. Clearly a knowledge of the isotropic functions $J^{abc}(\varrho)$ suffices in order to calculate the elastic constants. The anisotropic terms can be dealt with completely as shown in the expressions (5.41). The isotropic functions $J^{abc}(\varrho)$ depend on the form and strength of the intermolecular interactions and the temperature. Needless to say, a given set of elastic constants can be reproduced by an infinite number of different free energy densities.

As examples, the elastic constants will be given for to two different free energy densities. The choice of these free energies was motivated by their relationship to certain molecular models (see Part IV). The first density has the form

$$f(\boldsymbol{r}) = -\tfrac{1}{2} \int d\boldsymbol{\varrho}\, J(\varrho) [\boldsymbol{n}(\boldsymbol{r}) \cdot \boldsymbol{n}(\boldsymbol{r}+\boldsymbol{\varrho})]^2 \quad , \tag{5.42}$$

i.e. $a = b = 0$, $c = 2$. Consequently the elastic constants are given by

$$K_1 = K_2 = K_3 = (4\pi/3) \int_0^\infty d\varrho\, \varrho^4 J(\varrho) \quad . \tag{5.43}$$

The second example concerns a free energy density closely related to the induced dipole-dipole interaction:

$$\begin{aligned} f(\boldsymbol{r}) = & -\tfrac{1}{2}\int d\boldsymbol{\varrho}\, J(\varrho)\{[\boldsymbol{n}(\boldsymbol{r})\cdot\boldsymbol{n}(\boldsymbol{r}+\boldsymbol{\varrho})]^2 \\ & - 6[\boldsymbol{n}(\boldsymbol{r})\cdot\boldsymbol{u}][\boldsymbol{n}(\boldsymbol{r}+\boldsymbol{\varrho})\cdot\boldsymbol{u}][\boldsymbol{n}(\boldsymbol{r})\cdot\boldsymbol{n}(\boldsymbol{r}+\boldsymbol{\varrho})] \\ & + 9[\boldsymbol{n}(\boldsymbol{r})\cdot\boldsymbol{u}]^2[\boldsymbol{n}(\boldsymbol{r}+\boldsymbol{\varrho})\cdot\boldsymbol{u}]^2\} \end{aligned} \tag{5.44}$$

leading to

$$J^{abc}(\varrho) = J(\varrho)[\delta_{a0}\delta_{b0}\delta_{c2} - 6\delta_{a1}\delta_{b1}\delta_{c1} + 9\delta_{a2}\delta_{b2}\delta_{c0}] \quad . \tag{5.45}$$

Substitution of (5.45) into (5.41) yields

$$K_1 = K_3 = (5/11)K_2 = (4\pi/21)\int_0^\infty d\varrho\, \varrho^4 J(\varrho) \quad , \tag{5.46}$$

i.e. $K_1 : K_2 : K_3 = 5 : 11 : 5$ (Nehring and Saupe 1972).

In order to describe the chiral nematic phase, terms producing twist must be incorporated into the free energy density. In that case the term $k\boldsymbol{n}(\boldsymbol{r})\cdot \mathrm{curl}\, \boldsymbol{n}(\boldsymbol{r})$ appears in the expression for the Frank free energy density. The corresponding general term that must be added to the free energy density (5.37) is given by

$$\begin{aligned} -\frac{1}{2}\sum_{a,b,c}\int d\boldsymbol{\varrho}\;\; & G^{abc}(\varrho)[\boldsymbol{n}(\boldsymbol{r})\cdot\boldsymbol{u}]^a[\boldsymbol{n}(\boldsymbol{r}+\boldsymbol{\varrho})\cdot\boldsymbol{u}]^b \\ & \times [\boldsymbol{n}(\boldsymbol{r})\cdot\boldsymbol{n}(\boldsymbol{r}+\boldsymbol{\varrho})]^c[\boldsymbol{n}(\boldsymbol{r})\times\boldsymbol{n}(\boldsymbol{r}+\boldsymbol{\varrho})\cdot\boldsymbol{u}] \quad , \end{aligned} \tag{5.47}$$

where a, b and c run over zero and all natural numbers but such that $a+c$ and $b+c$ are odd. It is easily verified that this term does not contribute to f_0. Likewise the contribution to the elastic constants K_1, K_2 and K_3 is zero because of an integration over an odd number of components u_α. The constant k is determined by the term

$$(8\pi)^{-1}\sum_{a,b,c} G(a,b,c)\int d^2u[\boldsymbol{n}(\boldsymbol{r})\cdot\boldsymbol{u}]^{a+b}u_\alpha\varepsilon_{\alpha\beta\gamma}n_\beta(\boldsymbol{r})u_\delta\partial_\delta n_\gamma(\boldsymbol{r}) \tag{5.48}$$

with

$$G(a,b,c) = 4\pi\int_0^\infty d\varrho\, \varrho^3 G^{abc}(\varrho) \quad . \tag{5.49}$$

The expression (5.48) is obtained by expanding $\boldsymbol{n}(\boldsymbol{r}+\boldsymbol{\varrho})$. The easiest way to calculate k is by choosing $\boldsymbol{n}(\boldsymbol{r})$ along the z axis. Expression (5.48) then becomes

$$-\frac{1}{2}\sum_{a,b,c}\frac{G(a,b,c)}{(a+b+1)(a+b+3)}[\partial_y n_x(\boldsymbol{r})-\partial_x n_y(\boldsymbol{r})] \quad .$$

The corresponding term of the Frank free energy density (5.24) is given by

$$-k[\partial_y n_x(\boldsymbol{r})-\partial_x n_y(\boldsymbol{r})] \quad .$$

Consequently it holds that

$$k=\frac{1}{2}\sum_{a,b,c}\frac{G(a,b,c)}{(a+b+1)(a+b+3)} \quad . \tag{5.50}$$

A knowledge of the isotropic functions $G^{abc}(\varrho)$ clearly suffices in order to calculate the pitch of the helix.

5.5 Appendix

As already mentioned above (5.41) the integration over the solid angle can be performed quite easily by choosing the z axis of the coordinate system along $\boldsymbol{n}$. The following integrals are relevant

$$(4\pi)^{-1}\int d^2u\, u_z^{a+b+2}=(a+b+3)^{-1} \quad ,$$

$$\begin{aligned}(4\pi)^{-1}\int d^2u\, u_z^{a+b}u_x^2 &=(4\pi)^{-1}\int d^2u\, u_z^{a+b}u_y^2\\ &=(a+b+1)^{-1}(a+b+3)^{-1} \quad ,\end{aligned}$$

$$\begin{aligned}(4\pi)^{-1}\int d^2u\, u_z^{a+b-2}u_x^4 &=(4\pi)^{-1}\int d^2u\, u_z^{a+b-2}u_y^4\\ &=3(a+b-1)^{-1}(a+b+1)^{-1}(a+b+3)^{-1} \quad ,\end{aligned}$$

$$(4\pi)^{-1}\int d^2u\, u_z^{a+b-2}u_x^2u_y^2=(a+b-1)^{-1}(a+b+1)^{-1}(a+b+3)^{-1} \quad .$$

Consequently the contributions of the terms, that appear successively in the expression for the Frank free energy density (5.40), are

$$(4\pi)^{-1}\int d^2u\,u_z^{a+b-1}u_\alpha u_\beta u_\gamma \partial_\beta \partial_\gamma n_\alpha = (a+b+3)^{-1}\partial_z\partial_z n_z(\boldsymbol{r})$$
$$+(a+b+1)^{-1}(a+b+3)^{-1}[\partial_x\partial_x n_z(\boldsymbol{r}) + \partial_y\partial_y n_z(\boldsymbol{r})$$
$$+2\partial_x\partial_z n_x(\boldsymbol{r}) + 2\partial_y\partial_z n_y(\boldsymbol{r})] \quad ,$$

$$(4\pi)^{-1}\int d^2u\,u_z^{a+b-2}u_\alpha u_\beta u_\gamma u_\delta[\partial_\alpha n_\beta(\boldsymbol{r})][\partial_\gamma n_\delta(\boldsymbol{r})]$$
$$=(a+b+1)^{-1}(a+b+3)^{-1}\{[\partial_z n_x(\boldsymbol{r})]^2 + [\partial_z n_y(\boldsymbol{r})]^2\}$$
$$+(a+b-1)^{-1}(a+b+1)^{-1}(a+b+3)^{-1}\{3[\partial_x n_x(\boldsymbol{r})]^2$$
$$+3[\partial_y n_y(\boldsymbol{r})]^2 + [\partial_y n_x(\boldsymbol{r})]^2 + [\partial_x n_y(\boldsymbol{r})]^2$$
$$+2[\partial_x n_x(\boldsymbol{r})][\partial_y n_y(\boldsymbol{r})] + 2[\partial_x n_y(\boldsymbol{r})][\partial_y n_x(\boldsymbol{r})]\} \quad ,$$

$$(4\pi)^{-1}\int d^2u\,u_z^{a+b}u_\beta u_\gamma \partial_\beta\partial_\gamma n_z(\boldsymbol{r}) = (a+b+3)^{-1}\partial_z\partial_z n_z(\boldsymbol{r})$$
$$+(a+b+1)^{-1}(a+b+3)^{-1}[\partial_x\partial_x n_z(\boldsymbol{r}) + \partial_y\partial_y n_z(\boldsymbol{r})] \quad .$$

Making use of the fact that surface terms are irrelevant the following relations are derived:

$$\begin{aligned}
\partial_z\partial_x n_x + \partial_z\partial_y n_y + \partial_z\partial_z n_z &= -(\partial_x n_x)^2 - (\partial_y n_y)^2 - 2(\partial_x n_x)(\partial_y n_y)\\
&= -(\partial_x n_x)^2 - (\partial_y n_y)^2 - 2(\partial_y n_x)(\partial_x n_y) \quad ,\\
\partial_x\partial_x n_z + \partial_y\partial_y n_z + \partial_z\partial_z n_z &= -(\partial_x n_x)^2 - (\partial_y n_x)^2 - (\partial_z n_x)^2\\
&\quad - (\partial_x n_y)^2 - (\partial_y n_y)^2 - (\partial_z n_y)^2 \quad ,\\
\partial_z\partial_z n_z &= -(\partial_z n_x)^2 - (\partial_z n_y)^2 \quad .
\end{aligned}$$

Rewriting the Frank free energy density (5.24) in terms of second-order derivatives of the director field results, in the present coordinate system, in

$$\begin{aligned}
f_{\mathrm{d}}(\boldsymbol{r}) = &-\tfrac{1}{2}(K_1 - K_2)[\partial_z\partial_x n_x(\boldsymbol{r}) + \partial_z\partial_y n_y(\boldsymbol{r})]\\
&-\tfrac{1}{2}K_2[\partial_x\partial_x n_z(\boldsymbol{r}) + \partial_y\partial_y n_z(\boldsymbol{r})]\\
&-\tfrac{1}{2}(K_1 + K_3 - K_2)\partial_z\partial_z n_z(\boldsymbol{r}) \quad .
\end{aligned}$$

Finally a comparison between the coefficients in front of the second-order derivatives, as given by the original Frank free energy density and the general expression of the free energy (5.40), results in the following relations

$$K_1 - K_2 = \sum_{a,b,c}\frac{J(a,b,c)b}{(a+b+1)(a+b+3)}\left[1 - \frac{b-1}{(a+b-1)}\right] \quad ,$$

$$K_2 = \frac{1}{2}\sum_{a,b,c}\frac{J(a,b,c)}{(a+b+1)(a+b+3)}\left[b + c - \frac{b(b-1)}{(a+b-1)}\right] \quad ,$$

$$K_1 + K_3 - K_2 = \frac{1}{2} \sum_{a,b,c} \frac{J(a,b,c)}{(a+b+3)}$$
$$\times \left[b + c - \frac{2b(b-1)}{(a+b-1)(a+b+1)} - \frac{b(b-1)}{(a+b+1)} \right] .$$

After some simple algebra the expressions (5.41) are obtained.

6. Effects of External Fields

This chapter deals with the effects of external fields on the orientation of a nematic liquid crystal. First we consider the interaction between an applied magnetic or electric field and the director field. In this context some attention is also paid to the flexoelectric effect, whose origin is discussed briefly. Next we consider the important and technologically highly relevant Frederiks transition, which results from competing orienting effects of the boundaries of a given nematic layer and the applied field. At the transition a distortion of the director field sets in. This phenomenon occurs as soon as the strength of the applied field exceeds a certain threshold value. These threshold values are calculated for some particular combinations of boundary conditions and applied fields, chosen because of their relevance in the experimental determination of the elastic constants. Furthermore we deal with the unwinding of the helix of a chiral nematic by means of an external field applied perpendicular to the helix axis. This cholesteric-nematic transition also sets in at a well-defined threshold value of the applied field. Finally we briefly discuss the relation between the elastic constants and the structure of the molecules.

6.1 Interaction Between Applied Fields and Director Fields

The application of an external magnetic field $\boldsymbol{B}$ to a nematic results in a magnetization $\boldsymbol{M}$ given by

$$M_\alpha = \mu_0^{-1} \chi_{\alpha\beta} B_\beta \quad . \tag{6.1}$$

Choosing the uniaxial axis of the nematic along the basis vector $\overline{e}_3$ (i.e. $\boldsymbol{n} = \overline{e}_3$ or $\boldsymbol{n} = -\overline{e}_3$), the susceptibility tensor is given by

$$\tilde{\boldsymbol{\chi}} = \begin{pmatrix} \chi_\perp & 0 & 0 \\ 0 & \chi_\perp & 0 \\ 0 & 0 & \chi_\perp + \Delta\chi \end{pmatrix} \quad , \tag{6.2}$$

where $\Delta\chi = \chi_{\parallel} - \chi_{\perp}$. With respect to an arbitrary coordinate system spanned by the basis vectors $\boldsymbol{e}_\alpha$ the elements of the tensor are

$$\chi_{\alpha\beta} = \overline{R}_{\alpha\gamma}\overline{R}_{\beta\delta}\overline{\chi}_{\gamma\delta} \tag{6.3}$$

with $\overline{R}_{\alpha\beta} = \boldsymbol{e}_\alpha \cdot \overline{\boldsymbol{e}}_\beta$. This means that the components of $\tilde{\chi}$ with respect to an arbitrary coordinate system are given by

$$\chi_{\alpha\beta} = \chi_{\perp}\delta_{\alpha\beta} + \Delta\chi n_\alpha n_\beta \quad . \tag{6.4}$$

The diamagnetic susceptibilities will be discussed in more detail in Sect. 9.2. It appears that $\Delta\chi$ is positive for most nematics.

Application of a magnetic field gives the following contribution of the medium to the free energy density

$$-\int_0^B M'_\alpha dB'_\alpha = -\frac{1}{2}\mu_0^{-1}\chi_{\perp}B^2 - \frac{1}{2}\mu_0^{-1}\Delta\chi(n_\alpha B_\alpha)^2 \quad , \tag{6.5}$$

where B denotes the strength of the applied field $\boldsymbol{B}$. When only the influence of the magnetic field on the director field is considered, the first term is of no importance. The coupling between magnetic field and director field thus gives rise to a free energy density, f_{M}:

$$f_{\mathrm{M}} = -\tfrac{1}{2}\mu_0^{-1}\Delta\chi(n_\alpha B_\alpha)^2 \quad . \tag{6.6}$$

The interaction of a static electric field with a nematic is, in general, quite complicated because of the presence of charges. Even in the case of a nematic insulator two different types of coupling between the electric field and the director field are possible. The first type of coupling is analogous to the coupling with a magnetic field, whereas the second type of coupling is due to the fact that a spontaneous dielectric polarization may appear as a result of the distortion of the director field, the so-called flexoelectric effect. First the analogy with the magnetic case will be considered and then a brief discussion will be devoted to the flexoelectric effect.

Application of a static electric field $\boldsymbol{E}$ to a nematic gives rise to a polarization $\boldsymbol{P}$ of the form

$$P_\alpha = \varepsilon_0(\varepsilon_{\alpha\beta} - \delta_{\alpha\beta})E_\beta = D_\alpha - \varepsilon_0\delta_{\alpha\beta}E_\beta \quad , \tag{6.7}$$

where $\tilde{\varepsilon}$ and $\boldsymbol{D}$ denote the relative dielectric permittivity tensor and dielectric displacement, respectively. The components of $\tilde{\varepsilon}$ with respect to an arbitrary coordinate system, analogous to (6.4), are

$$\varepsilon_{\alpha\beta} = \varepsilon_{\perp}\delta_{\alpha\beta} + \Delta\varepsilon n_{\alpha} n_{\beta} \quad , \tag{6.8}$$

where $\Delta\varepsilon = \varepsilon_{\|} - \varepsilon_{\perp}$ denotes the dielectric anisotropy. The electric field gives rise to the following free energy density

$$-\int_0^E D'_{\alpha} dE'_{\alpha} = -\frac{1}{2}\varepsilon_0 \varepsilon_{\perp} E^2 - \frac{1}{2}\varepsilon_0 \Delta\varepsilon (n_{\alpha} E_{\alpha})^2 \quad , \tag{6.9}$$

where E denotes the strength of the applied field $\boldsymbol{E}$. Again only the last term on the right-hand side is relevant for the description of the interaction between an electric field and a director field. Depending on the sign of $\Delta\varepsilon$ either a parallel alignment ($\Delta\varepsilon>0$) of the electric field and the director field is favoured or a mutually perpendicular orientation ($\Delta\varepsilon<0$) of the fields.

As is well known, the application of stress to a solid results in a strain, which in some cases, e.g. quartz, can induce a polarization. This phenomenon is called piezoelectricity. An analogous effect can be observed in liquid crystals, where the distortion of the director field takes the place of the ordinary strain in a solid. In this effect, called flexoelectricity, the appearing curvature strain induces a polarization $\boldsymbol{P}$. This polarization is proportional to the first-order spatial derivatives of the components $n_{\alpha} n_{\beta}$ of the orientational tensor order parameter. This means that

$$P_{\mu}(\boldsymbol{r}) = K_{\mu\alpha\beta\gamma}(\boldsymbol{r}) n_{\alpha}(\boldsymbol{r}) \partial_{\beta} n_{\gamma}(\boldsymbol{r}) \quad , \tag{6.10}$$

where $K_{\mu\alpha\beta\gamma}(\boldsymbol{r})$ is some tensor of the fourth rank consisting of all possible combinations of $n_{\alpha}(\boldsymbol{r}) n_{\beta}(\boldsymbol{r})$, the Kronecker delta tensor and the Levi-Civita tensor. Only two tensors give an essentially different and non-zero contribution to the polarization, namely the forms

$$e_1 \delta_{\mu\alpha} \delta_{\beta\gamma} \quad \text{and} \quad e_3 \delta_{\mu\gamma} \delta_{\alpha\beta} \quad , \tag{6.11}$$

where the flexoelectric coefficients e_1 and e_3 are functions of the order parameter. Consequently the general form of the polarization is

$$P_{\mu}(\boldsymbol{r}) = e_1 n_{\mu}(\boldsymbol{r}) \partial_{\alpha} n_{\alpha}(\boldsymbol{r}) + e_3 n_{\alpha}(\boldsymbol{r}) \partial_{\alpha} n_{\mu}(\boldsymbol{r}) \tag{6.12a}$$

or, in vector notation,

$$\begin{aligned} \boldsymbol{P}(\boldsymbol{r}) &= e_1 \boldsymbol{n}(\boldsymbol{r})[\nabla \cdot \boldsymbol{n}(\boldsymbol{r})] + e_3 [\boldsymbol{n}(\boldsymbol{r}) \cdot \nabla] \boldsymbol{n}(\boldsymbol{r}) \\ &= e_1 \boldsymbol{n}(\boldsymbol{r})[\nabla \cdot \boldsymbol{n}(\boldsymbol{r})] + e_3 [\operatorname{curl} \boldsymbol{n}(\boldsymbol{r})] \times \boldsymbol{n}(\boldsymbol{r}) \quad . \end{aligned} \tag{6.12b}$$

The terms $e_1 \boldsymbol{n}(\boldsymbol{r})[\nabla \cdot \boldsymbol{n}(\boldsymbol{r})]$ and $e_3[\operatorname{curl} \boldsymbol{n}(\boldsymbol{r})] \times \boldsymbol{n}(\boldsymbol{r})$ are correlated with a splay and bend distortion, respectively. Both terms influence the distortions that result from the application of an electric field.

The coupling of the elastic distortions to an electric field gives rise to an interaction energy density of the form

$$f_{\mathrm{f}} = -e_1[\boldsymbol{n}(\boldsymbol{r})\nabla \cdot \boldsymbol{n}(\boldsymbol{r})] \cdot \boldsymbol{E} - e_3\{[\mathrm{curl}\, \boldsymbol{n}(\boldsymbol{r})] \times \boldsymbol{n}(\boldsymbol{r})\} \cdot \boldsymbol{E} \quad . \tag{6.13}$$

It should be mentioned here that in addition to this coupling, called dipolar flexoelectricity, the coupling of the director field to the spatial derivatives of the electric field can be considered as well. This type of interaction is known as quadrupolar flexoelectricity (Prost and Marcerou 1977; Marcerou and Prost 1980). The molecular origin of the appearance of dipolar flexoelectricity may be attributed to the particular shape (e.g. resembling a wedge or banana) of the molecules in question, that in addition must carry a permanent dipole moment (Meyer 1969; Straley 1976).

6.2 Frederiks Transitions

Consider a nematic cell with given boundary conditions, such that a uniform director field results. Next apply a magnetic or electric field. When the orienting effect of the boundaries conflicts with the orienting effect of the applied field, a distortion of the original director field occurs. In the case where the external field is perpendicular to the original director the distortion normally sets in only if the strength of the applied field exceeds a certain well-defined threshold value. This type of transition is called a Frederiks transition. The existence of a threshold value of the field can be easily understood by considering a small spontaneous local distortion of the director field. This fluctuation will experience a stabilizing generalized elastic force due to the boundary conditions. In addition, depending on the sign of the anisotropy, a destabilizing effect can be present due to the applied field. The amount of destabilization depends, of course, on the strength of the field. The threshold for the appearance of a distortion in the equilibrium situation is then reached if the strength of the destabilizing interaction with the applied field equals the stabilizing generalized elastic force.

Frederiks transitions can be used to measure directly the elastic constants, provided the geometries, i.e. the boundary conditions and the direction of the applied field, are chosen in the appropriate way. Attention will be paid here to the calculation of the threshold values belonging to four different geometries (Fig. 6.1). The first three geometries give rise to expressions that directly relate the elastic constants for twist, splay and bend, respectively, to the strength of the applied field. The fourth geometry concerns the twisted nematic cell, and is important because the change in optical properties, associated with the field-induced distortion, is widely used in liquid crystal displays (LCD's, see for example, Raynes 1983).

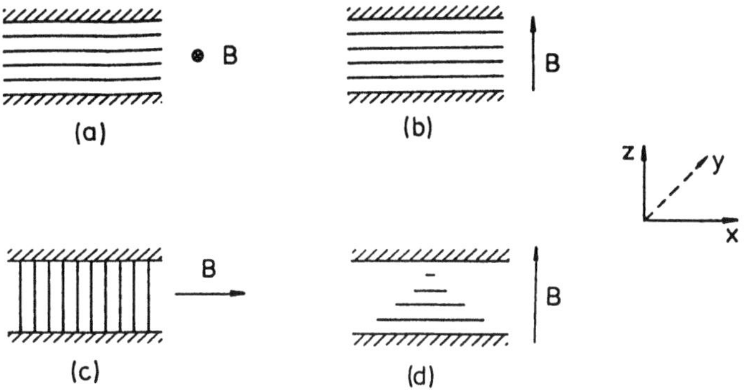

Fig. 6.1. **(a)** Twist geometry; **(b)** splay geometry; **(c)** bend geometry; **(d)** geometry of the twisted nematic layer

The starting point for the calculations is the free energy density, $f(\boldsymbol{r})$, consisting of the Frank free energy and the field energy density. Here a magnetic field will be considered. The following form of the free energy density appears to be the most convenient:

$$
\begin{aligned}
f(\boldsymbol{r}) = \tfrac{1}{2}(K_1 - K_2)[\partial_\alpha n_\alpha(\boldsymbol{r})][\partial_\beta n_\beta(\boldsymbol{r})] + \tfrac{1}{2}K_2[\partial_\alpha n_\beta(\boldsymbol{r})][\partial_\alpha n_\beta(\boldsymbol{r})] \\
+ \tfrac{1}{2}(K_3 - K_2) n_\alpha(\boldsymbol{r}) n_\beta(\boldsymbol{r})[\partial_\alpha n_\gamma(\boldsymbol{r})][\partial_\beta n_\gamma(\boldsymbol{r})] \\
- \tfrac{1}{2}\mu_0^{-1}\Delta\chi[n_\alpha(\boldsymbol{r})B_\alpha]^2 \quad . \qquad (6.14)
\end{aligned}
$$

The free energy density f_0, relating to the state of uniform parallel alignment, is disregarded here because it is merely an additive constant.

(1) *The Twist Geometry.* Consider a nematic layer of thickness d with uniform planar boundary conditions (see Fig. 6.1a). The undistorted state is described by the director field $\boldsymbol{n} = (1,0,0)$. Next a uniform magnetic field is applied in the direction of the y axis. Then the director field, because of symmetry, is given by

$$\boldsymbol{n} = [\cos\phi(z),\ \sin\phi(z),\ 0] \quad , \qquad (6.15)$$

where $\phi(z)$ must satisfy the boundary conditions

$$\phi(0) = \phi(d) = 0 \quad . \qquad (6.16)$$

Substitution of (6.15) into (6.14) leads to the following free energy per unit surface of the nematic slab

$$F = \frac{1}{2}\int_0^d dz(K_2\phi_z^2 - \mu_0^{-1}\Delta\chi B^2 \sin^2\phi) \quad , \qquad (6.17)$$

where ϕ_z is shorthand for $d\phi/dz$. The function $\phi(z)$ is determined by the requirement that the energy F must be stationary. This means that $\phi(z)$ must be such that a replacement of $\phi(z)$ by $\phi(z)+\alpha\xi(z)$ does not change the value of the integral, i.e. F, up to order α. The function $\xi(z)$ is arbitrary except that $\xi(0)=\xi(d)=0$, whereas α is a small parameter. Carrying out the replacement yields

$$F(\alpha)=F(0)+\alpha\int_0^d dz(K_2\xi_z\phi_z-\mu_0^{-1}\Delta\chi B^2\xi\sin\phi\cos\phi)+0(\alpha^2) \quad .$$

Partial integration gives

$$\int_0^d dz\,\xi_z\phi_z=\xi(z)\phi_z(z)\Big|_0^d-\int_0^d dz\,\xi\phi_{zz}=-\int_0^d dz\,\xi\phi_{zz} \quad ,$$

where ϕ_{zz} is shorthand for $d^2\phi/dz^2$. Consequently the requirement, that F must not change up to order α, boils down to requiring

$$\int_0^d dz\,\xi(K_2\phi_{zz}+\mu_0^{-1}\Delta\chi B^2\sin\phi\cos\phi)=0 \quad .$$

This requirement must hold for arbitrary $\xi(z)$. This means

$$K_2\phi_{zz}+\mu_0^{-1}\Delta\chi B^2\sin\phi\cos\phi=0. \tag{6.18}$$

This differential equation is the Euler-Lagrange equation. The equation (6.18) has the trivial solution $\phi(z)=0$. A non-trivial solution appears as soon as B exceeds a certain threshold value. This value can be easily obtained by realizing that $\phi(z)$ is very small just above the threshold value. Then $\sin\phi\cos\phi$ can be linearized. The resulting linear differential equation

$$K_2\phi_{zz}=-\mu_0^{-1}\Delta\chi B^2\phi \tag{6.19}$$

can easily be solved. Its general solution reads

$$\phi(z)=A_1\cos\left[B\left(\frac{\Delta\chi}{\mu_0K_2}\right)^{1/2}z\right]+A_2\sin\left[B\left(\frac{\Delta\chi}{\mu_0K_2}\right)^{1/2}z\right] \quad .$$

The boundary conditions (6.16) imply immediately $A_1=0$ and

$$B(\Delta\chi/\mu_0K_2)^{1/2}d=n\pi, \quad n=0,1,2,\ldots \ .$$

Consequently a non-trivial solution of (6.19) and the original equation (6.18) is obtained as soon as B exceeds a critical value B_c given by

$$B_c = \frac{\pi}{d}\left(\frac{\mu_0 K_2}{\Delta\chi}\right)^{1/2} \quad . \tag{6.20}$$

The Euler-Lagrange equation (6.18) is solved by multiplying by $\frac{1}{2}\phi_z$ and then integrating. This results in

$$\phi_z^2 = C - \frac{\Delta\chi B^2}{\mu_0 K_2}\sin^2\phi \quad . \tag{6.21}$$

Because of symmetry $\phi(z)$ reaches its maximal value ϕ_m at $z = \frac{1}{2}d$, i.e. $\phi_z(\frac{1}{2}d) = 0$. This means

$$\phi_z^2 = \frac{\Delta\chi B^2}{\mu_0 K_2}(\sin^2\phi_m - \sin^2\phi) \quad . \tag{6.22}$$

The function $\phi(z)$ can be simply obtained by integration. Using $\phi(z) = \phi(d-z)$ and (6.20) one obtains the result

$$\frac{z}{d}\frac{B}{B_c} = \pi^{-1}\int_0^{\phi(z)} \frac{d\phi'}{(\sin^2\phi_m - \sin^2\phi')^{1/2}} \quad , \quad 0 \leqq z \leqq \tfrac{1}{2}d \quad . \tag{6.23}$$

The constant ϕ_m follows directly from $\phi(\frac{1}{2}d) = \phi_m$. Rewriting the resulting equation for ϕ_m, in terms of a new variable ψ defined by

$$\sin\psi = \frac{\sin\phi'}{\sin\phi_m} \quad ,$$

gives the following equation for ϕ_m

$$\frac{B}{B_c} = \frac{2}{\pi}\int_0^{\frac{\pi}{2}} \frac{d\psi}{(1 - \sin^2\phi_m \sin^2\psi)^{1/2}} = K(\sin\phi_m) \quad . \tag{6.24}$$

Here $K(\sin\phi_m)$ is the complete elliptic integral of the first kind.

The distorted director field (6.15) is obtained by solving (6.23) and (6.24). Examples of solutions are shown in Fig. 6.2. The effect of the field strength, however, can also be reasonably estimated with hardly any effort by a simple approximation. For that purpose $\phi(z)$ is taken to be proportional

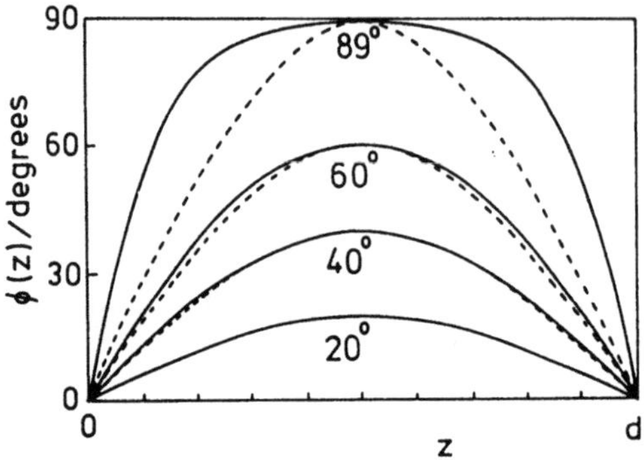

Fig. 6.2. The distorted director field for the twist geometry. The appearing tilt angle ϕ is plotted as a function of position for different magnetic fields as indicated by the value of ϕ_m. The broken lines represent a sine function with amplitude ϕ_m

to its first Fourier coefficient, i.e.

$$\phi(z) = b \sin(\pi z/d) \quad , \tag{6.25}$$

and the amplitude b is determined in the middle of the layer, i.e. at $z = \frac{1}{2}d$, using the Euler-Lagrange equation (6.18). Then the following equation for b results

$$K_2(\pi/d)^2 b = \tfrac{1}{2}\mu_0^{-1}\Delta\chi B^2 \sin(2b) \quad . \tag{6.26}$$

Clearly this relation determines the exact value of B_c. Using (6.20) and putting $b = \frac{\pi}{2}(1-\varepsilon)$ Eq. (6.26) can be written as

$$(B_c/B)^2\pi(1-\varepsilon) = \sin(\pi\varepsilon) \quad . \tag{6.27}$$

This equation for ε can easily be solved if $B \gg B_c$; $\sin(\pi\varepsilon)$ can be approximated by $\pi\varepsilon$ and ε is given by

$$\varepsilon = \frac{B_c^2}{(B^2 + B_c^2)} \quad .$$

For large fields, i.e. $B \gg B_c$, the distortion can be approximated by

$$\phi(z) = \frac{\pi}{2}\frac{B^2}{B^2 + B_c^2} \sin(\pi z/d) \quad . \tag{6.28}$$

This approximation already approaches the exact value of ϕ_m, i.e. the distortion in the middle of the layer, to within 10 % as soon as $B = 2B_c$ and it converges rapidly to the exact value with increasing field strength. Expression (6.20) relates the elastic constant for twist with a threshold value of a magnetic field for a given configuration of the director field, as imposed by the boundary conditions, and the applied field. Similar expressions hold for the elastic constants for splay and bend as will be discussed briefly.

(2) *The Splay Geometry.* Again consider a nematic layer of thickness d with uniform planar boundary conditions. At the boundaries the director is taken to be along the x axis (Fig. 6.1b). In order to express the elastic constant for splay, K_1, in terms of a threshold value of an applied magnetic field this field must now be applied along the z axis, i.e. perpendicular to the layer. Because of symmetry, the resulting distorted director field is given by

$$\boldsymbol{n} = [\cos\theta(z),\ 0,\ \sin\theta(z)] \quad , \tag{6.29}$$

where $\theta(z)$ must again satisfy the boundary conditions $\theta(0) = \theta(d) = 0$. Clearly the undistorted state is described by $\theta(z) = 0$. Substitution of (6.29) into expression (6.14) gives the following free energy per unit surface of the layer

$$F = \frac{1}{2}\int_0^d dz[(K_1 \cos^2\theta + K_3 \sin^2\theta)\theta_z^2 - \mu_0^{-1}\Delta\chi B^2 \sin^2\theta] \quad . \tag{6.30}$$

The function $\theta(z)$ is determined by the requirement that the energy F must be stationary. Consequently it satisfies the Euler-Lagrange equation

$$\begin{aligned}&(K_1 \cos^2\theta + K_3 \sin^2\theta)\theta_{zz} + (K_3 - K_1)\theta_z^2 \sin\theta \cos\theta \\ &\quad +\mu_0^{-1}\Delta\chi B^2 \sin\theta \cos\theta = 0 \quad ,\end{aligned} \tag{6.31}$$

as can be easily verified using the calculus of variations discussed before. Just above threshold $\theta(z)$ is very small, i.e. the Euler-Lagrange equation can be linearized to

$$K_1\theta_{zz} = -\mu_0^{-1}\Delta\chi B^2\theta \tag{6.32}$$

giving rise to a threshold value

$$B_c = \frac{\pi}{d}\left(\frac{\mu_0 K_1}{\Delta\chi}\right)^{1/2} \quad . \tag{6.33}$$

The Euler-Lagrange equation (6.31) can be solved by multiplying by $\frac{1}{2}\theta_z$ and integrating. This yields

$$(K_1 \cos^2\theta + K_3 \sin^2\theta)\theta_z^2 = C - \mu_0^{-1}\Delta\chi B^2 \sin^2\theta \quad . \tag{6.34}$$

Because of symmetry $\theta(z)$ reaches its maximum at $z = \frac{1}{2}d$, i.e. $\theta_z(\frac{1}{2}d) = 0$.

Putting $\theta(\frac{1}{2}d) = \theta_m$ it follows that

$$\theta_z^2 = \frac{\Delta\chi B^2}{\mu_0 K_1}\left[\frac{\sin^2\theta_m - \sin^2\theta}{\cos^2\theta + (K_3/K_1)\sin^2\theta}\right] \quad . \tag{6.35}$$

The function $\theta(z)$ is then determined by

$$\frac{z}{d}\frac{B}{B_c} = \pi^{-1}\int_0^{\theta(z)} d\theta' \left[\frac{\cos^2\theta' + (K_3/K_1)\sin^2\theta'}{\sin^2\theta_m - \sin^2\theta'}\right]^{1/2} \quad , \quad 0 \leqq z \leqq \tfrac{1}{2}d \tag{6.36a}$$

$$\theta(z) = \theta(d-z) \quad , \quad \tfrac{1}{2}d \leqq z \leqq d \quad . \tag{6.36b}$$

Substituting $\kappa = (K_3 - K_1)/K_1$ and using the same change of variables as introduced before [see (6.24)] the constant θ_m is given by the equation

$$\frac{B}{B_c} = \frac{2}{\pi}\int_0^{\frac{\pi}{2}} d\psi \left(\frac{1 + \kappa\sin^2\theta_m \sin^2\psi}{1 - \sin^2\theta_m \sin^2\psi}\right)^{1/2} \quad . \tag{6.37}$$

The threshold value (6.33) also follows directly from this equation. Approximate expressions for the distorted director field can be derived by replacing $\theta(z)$ by its first Fourier component.

(3) *The Bend Geometry.* Consider a nematic layer of thickness d with homeotropic boundary conditions, i.e. the orientation of the nematic at the boundaries is parallel to the normal to the layer (Fig. 6.1c). Choosing the normal to the layer along the z axis and the boundaries of the layer at $z = 0$ and $z = d$, the undistorted director field is described by $\boldsymbol{n} = (0, 0, 1)$. Next a magnetic field is applied parallel to the layer, say in the x direction. Because of symmetry the distorted director field must be given by

$$\boldsymbol{n} = [\sin\theta(z),\ 0,\ \cos\theta(z)] \quad , \tag{6.38}$$

where $\theta(z)$ must satisfy the boundary conditions $\theta(0) = \theta(d) = 0$. Substitution of (6.38) into (6.14) gives the following free energy per unit surface of the layer

$$F = \frac{1}{2}\int_0^d dz[(K_1\sin^2\theta + K_3\cos^2\theta)\theta_z^2 - \mu_0^{-1}\Delta\chi B^2\sin^2\theta] \quad . \tag{6.39}$$

Interchanging K_1 and K_3 in this expression gives rise to the expression

(6.30). Consequently the threshold value of the applied field and the expression for the distorted field can simply be obtained by interchanging K_1 and K_3 in the expressions (6.33), (6.36) and (6.37).

(4) *The Geometry of the Twisted Nematic Cell.* Consider a nematic layer of thickness d with planar boundary conditions. At the boundaries, defined by the planes $z = 0$ and $z = d$, the director is taken to be in different directions, namely along the x axis and in a direction, that has an angle ϕ_0 with respect to the x axis, respectively (see Fig. 6.1d for $\phi_0 = \frac{\pi}{2}$). Next a magnetic field is applied perpendicular to the axis (Leslie 1970). Because of symmetry the distorted director field must be

$$\boldsymbol{n} = [\cos\theta(z)\cos\phi(z),\ \cos\theta(z)\sin\phi(z),\ \sin\theta(z)] \quad , \tag{6.40}$$

where $\theta(z)$ and $\phi(z)$ must satisfy the boundary conditions

$$\theta(0) = \theta(d) = 0 \quad , \tag{6.41a}$$

$$\phi(0) = 0 \quad , \quad \phi(d) = \phi_0 \quad . \tag{6.41b}$$

Substitution of (6.40) into (6.14) gives the following free energy per unit surface of the layer

$$\begin{aligned} F = \frac{1}{2}\int_0^d dz[(K_1\cos^2\theta + K_3\sin^2\theta)\theta_z^2 \\ + \cos^2\theta(K_2\cos^2\theta + K_3\sin^2\theta)\phi_z^2 - \mu_0^{-1}\Delta\chi B^2\sin^2\theta] \quad . \end{aligned} \tag{6.42}$$

The functions $\theta(z)$ and $\phi(z)$ are determined by the requirement that the energy F must be stationary. Consequently they must satisfy the Euler-Lagrange equations

$$\frac{d}{dz}[\cos^2\theta(K_2\cos^2\theta + K_3\sin^2\theta)\phi_z] = 0 \quad , \tag{6.43a}$$

$$\begin{aligned} &(K_1\cos^2\theta + K_3\sin^2\theta)\theta_{zz} + (K_3 - K_1)\theta_z^2\sin\theta\cos\theta \\ &\quad + \mu_0^{-1}\Delta\chi B^2\sin\theta\cos\theta \\ &\quad + \sin\theta\cos\theta[2K_2\cos^2\theta + K_3(\sin^2\theta - \cos^2\theta)]\phi_z^2 = 0 \quad . \end{aligned} \tag{6.43b}$$

When the magnetic field is absent, the director field is described by

$$\theta(z) = 0 \quad ; \quad \phi(z) = \phi_0 z/d \quad . \tag{6.44}$$

The threshold value is obtained by linearizing the Euler-Lagrange equations after inserting (6.44). This yields

$$\phi_{zz} = 0 \quad , \tag{6.45a}$$

$$K_1\theta_{zz} + \mu_0^{-1}\Delta\chi B^2\theta + (2K_2 - K_3)\theta\phi_z^2 = 0 \quad . \tag{6.45b}$$

The non-trivial solution $\theta(z) = A_1 \sin(\pi z/d)$ is obtained as soon as B reaches the threshold value B_c given by

$$-K_1\left(\frac{\pi}{d}\right)^2 + \mu_0^{-1}\Delta\chi B_c^2 + (2K_2 - K_3)\left(\frac{\phi_0}{d}\right)^2 = 0$$

or

$$B_c = \frac{\pi}{d}\left\{\frac{\mu_0}{\Delta\chi}\left[K_1 + (K_3 - 2K_2)\left(\frac{\phi_0}{\pi}\right)^2\right]\right\}^{1/2} \quad . \tag{6.46}$$

In practice, the twist angle ϕ_0 cannot achieve values larger than $\frac{\pi}{2}$. This ultimate situation is of considerable practical interest. For fields $B \gg B_c$ the optical properties of such a sample change drastically compared to the zero-field situation (see Sect. 7.4), and this behaviour makes the effect suitable for display applications. In practice of course, an electric field will be used instead of a magnetic field (Schadt and Helfrich 1971).

From the discussion above it is clear that the elastic constants of a particular nematogenic compound can be determined by measuring the threshold values for the three appropriate geometries, provided that the anisotropy $\Delta\chi$ is known. The average state of alignment can be found in principle by making use of any anisotropic property, such as the birefringence, dielectric permittivity, or the electric or thermal conductivity. The distortion of the director pattern is most accurately detected optically by monitoring the phase difference between an ordinary and extraordinary ray (see for example, Deuling 1978). Light incident normal to the nematic layer allows determination of only two elastic constants, namely the constants for splay and bend. The twist constant cannot be obtained in this way, because the sample can be divided into two halves that give equal contributions to the total effect, but of opposite sign. The twist transition can however be detected by making use of light incident at an angle to the normal to the layer. The most elegant method to achieve this is conoscopy. The characteristic conoscopic figure of a uniform planar layer, namely a set of hyperbolae with one axis parallel to $\boldsymbol{n}$, starts to rotate as soon as B exceeds the threshold value B_c (Cladis 1972).

As follows from the equations for the thresholds in the various geometries, B_c is inversely proportional to d, the thickness of the nematic layer. In the derivation of these formulae strong anchoring at the boundaries is assumed. If this is not the case, the threshold value of B_c will be lower, which can be translated into the onset of a distortion in a layer of thickness $d + d'$. The extrapolation length d' is a measure of the surface energy and can be determined from measurements at various thicknesses. It should be remarked here that in the early literature the importance of different surface treatments was not fully appreciated. In this respect one can say that a high value of a specific elastic constant (corresponding to a high threshold) is in general more reliable than a lower one. Experimental results for elastic constants will be discussed in Sect. 6.4.

6.3 The Cholesteric-Nematic Transition

A particularly interesting transition concerns the unwinding of the helix of a chiral nematic by means of an applied magnetic field, which is directed perpendicular to the helix axis (De Gennes 1968a). As soon as the field strength reaches a certain threshold value the system becomes a nematic, i.e. the period p of the helix diverges at this critical field strength. Without loss of generality the helix axis may be taken along the z axis and the applied field along the y axis. The director field is now given by

$$\boldsymbol{n} = [\cos\theta(z),\ \sin\theta(z),\ 0] \quad , \tag{6.47}$$

where it can be assumed that $\theta(0) = 0$ and $\theta[p(B)] = 2\pi$. Consequently the period $p(B)$ is determined by

$$p(B) = \int_0^{p(B)} dz = \int_0^{2\pi} d\theta \, \frac{dz}{d\theta} \quad . \tag{6.48}$$

In order to study the behaviour of the period p as a function of B the function $\theta(z)$ must be calculated first. This function $\theta(z)$ is obtained by minimizing the free energy per unit volume. Substituting (6.47) into (6.14) and taking into account the additional twist term $k\varepsilon_{\alpha\beta\gamma}n_\gamma(\boldsymbol{r})\partial_\alpha n_\beta(\boldsymbol{r})$ [see also (5.24)] gives the following free energy per unit volume

$$F = [2p(B)]^{-1} \int_0^{p(B)} dz[K_2(\theta_z - t_0)^2 - \mu_0^{-1}\Delta\chi B^2 \sin^2\theta] - \frac{1}{2}K_2 t_0^2 \quad . \tag{6.49}$$

The function $\theta(z)$ follows from the requirement that the energy F must be stationary. Consequently $\theta(z)$ must satisfy the Euler-Lagrange equation

$$\xi^2 \theta_{zz} + \sin\theta \cos\theta = 0 \tag{6.50}$$

with $\xi^2 = \mu_0 K_2/(\Delta\chi B^2)$. Multiplying (6.50) by θ_z and integrating yields

$$\xi^2 \theta_z^2 = C - \sin^2\theta \quad . \tag{6.51}$$

This means

$$z = \xi \int_0^{\theta} d\theta' (C - \sin^2\theta')^{-1/2} \quad . \tag{6.52}$$

The dependence of the period $p(B)$ on the integration constant C is given by

$$p(B) = \xi \int_0^{2\pi} d\theta \, (C - \sin^2\theta)^{-1/2} \quad . \tag{6.53}$$

Expressions (6.52) and (6.53) imply immediately

$$\theta(z + p) = \theta(z) + 2\pi$$

provided that $C>1$. The value $C = 1$ is a critical value, because then the period diverges. The only problem left is the calculation of the integration constant C as a function of B. For that purpose the free energy per unit volume is expressed as a function of C. Using (6.48) and (6.51) it follows that

$$\begin{aligned} F &= \frac{K_2 t_0^2}{2p(B)} \int_0^{2\pi} d\theta \left[\frac{2}{t_0^2}\theta_z^2 - \frac{2}{t_0}\theta_z + 1 - \frac{C}{\xi^2 t_0^2} \right] \frac{dz}{d\theta} - \frac{1}{2} K_2 t_0^2 \\ &= \frac{K_2}{\xi p(B)} \int_0^{2\pi} d\theta (C - \sin^2\theta)^{1/2} - \frac{2\pi K_2 t_0}{p(B)} - \frac{K_2 C}{2\xi^2} \quad , \end{aligned} \tag{6.54}$$

where use has also been made of (6.53). Minimization of F with respect to C gives the condition $dF/dC = 0$, or

$$\frac{K_2}{\xi p^2(B)} \left[\int_0^{2\pi} d\theta (C - \sin^2\theta)^{1/2} - 2\pi \xi t_0 \right] \frac{dp}{dC} = 0 \quad . \tag{6.55}$$

This means that the dependence of C on B is obtained by solving

$$\int_0^{\frac{\pi}{2}} d\theta (C - \sin^2\theta)^{1/2} = \frac{\pi t_0}{2B}\left(\frac{\mu_0 K_2}{\Delta\chi}\right)^{1/2} \quad . \tag{6.56}$$

This equation determines a threshold value for the applied field, because it does not allow a solution for C as soon as the field strength exceeds a critical value. The lowest possible value for C is $C = 1$. Consequently the critical field value is given by

$$\int_0^{\frac{\pi}{2}} d\theta (1 - \sin^2\theta)^{1/2} = \int_0^{\frac{\pi}{2}} d\theta \cos\theta = 1 = \frac{\pi t_0}{2B_c}\left(\frac{\mu_0 K_2}{\Delta\chi}\right)^{1/2}$$

or

$$B_c = \frac{\pi t_0}{2}\left(\frac{\mu_0 K_2}{\Delta\chi}\right)^{1/2} \quad . \tag{6.57}$$

At this value of B the period of the helix becomes infinite, i.e. the cholesteric-nematic transition occurs. The behaviour of the pitch in a magnetic field is shown in Fig. 6.3.

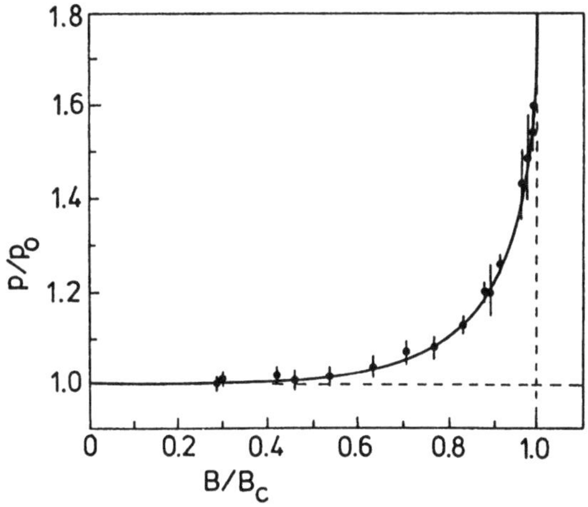

Fig. 6.3. The behaviour of the pitch of a chiral nematic in a magnetic field (Durand et al. 1969)

6.4 Elastic Constants and Molecular Structure

The study of the Frederiks transition in magnetic as well as electric fields has provided a large number of experimental data on the elastic constants. The available information can be summarized as follows (De Jeu 1981):

(i) The elastic constants are of the order of 10^{-11} N. In all cases K_2 is the smallest of the three. For the rod-like nematics studied so far one finds approximately

$$0.5 < K_3/K_1 < 3.0 \quad , \quad 0.5 < K_2/K_1 < 0.8 \quad .$$

(ii) Until now little or no evidence has been found for a noticeable influence of the polarizability and the dipole moment of the constituent molecules on the elastic behaviour. A relevant parameter seems to be the molecular length-to-width ratio L/W. It appears that compounds with different bridging or end groups can have similar elastic constants provided that the molecular dimensions are about the same. This observation has led to the speculation that there must be a relation between the ratios K_3/K_1 and L/W. The relevance of this speculation, however, is not clear. It appears that the dependence of K_3/K_1 on L/W is rather different for molecules with and without alkyl chains. The introduction of relatively rigid end or side groups in molecules without alkyl chains changes K_3/K_1 in the same direction as L/W (see Table 6.1). On the other hand, if the length of alkyl chains is increased in a homologous series, K_3/K_1 is found to decrease. This trend is shown in Fig. 6.4, and also in the last example of Table 6.1. However, several observations have also been made where there is a

Table 6.1. Ratio K_3/K_1 for some compounds with a different length-to-width ratio at constant reduced temperature $T/T_{NI} = 0.96$ (Leenhouts and Dekker 1981)

Compound	K_3/K_1
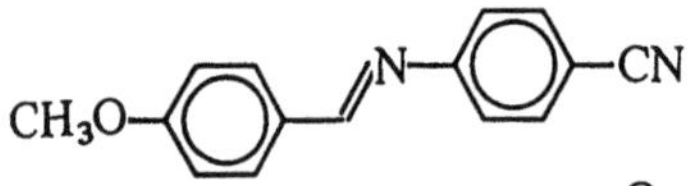	1.9
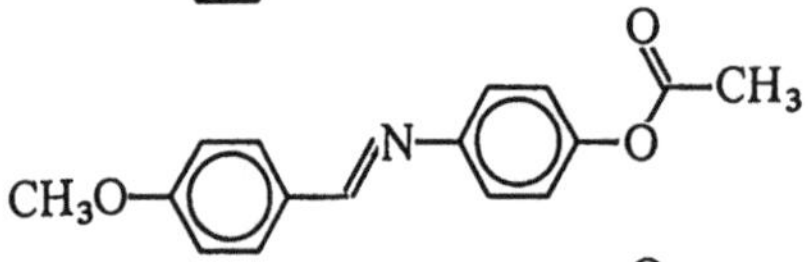	2.1
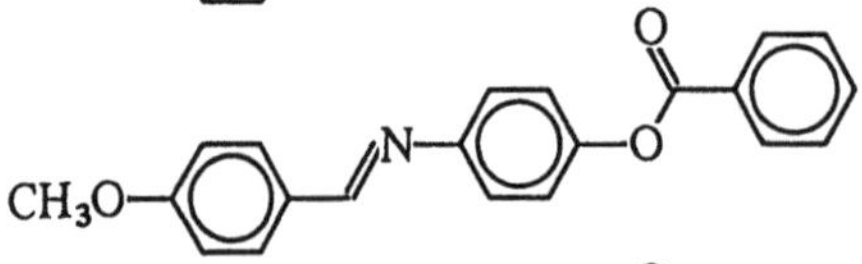	2.4
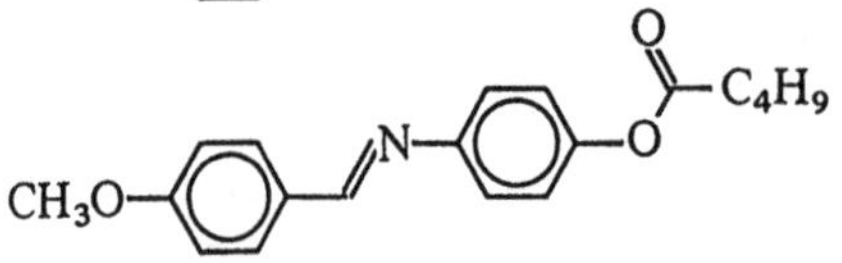	1.3

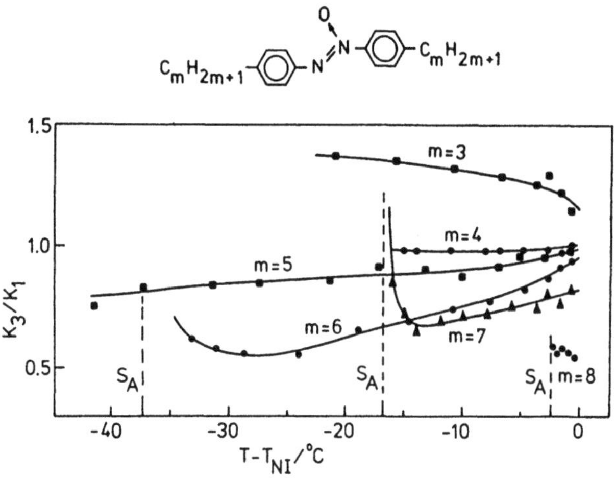

Fig. 6.4. Variation of the elastic ratio K_3/K_1 for a homologous series (De Jeu and Claassen 1977)

change in K_3/K_1 without a noticeable change in L/W. This is particularly true when a benzene ring is replaced by a saturated ring, or when hetero atoms are introduced (Schadt and Gerber 1982). Besides evidence is available that differences in short-range order have a strong influence on the elastic properties (Bradshaw et al. 1984).

(iii) As far as the temperature dependence of the elastic constants is concerned a first approximation gives $K_i \sim S^2$ $(i = 1, 2, 3)$, where S denotes the order parameter. In this case the ratio of two elastic constants should be independent of the temperature. It directly follows from Fig. 6.4 that the given proportionality between K_i and S^2 only holds approximately, unless K_1 and K_3 are about equal. For nematics that show a smectic phase at lower temperatures, a presmectic stiffening of K_2 and K_3 is observed, as discussed in Sect. 5.3, when the nematic-smectic transition is approached. This can be seen in Fig. 6.4 for $m = 6$ and $m = 7$.

The experimental results can easily be described in terms of the tensor fields discussed in Sect. 5.4. The interpretation of the elastic constants in terms of molecular theories, however, still constitutes one of the outstanding and important problems in the field of liquid crystals. For a discussion of some models the reader is referred to Chap. 13.

7. Applications of Continuum Theory

In this chapter we treat some important physical phenomena in terms of continuum theory. First of all we pay attention to the observed textures, for which the orientational field around the appearing discontinuities can be calculated. Two relevant discontinuities are dealt with, namely the orientational field around an axial and around a perpendicular disclination line. Then the thermal fluctuations of the orientational field are discussed and the quantitative validity of the present formulation of the fluctuation theory is examined. Next the strong scattering of light in the nematic phase is explained in terms of the thermal fluctuations of the components of the orientational tensor order parameter. Then we discuss the optical properties of chiral nematics. The observed Bragg reflection, the polarization of the reflected light and the rotary power are dealt with in some detail. Finally the absence of true long-range correlation in the smectic A phase is considered.

7.1 Disclinations

In continuum theory liquid crystals are described by director fields. The basic assumption of this theory concerns the smoothness of the director field $\boldsymbol{n}(\boldsymbol{r})$, i.e. singularities in the field are excluded. Without this assumption the distortion free energy density cannot be defined throughout the whole medium. In practice, however, textures are observed originating from singularities in the orientational field, e.g. in the centre of a schlieren texture. These discontinuities in orientation are called disclinations (Frank 1958). Although the energy of the disclinations themselves is unknown, the energy of the surrounding distorted director field can be calculated by making use of the continuum theory, i.e. using the expression for the distortion free energy density.

In principle the singularities which appear in the director field may be zero-, one- or two-dimensional and they are referred to as point, line or sheet disclinations, respectively. In practice, however, the sheet disclinations appear to be unstable: the corresponding energy can be lowered by smearing out the discontinuity over a large distance giving rise to so-called walls. Examples of such a director pattern are the alignment-inversion walls

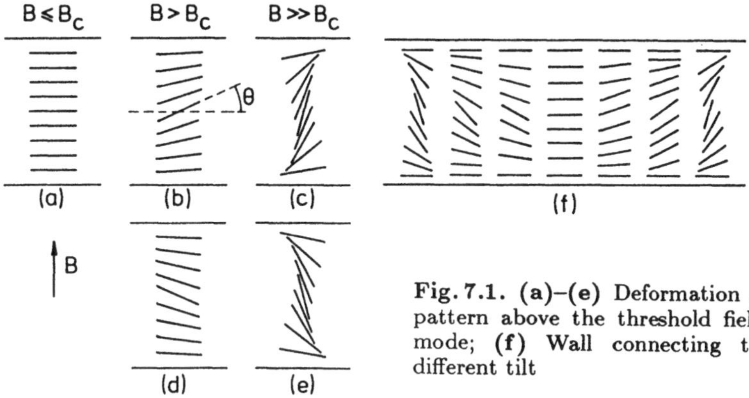

Fig. 7.1. (a)–(e) Deformation of the director pattern above the threshold field in the splay mode; **(f)** Wall connecting two regions of different tilt

associated with Frederiks transitions. The creation of such a wall is sketched in Fig. 7.1. For a magnetic field $B>B_c$ the uniform planar structure (a) is unstable and the system jumps into one of the two possible states (b) or (d). If the field is increased further these states develop into (c) and (e), respectively. In practice, the system may choose (b) in one domain and (d) in another. The transition between such domains is an alignment-inversion wall that can be observed using a polarizing microscope. In the wall the orientation of the field rotates continuously from $+\theta$ to $-\theta$ (Fig. 7.1f). Sometimes closed domains are generated. The equilibrium shape of a wall surrounding such a domain is an ellipse, with an axial ratio equal to $(K_1/K_2)^{1/2}$ and $(K_3/K_2)^{1/2}$ in the splay and bend geometry, respectively (Brochard 1972). A closed wall will collapse spontaneously, but its shape will remain an ellipse of fixed ellipticity during this collapsing process as long as the size of the axes is larger than the wall thickness.

In the following the discussion of defects is restricted to the disclination lines in nematics. The singularities are experimentally observed as a system of dark, flexible filaments. In this context it should be remarked that the name "nematic" stems from this observed thread-like structure (the Greek word $\nu\eta\mu\alpha$ means thread). The study of defects is a field of interest in itself. In view of the general nature of this book the discussion here is restricted to the treatment of only two types of disclination lines, namely the axial and the perpendicular disclination, the latter being relevant for a discussion of the Grandjean texture. These examples suffice to demonstrate the way defects can be dealt with within the framework of the continuum theory. A more detailed introduction can be found in Chap. 4 of the book of De Gennes (1974). For full information on experimental and theoretical aspects of disclinations the reader is referred to Kleman (1983).

7.1.1 The Axial Disclination Line

This type of singularity, first considered by Oseen (1933), is a straight line and the director field is such that the director is perpendicular to the direction of the disclination. Without loss of generality this disclination line may be supposed to lie along the z axis, i.e. the director field is given by

$$\boldsymbol{n}(\boldsymbol{r}) = [\cos\phi(x,y),\ \sin\phi(x,y),\ 0] \quad , \tag{7.1}$$

where $\phi(x,y)$ is the angle between the director and the x axis. Next consider the distortion free energy density. In order to avoid unnecessarily complicated mathematics the one-constant approximation is used, i.e. $K_1 = K_2 = K_3 = K$. As will be seen later, the price to be paid for this approximation consists of the introduction of a small artefact in the mathematical description of the disclination. In the one-constant approximation substitution of the director field (7.1) in (5.23) or (5.24) gives rise to

$$f_{\mathrm{d}}(\boldsymbol{r}) = \tfrac{1}{2}K(\phi_x^2 + \phi_y^2) \quad . \tag{7.2}$$

The singularity is now introduced into the description, by changing over to cylindrical coordinates (ϱ, ψ, z) where

$$x = \varrho\cos\psi \quad , \quad y = \varrho\sin\psi \quad . \tag{7.3}$$

Using

$$\partial_x = \cos\psi\,\partial_\varrho - \frac{\sin\psi}{\varrho}\partial_\psi \quad , \quad \partial_y = \sin\psi\,\partial_\varrho + \frac{\cos\psi}{\varrho}\partial_\psi$$

the distortion free energy density in cylindrical coordinates is given by

$$f_{\mathrm{d}}(\varrho,\psi,z) = \frac{1}{2}K\left(\phi_\varrho^2 + \frac{1}{\varrho^2}\phi_\psi^2\right) \quad . \tag{7.4}$$

Clearly this expression diverges at $\varrho = 0$ provided that ϕ_ψ has a non-zero value there. Alternatively stated, the divergence is caused by a discontinuity in orientation at $\varrho = 0$.

The functional dependence of ϕ on ϱ and ψ is determined by requiring that the free energy of a distorted region, excluding the axial disclination itself, must be stationary with respect to variations of ϕ. This means that ϕ must satisfy the Euler-Lagrange equation

$$\phi_{\varrho\varrho} + \frac{1}{\varrho}\phi_\varrho + \frac{1}{\varrho^2}\phi_{\psi\psi} = 0 \quad . \tag{7.5}$$

The axial disclinations are represented by the general solution

$$\phi = A\psi + \phi_0 \quad , \tag{7.6}$$

where ϕ_0 is a constant and A is determined by the requirement that the director field must be single valued. This means that changing ψ by 2π must give rise to the same director field. A further restriction is obtained from the fact that the states described by $\boldsymbol{n}$ or $-\boldsymbol{n}$ are indistinguishable (head-tail symmetry). Consequently A must be equal to m, where $2m$ is an integer. Thus the axial disclinations are described by

$$\phi = m\psi + \phi_0 \quad , \quad m = \pm\tfrac{1}{2},\ \pm 1,\ \pm\tfrac{3}{2},\ \pm 2, \dots \quad . \tag{7.7}$$

The solution $m = 0$ must be excluded because it represents the state with uniform parallel alignment (no disclination present). A variation of ψ from 0 to 2π corresponds to a variation of the angle ϕ (the angle between the director and the x axis) from ϕ_0 to $\phi_0 + 2\pi m$.

The distortion free energy per unit length of the disclination is calculated by considering a region consisting of two concentric cylinders with the disclination as axis. The radii of both cylinders are given by the core size a_m of the disclination with index m and the size $\varrho_{\max}$ of the sample. Thus the distortion free energy per unit length is given by

$$F = \frac{1}{2}K \int_0^{2\pi} d\psi \int_{a_m}^{\varrho_{\max}} \varrho \, d\varrho \left(\phi_\varrho^2 + \frac{1}{\varrho^2}\phi_\psi^2 \right) = \pi m^2 K \ln\left(\varrho_{\max}/a_m\right) \quad . \tag{7.8}$$

Clearly the elastic energy increases with m^2. Accordingly the formation of disclinations with a large index m seems highly unfavourable, although it must be admitted that this analysis is not complete because the core energy of a disclination with index m is unknown.

Analogous to an electromagnetic field, a director field can be described in terms of a set of field or flux lines. This set of flux lines, which will usually be curved, is drawn in such a way that at any point on one of these lines the direction of $\boldsymbol{n}$ coincides with the tangent to the line. Since the direction of the line is the same as the direction of $\boldsymbol{n}$ the lines must be defined in the underlying case by

$$\frac{dy}{dx} = \frac{n_y}{n_x} = \tan\phi(x, y) \quad , \tag{7.9}$$

where (7.1) has been used. In order to rewrite Eq. (7.9) in cylindrical coordinates use is made of

$$dx = \cos\psi \, d\varrho - \varrho \sin\psi \, d\psi \quad , \tag{7.10a}$$

$$dy = \sin\psi\, d\varrho + \varrho \cos\psi\, d\psi \quad . \tag{7.10b}$$

Substitution of (7.10) into (7.9) gives

$$\frac{\sin\psi\, d\varrho + \varrho\cos\psi\, d\psi}{\cos\psi\, d\varrho - \varrho\sin\psi\, d\psi} = \tan\phi \quad \text{or}$$

$$\frac{1}{\varrho}\frac{d\varrho}{d\psi} = \frac{d}{d\psi}\ln\varrho = \cot(\phi - \psi) \quad . \tag{7.11}$$

Using the solutions (7.7), the axial disclinations are represented by flux lines, which are the solutions of the differential equation

$$\frac{d}{d\psi}\ln\varrho = \cot[(m-1)\psi + \phi_0] \quad . \tag{7.12}$$

Clearly this differential equation contains singularities determined by the relation

$$(m-1)\psi + \phi_0 = k\pi \quad , \tag{7.13}$$

where k is an integer. An exceptional case, however, arises for $m = 1$, because here singularities appear only for special values of the constant ϕ_0, namely $\phi_0 = k\pi$, whereas in the general case $(m \neq 1)$, values of ψ can always be found that satisfy the relation (7.13).

The solution of the differential equation (7.12) can easily be obtained for the case $m = 1$. The flux lines are described by

$$\varrho = C\exp(\psi\cot\phi_0) \quad , \tag{7.14}$$

where C is a positive constant. The flux lines of the general case, $m \neq 1$, are, apart from the singularities, described by

$$\varrho = C\{\sin[(m-1)\psi + \phi_0]\}^{1/(m-1)} \quad , \tag{7.15}$$

where C is some positive constant. In contrast to the $m = 1$ situation the form of the flux lines does not depend here on the value of the constant ϕ_0. A change of ϕ_0 only results in a rotation of the pattern of flux lines as follows directly from (7.15). No real significance, however, should be attributed to the dependence on ϕ_0 of a pattern of flux lines belonging to axial disclinations of index 1, because this dependence is merely an artefact of the one-constant approximation (Dzyaloshinskii 1970); as soon as $K_1 \neq K_3$ only the solutions $\phi_0 = 0$ and $\phi_0 = \frac{\pi}{2}$ are possible in the case $m = 1$.

As an illustration of the foregoing discussion, three different patterns of flux lines are shown in Fig. 7.2. The first pattern belongs to an axial

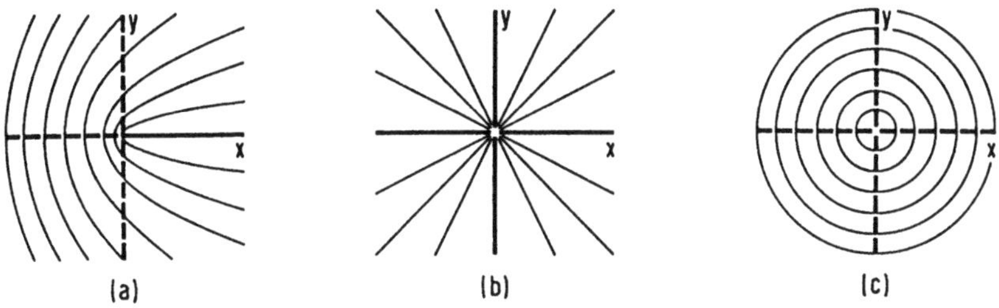

Fig. 7.2a–c. Pattern of flux lines of three different axial disclinations. **(a)** The parabolic pattern of the $m = \frac{1}{2}$ disclination with $\phi_0 = 0$. **(b)** The radial pattern of the $m = 1$ disclination with $\phi_0 = 0$. **(c)** The circular pattern of the $m = 1$ disclination with $\phi_0 = \pi/2$

disclination of index $m = \frac{1}{2}$ $(\phi_0 = 0)$ and is, according to (7.15), described by

$$\varrho = C\frac{1}{\sin^2(\frac{1}{2}\psi)} = \frac{2C}{1 - \cos\psi} \quad \text{or}$$

$$y^2 = 4C(C + x) \quad ,$$

i.e. an axial disclination of Frank index $\frac{1}{2}$ corresponds to a parabolic pattern of flux lines. The second and third patterns of flux lines belong to an axial disclination of index 1 with $\phi_0 = 0$ and $\phi_0 = \frac{\pi}{2}$, respectively. Consequently the second pattern is described by

$$\frac{d}{d\psi}\ln\varrho = \infty \quad ,$$

or ψ is a constant. The third pattern is given by

$$\varrho = C \quad .$$

This means that the second pattern consists of radial flux lines, whereas the flux lines of the third pattern are concentric circles.

Experimentally the situation of an axial disclination line with $m = 1$ and $\phi_0 = 0$ (see Fig. 7.2b) could in principle be realized in a cylindrical tube with homeotropic boundary conditions as denoted in Fig. 7.3a. In practice such a disclination often turns out to be unstable: the total energy can be lowered by a so-called "escape into the third dimension" (see Fig. 7.3b) leaving a smooth curved structure and a point disclination. This is in fact a possibility for all lines of integral strength $(m = \pm 1, \pm 2, \ldots)$.

A disclination line closely related to the one described appears in chiral nematics. This disclination line is parallel to the twist axis and gives rise to director fields as shown in Fig. 7.2 in a plane perpendicular to the twist axis. However, due to chirality, the director fields in successive planes are slightly rotated with respect to each other. This type of disclination line in cholesterics is called a χ type. Clearly both disclinations are identical in the limit of an infinite pitch.

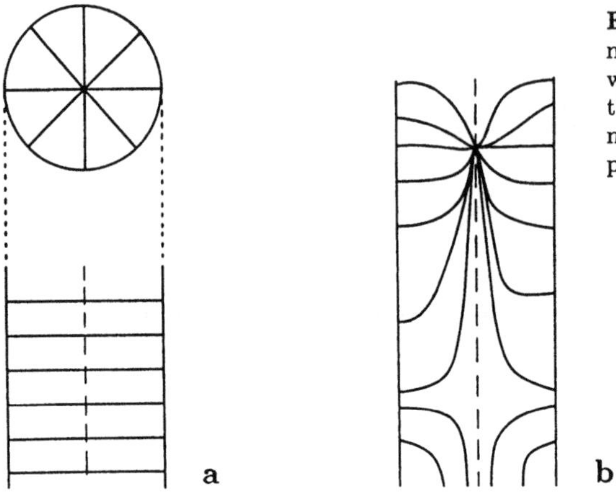

Fig. 7.3. (**a**) Unstable axial disclination of strength +1 in a cylinder with homeotropic boundary conditions. (**b**) Escape into the third dimension with two types of singular points

7.1.2 The Perpendicular Disclination Line

This type of singularity (De Gennes 1968b) is a straight line and the corresponding director field is such that the director only has components along the disclination and in one fixed direction perpendicular to the disclination. Without loss of generality the disclination may be assumed to lie along the z axis. Choosing the fixed direction along the x axis the director field is given by

$$\boldsymbol{n}(\boldsymbol{r}) = [\cos\phi(x,y),\ \ 0,\ \ \sin\phi(x,y)] \quad . \tag{7.16}$$

An example of such a situation is given in Fig. 7.4. It is of practical interest because of the importance of a twisted nematic layer in display applications (Geurst et al. 1975). The mathematical treatment of this type of singularity proceeds entirely by analogy with the axial disclination line. It is easily seen that the perpendicular disclinations are also described by the solutions (7.7) and that the distortion free energy per unit length is given by expression (7.8).

The set of flux lines belonging to a particular perpendicular disclination is analogous to (7.9) and described by the differential equation

$$\begin{aligned} \frac{dz}{dx} &= \tan\phi(x,y) = \tan(m\psi + \phi_0) \\ &= \tan\left[m \arctan\left(\frac{y}{x}\right) + \phi_0\right] \quad . \end{aligned} \tag{7.17}$$

This differential equation can be solved by simple quadrature. The pattern of flux lines appears to be very simple in the plane $y = 0$. It follows directly

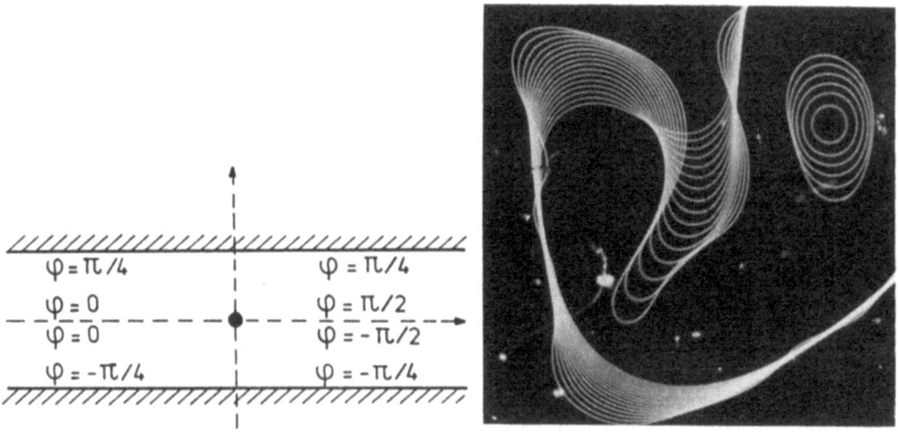

Fig. 7.4. **(a)** Boundary values for the director angle ϕ in a cross section of a $\frac{\pi}{2}$-twisted nematic layer perpendicular to the disclination line. **(b)** Perpendicular disclination lines separating regions of opposite twist. This is a multi-exposure photograph showing how the lines move due to their curvature while one twist region grows at the expense of the other (courtesy of C.J. Gerritsma)

from (7.17) that the pattern of flux lines in this plane is described by

$$\frac{dz}{dx} = \tan\phi_0 \quad , \quad x>0 \quad ;$$
$$\frac{dz}{dx} = \frac{-1}{\tan\phi_0} \quad , \quad x<0 \quad , \quad 2m \text{ odd} \quad ;$$
$$\frac{dz}{dx} = \tan\phi_0 \quad , \quad x<0 \quad , \quad 2m \text{ even} \quad .$$

This means that in the case $\phi_0 = 0$ the flux lines are perpendicular to the disclination line in the half-plane $x>0$ and parallel ($2m$ odd) or perpendicular ($2m$ even) to the disclination line in the half-plane $x<0$.

As an example, the full analytical expression for the pattern of flux lines of the perpendicular disclination with Frank index $m = 1$ and $\phi_0 = 0$ will be given. This pattern satisfies the equation

$$\frac{dz}{dx} = \frac{y}{x}$$

and is described accordingly by

$$z = y \ln|x| + C \quad ,$$

where C is an arbitrary constant. Here a given flux line lies in a plane which is perpendicular to the y axis.

Finally it should be remarked that the perpendicular disclination line appears in chiral nematics as well. In contrast to the axial disclination line the perpendicular disclination line induces distortions which are partly of twist character. This difference between the two types can easily be verified by substituting the accompanying director fields in the expression (5.24) for the distortion free energy. It is found that an axial disclination gives rise to distortions that have only splay and bend character.

7.2 Fluctuations in the Nematic Phase

The nematic-isotropic transition temperature is often referred to as the clearing point because of the turbidity of the nematic state and the transparency of the isotropic state. This means that light is strongly scattered in the nematic state. Any influence of the density fluctuations must be small as these also contribute to the scattering by the isotropic fluid. The turbidity of nematics must be attributed to the orientational fluctuations of the molecules, which in turn give rise to fluctuations in the elements of the optical dielectric tensor, leading to the strong scattering of light. The calculation of the orientational fluctuations belongs to the realm of statistical mechanics. A prerequisite for such a calculation is the knowledge of the intermolecular interactions, which are largely unknown. For that reason a hybrid approach is adopted, in which the macroscopic theory plays a prominent part. The relevant fluctuations are now conceived as the fluctuations of the components $Q_{\alpha\beta}$ of the macroscopic tensor order parameter with respect to their equilibrium values, i.e. the fluctuations of average quantities are introduced. The magnitude of these fluctuations is then calculated with the aid of the microscopic theory of fluctuations starting from the macroscopic Frank free energy associated with the fluctuations. In this section some details are given of this frequently used theory because the calculation of fluctuations is indispensable for a discussion of the light scattering in nematics.

The starting point of the fluctuation theory is the assumption that the components $Q_{\alpha\beta}(\boldsymbol{r})$ of the local tensor order parameter are given by

$$Q_{\alpha\beta}(\boldsymbol{r}) = S(T)[N_{\alpha\beta}(\boldsymbol{r}) - \tfrac{1}{3}\delta_{\alpha\beta}] \tag{7.18}$$

and that the components of the tensor $\tilde{\boldsymbol{N}}(\boldsymbol{r})$ are the relevant fluctuating quantities. In order to calculate the thermal behaviour of the fluctuations a knowledge of the energy of these fluctuations is required. For that purpose the free energy density $f(\boldsymbol{r})$ belonging to a field $\tilde{\boldsymbol{N}}(\boldsymbol{r})$ is expanded about the state of uniform parallel alignment. Assuming the spatial derivatives of the field to be small, the free energy density may be approximated by $f(\boldsymbol{r}) = f_0 + f_{\mathrm{d}}(\boldsymbol{r})$, where f_0 is the free energy density of the state of uniform parallel alignment and $f_{\mathrm{d}}(\boldsymbol{r})$ the distortion free energy density (5.24). Rewriting $f_{\mathrm{d}}(\boldsymbol{r})$ in terms of $\boldsymbol{N}$, the following expression results

$$
\begin{aligned}
f_{\mathrm{d}}(\boldsymbol{r}) = & \tfrac{1}{2}K_1 N_{\alpha\beta}(\boldsymbol{r})[\partial_\gamma N_{\alpha\gamma}(\boldsymbol{r})][\partial_\mu N_{\beta\mu}(\boldsymbol{r})] \\
& + \tfrac{1}{2}K_2\{N_{\alpha\beta}(\boldsymbol{r})[\partial_\gamma N_{\alpha\mu}(\boldsymbol{r})][\partial_\gamma N_{\beta\mu}(\boldsymbol{r})] \\
& - N_{\alpha\beta}(\boldsymbol{r})[\partial_\gamma N_{\alpha\mu}(\boldsymbol{r})][\partial_\mu N_{\beta\gamma}(\boldsymbol{r})] \\
& - N_{\alpha\beta}(\boldsymbol{r})[\partial_\alpha N_{\beta\gamma}(\boldsymbol{r})][\partial_\mu N_{\mu\gamma}(\boldsymbol{r})]\} \\
& + \tfrac{1}{2}K_3 N_{\alpha\beta}(\boldsymbol{r})[\partial_\alpha N_{\beta\gamma}(\boldsymbol{r})][\partial_\mu N_{\mu\gamma}(\boldsymbol{r})] \\
& - \tfrac{1}{2}\mu_0^{-1}\Delta\chi N_{\alpha\beta}(\boldsymbol{r})B_\alpha B_\beta \quad .
\end{aligned}
\tag{7.19}
$$

It should be remarked here that, from a statistical mechanical point of view, the outlined procedure is only allowed for the calculation of fluctuations at low temperatures. According to statistical mechanics the energy must be expanded around the energy of the ground state, i.e. the perfectly aligned state existing at $T = 0$. For the calculation of the low temperature properties of the system it suffices to consider only terms, that are quadratic in the fluctuations. At higher temperatures, however, and certainly near the NI transition temperature the energy expansion must contain higher order terms as well, i.e. the interaction between the fluctuations must be taken into account. Clearly an expansion of the energy around the energy of the ground state does not involve temperature-dependent coupling constants. On the other hand the usual procedure to calculate the fluctuations is to start from a perfectly aligned state at an arbitrary temperature. Next the distortion free energy is expanded up to quadratic terms in the distortions giving rise to temperature-dependent coupling constants (the elastic constants depending on S). The fluctuations are then identified with the distortions of the director field. From a microscopic point of view, however, the molecules fluctuate around their average orientation, i.e. the director, even in the distorted state. Clearly the theory lacks a proper justification from the statistical mechanical point of view and the quantitative validity of this rather hybrid approach is not directly obvious.

Realizing the approximations involved, the fluctuations in the nematic phase are calculated in the following way. First of all a magnetic field is imposed along the z axis for reasons of convenience, i.e. the optical axis of the nematic is fixed along the z axis. Next the fluctuations are assumed to be small. This means in terms of a small parameter α that $1 - N_{zz}(\boldsymbol{r})$, $N_{xx}(\boldsymbol{r})$, $N_{yy}(\boldsymbol{r})$ and $N_{xy}(\boldsymbol{r})$ are of the order α^2 and can be disregarded compared with $N_{xz}(\boldsymbol{r})$ and $N_{yz}(\boldsymbol{r})$ that are of order α. Next the distortion free energy is considered up to order α^2. In terms of the fluctuating quantities $N_{xz}(\boldsymbol{r})$ and $N_{yz}(\boldsymbol{r})$ this expression is given by

$$
\begin{aligned}
f_{\mathrm{d}}(\boldsymbol{r}) = {} & \tfrac{1}{2}K_1[\partial_x N_{xz}(\boldsymbol{r}) + \partial_y N_{yz}(\boldsymbol{r})]^2 + \tfrac{1}{2}K_2[\partial_y N_{xz}(\boldsymbol{r}) - \partial_x N_{yz}(\boldsymbol{r})]^2 \\
& + \tfrac{1}{2}K_3\{[\partial_z N_{xz}(\boldsymbol{r})]^2 + [\partial_z N_{yz}(\boldsymbol{r})]^2\} \\
& + \tfrac{1}{2}\mu_0^{-1}\Delta\chi B^2[N_{xz}^2(\boldsymbol{r}) + N_{yz}^2(\boldsymbol{r})] \quad , \qquad (7.20)
\end{aligned}
$$

where use is made of $N_{\alpha\beta}(\boldsymbol{r}) = N_{\alpha\gamma}(\boldsymbol{r})N_{\gamma\beta}(\boldsymbol{r})$. Next the local fluctuations are expanded in a Fourier series

$$
N_{\alpha\beta}(\boldsymbol{r}) = \sum_{\boldsymbol{k}} N_{\alpha\beta}(\boldsymbol{k})\exp(\mathrm{i}\boldsymbol{k}\cdot\boldsymbol{r}) \quad , \quad \text{where} \qquad (7.21)
$$

$$
N_{\alpha\beta}(\boldsymbol{k}) = \Omega^{-1}\int d\boldsymbol{r}\, N_{\alpha\beta}(\boldsymbol{r})\exp(-\mathrm{i}\boldsymbol{k}\cdot\boldsymbol{r}) \qquad (7.22)
$$

with Ω denoting the volume of the system. Since $N_{\alpha\beta}(\boldsymbol{r})$ is real it obeys

$$
N_{\alpha\beta}^*(\boldsymbol{k}) = N_{\alpha\beta}(-\boldsymbol{k}) \quad , \qquad (7.23)
$$

where $N_{\alpha\beta}^*(\boldsymbol{k})$ is the complex conjugate of $N_{\alpha\beta}(\boldsymbol{k})$. Substitution of (7.21) into (7.20) and integration of the resulting expression over the volume of the system gives the following distortion free energy up to order α^2,

$$
\begin{aligned}
F_{\mathrm{d}} = {} & \tfrac{1}{2}\Omega\sum_{\boldsymbol{k}}\{(K_1k_x^2 + K_2k_y^2 + K_3k_z^2 + \mu_0^{-1}\Delta\chi B^2)|N_{xz}(\boldsymbol{k})|^2 \\
& + (K_1k_y^2 + K_2k_x^2 + K_3k_z^2 + \mu_0^{-1}\Delta\chi B^2)|N_{yz}(\boldsymbol{k})|^2 \\
& + (K_1 - K_2)k_xk_y[N_{xz}^*(\boldsymbol{k})N_{yz}(\boldsymbol{k}) + N_{xz}(\boldsymbol{k})N_{yz}^*(\boldsymbol{k})]\} \quad . \qquad (7.24)
\end{aligned}
$$

The value of F_{d} depends strongly on the values of the amplitudes $N_{\alpha\beta}(\boldsymbol{k})$. Conversely the appearance of a given set of amplitudes $\{N_{\alpha\beta}(\boldsymbol{k})\}$ is described by a probability distribution of the form

$$
P[\{N_{\alpha\beta}(\boldsymbol{k})\}] = Z^{-1}\exp(-\beta F_{\mathrm{d}}) \quad , \qquad (7.25)
$$

where $\beta = (k_{\mathrm{B}}T)^{-1}$ with k_{B} Boltzmann's constant and T the temperature. Z is a normalization constant such that the probability distribution is normalized to one. The amplitudes $N_{\alpha\beta}(\boldsymbol{k})$ are complex and can be written as

$$
N_{\alpha\beta}(\boldsymbol{k}) = r_{\alpha\beta}(\boldsymbol{k}) + \mathrm{i}s_{\alpha\beta}(\boldsymbol{k}) \qquad (7.26)
$$

with $r_{\alpha\beta}(\boldsymbol{k})$ and $s_{\alpha\beta}(\boldsymbol{k})$ being real variables. Next it should be remarked that the number of independent variables is reduced considerably because of the condition (7.23), which implies directly

$$r_{\alpha\beta}(\boldsymbol{k}) = r_{\alpha\beta}(-\boldsymbol{k}) \quad ; \quad s_{\alpha\beta}(\boldsymbol{k}) = -s_{\alpha\beta}(-\boldsymbol{k}) \quad . \tag{7.27}$$

Substituting (7.27) into (7.24) and noting that $s_{\alpha\beta}(0) = 0$ the following probability distribution of truly independent variables is obtained

$$P = Z^{-1} \exp[-\beta F_{\mathrm{d}}(0)] \prod_{\boldsymbol{k}}{}' \exp[-\beta F_{\mathrm{d}}(\boldsymbol{k})] \tag{7.28}$$

with $\prod_{\boldsymbol{k}}'$ being a product over all $\boldsymbol{k}$ values except $\boldsymbol{k} = 0$, that label independent variables (meaning that the vectors $-\boldsymbol{k}$ are excluded), whereas

$$F_{\mathrm{d}}(0) = \tfrac{1}{2}\Omega\mu_0^{-1}\Delta\chi B^2[r_{xz}^2(0) + r_{yz}^2(0)] \quad \text{and} \tag{7.29}$$

$$\begin{aligned} F_{\mathrm{d}}(\boldsymbol{k}) = {} & \Omega(K_1k_x^2 + K_2k_y^2 + K_3k_z^2 + \mu_0^{-1}\Delta\chi B^2)[r_{xz}^2(\boldsymbol{k}) + s_{xz}^2(\boldsymbol{k})] \\ & + \Omega(K_1k_y^2 + K_2k_x^2 + K_3k_z^2 + \mu_0^{-1}\Delta\chi B^2)[r_{yz}^2(\boldsymbol{k}) + s_{yz}^2(\boldsymbol{k})] \\ & + 2\Omega(K_1 - K_2)k_xk_y[r_{xz}(\boldsymbol{k})r_{yz}(\boldsymbol{k}) + s_{xz}(\boldsymbol{k})s_{yz}(\boldsymbol{k})] \quad . \end{aligned} \tag{7.30}$$

In order to calculate the thermal averages $\langle|N_{\alpha\beta}(\boldsymbol{k})|^2\rangle = \langle r_{\alpha\beta}^2(\boldsymbol{k})\rangle + \langle s_{\alpha\beta}^2(\boldsymbol{k})\rangle$ use is made of the fact that a Gaussian distribution of the form

$$P = Z^{-1} \exp\left(-\frac{1}{2}\sum_{i,j} A_{ij}\sigma_i\sigma_j\right) \quad , \tag{7.31}$$

where σ_i represents a real-valued stochastic variable and Z is a normalization constant, gives rise to the following expression for the thermal averages in question

$$\langle\sigma_i\sigma_j\rangle = A_{ij}^{-1} \tag{7.32}$$

with A_{ij}^{-1} being an element of the matrix inverse to A_{ij} (Landau and Lifshitz 1980). In the present case the stochastic variables are $r_{\alpha\beta}(\boldsymbol{k})$ and $s_{\alpha\beta}(\boldsymbol{k})$ and

$$A_{11}(\boldsymbol{k}) = 2\beta\Omega(K_1k_x^2 + K_2k_y^2 + K_3k_z^2 + \mu_0^{-1}\Delta\chi B^2) \quad , \tag{7.33a}$$

$$A_{12}(\boldsymbol{k}) = A_{21}(\boldsymbol{k}) = 2\beta\Omega(K_1 - K_2)k_xk_y \quad , \tag{7.33b}$$

$$A_{22}(\boldsymbol{k}) = 2\beta\Omega(K_1k_y^2 + K_2k_x^2 + K_3k_z^2 + \mu_0^{-1}\Delta\chi B^2) \quad . \tag{7.33c}$$

Now (7.32) leads to

$$\langle |N_{xz}(\boldsymbol{k})|^2 \rangle = 2A_{11}^{-1} \quad , \tag{7.34a}$$

$$\langle |N_{yz}(\boldsymbol{k})|^2 \rangle = 2A_{22}^{-1} \quad , \tag{7.34b}$$

$$\begin{aligned} &\langle N_{xz}^*(\boldsymbol{k}) N_{yz}(\boldsymbol{k}) + N_{xz}(\boldsymbol{k}) N_{yz}^*(\boldsymbol{k}) \rangle \\ &\quad = 2\langle r_{xz}(\boldsymbol{k}) r_{yz}(\boldsymbol{k}) \rangle + 2\langle s_{xz}(\boldsymbol{k}) s_{yz}(\boldsymbol{k}) \rangle = 4A_{12}^{-1} \quad . \end{aligned} \tag{7.34c}$$

Introducing

$$L(\boldsymbol{k}) = A_{11}(\boldsymbol{k}) A_{22}(\boldsymbol{k}) - A_{12}^2(\boldsymbol{k}) \tag{7.35}$$

the elements of the inverse matrix A^{-1} are given by

$$\begin{aligned} A_{11}^{-1}(\boldsymbol{k}) &= \frac{A_{22}(\boldsymbol{k})}{L(\boldsymbol{k})} \quad ; \\ A_{21}^{-1}(\boldsymbol{k}) = A_{12}^{-1}(\boldsymbol{k}) &= \frac{A_{12}(\boldsymbol{k})}{L(\boldsymbol{k})} \quad ; \\ A_{22}^{-1}(\boldsymbol{k}) &= \frac{A_{11}(\boldsymbol{k})}{L(\boldsymbol{k})} \quad . \end{aligned} \tag{7.36}$$

The Eqs. (7.34) are the main result of the fluctuation theory for nematics. They allow the calculation of a number of physical quantities, e.g. the correlation length and the correlation between the components of the tensor $\tilde{\boldsymbol{N}}$ taken at different points $\boldsymbol{r}_1$ and $\boldsymbol{r}_2$, i.e.

$$\langle N_{\alpha\beta}(\boldsymbol{r}_1) N_{\alpha\beta}(\boldsymbol{r}_2) \rangle = \sum_{\boldsymbol{k}} \langle |N_{\alpha\beta}(\boldsymbol{k})|^2 \rangle \exp\left[\mathrm{i}\boldsymbol{k}(\boldsymbol{r}_2 - \boldsymbol{r}_1) \right] \quad . \tag{7.37}$$

Here, in order to avoid unnecessarily complicated mathematics, the discussion is confined to the one-constant approximation: $K_1 = K_2 = K_3 = K$. Now the relevant thermal averages are given by

$$\langle |N_{xz}(\boldsymbol{k})|^2 \rangle = \langle |N_{yz}(\boldsymbol{k})|^2 \rangle = \frac{k_\mathrm{B} T}{\Omega (K k^2 + \mu_0^{-1} \Delta\chi B^2)} \quad . \tag{7.38}$$

In order to calculate the correlation length a summation over $\boldsymbol{k}$ must be carried out. However, only a finite number of Fourier components can be independent, because the system consists of a finite number of molecules, i.e. a finite number of degrees of freedom. Assuming that the fluid consists of N molecules, only N $\boldsymbol{k}$ vectors can be chosen to be independent. These vectors are selected according to a procedure of Debye. This means that the $\boldsymbol{k}$-mode spectrum is cut off and limited to a sphere of radius k_D given by

$$k_\mathrm{D} = (6\pi^2 \varrho)^{1/3} \tag{7.39}$$

with $\varrho = N/\Omega$ being the density of the system. With $R = |\boldsymbol{r}_2 - \boldsymbol{r}_1|$ the relevant correlation functions are now given by

$$\begin{aligned} \langle N_{xz}(0) N_{xz}(R) \rangle &= \langle N_{yz}(0) N_{yz}(R) \rangle \\ &= \sum_{\substack{\boldsymbol{k} \\ |\boldsymbol{k}| \leqq k_\mathrm{D}}} \frac{\exp(\mathrm{i}\boldsymbol{k} \cdot \boldsymbol{R})}{Kk^2 + \mu_0^{-1} \Delta\chi B^2} \quad . \end{aligned} \tag{7.40}$$

Replacing

$$\sum_{\boldsymbol{k}} \quad \text{by} \quad \frac{\Omega}{(2\pi)^3} \int d\boldsymbol{k}$$

and changing over to polar coordinates it is found that

$$\begin{aligned} \langle N_{xz}(0) N_{xz}(R) \rangle &= \langle N_{yz}(0) N_{yz}(R) \rangle \\ &= \frac{k_\mathrm{B} T}{2\pi^2 R} \int_0^{k_\mathrm{D}} \frac{k \sin(kR)}{Kk^2 + \mu_0^{-1} \Delta\chi B^2} \end{aligned} \tag{7.41}$$

or replacing k_D by ∞

$$\langle N_{xz}(0) N_{xz}(R) \rangle = \langle N_{yz}(0) N_{yz}(R) \rangle = \frac{k_\mathrm{B} T}{4\pi K R} \exp\left(-\frac{R}{\xi}\right) \quad . \tag{7.42}$$

The quantity ξ is called the magnetic coherence length and is given by

$$\xi = \left[\frac{K}{\mu_0^{-1} \Delta\chi B^2}\right]^{1/2} \quad . \tag{7.43}$$

In the case of zero field ξ is infinite, and the correlation decreases slowly as $1/R$.

In the literature the fluctuation theory is usually discussed in terms of the Fourier components of $n_x(\boldsymbol{r})$ and $n_y(\boldsymbol{r})$. Then the nematic is conceived as a vector field instead of a second rank tensor field. Consequently the right-hand side of (7.42) is equated for example to the correlation function $\langle n_x(0) n_x(R) \rangle$. This is fundamentally incorrect since $\langle n_x(0) n_x(R) \rangle = 0$ due to the local head-tail symmetry.

In order to test the validity of the final results of the fluctuation theory, $\langle N_{xz}^2(\boldsymbol{r})\rangle$ will now be calculated. This quantity must be small in order to justify the use of the expansion (7.19), i.e. the continuum theory. Substituting $R=0$ into (7.40) leads to

$$\begin{aligned}\langle N_{xz}^2(0)\rangle &= \frac{k_\mathrm{B}T}{2\pi^2}\int_0^{k_\mathrm{D}} \frac{k^2\,dk}{Kk^2+\mu_0^{-1}\Delta\chi B^2} \\ &= \frac{k_\mathrm{B}Tk_\mathrm{D}}{2\pi^2 K}\left[1-(\xi k_\mathrm{D})^{-1}\arctan(\xi k_\mathrm{D})\right] \quad . \end{aligned} \tag{7.44}$$

Using $k_\mathrm{D}\approx 0.5\times 10^{10}\,\mathrm{m}^{-1}$ and $K\approx 5\times 10^{-12}\,\mathrm{N}$ one obtains as a typical lower bound $\langle N_{xz}^2(0)\rangle \geq 0.24$ at a clearing point of 350 K in zero magnetic field meaning $\langle N_{zz}^2(0)\rangle \leq 0.52$. This strongly contradicts the assumption that $1-N_{zz}^2(0)$ must be small. In order to have small fluctuations the right-hand side of (7.44) must be small, i.e.

$$\mu_0^{-1}\Delta\chi B^2 \gg Kk_\mathrm{D}^2 \quad .$$

However, this condition is not satisfied experimentally.

One must conclude that the assumptions of the present fluctuation theory for nematics do not guarantee a quantitatively correct description near the clearing point. Care has to be exercised when interpreting experimental data in terms of this hybrid approach.

7.3 Light Scattering by Nematics

The strong light scattering by nematics is caused by the thermal fluctuations of the tensor order parameter. These orientational fluctuations give rise to fluctuations in the components of the dielectric tensor, which in turn scatter light strongly, resulting in the turbid appearance of the nematic. Before discussing this phenomenon, however, the basic formula is derived for the scattering of light by fluctuations of the components of the dielectric tensor (Jackson 1975).

The appropriate quantity for the description of the scattering of light is the differential cross-section per unit solid angle, which is derived in the following way. The starting point is the Maxwell equations in the absence of sources

$$\begin{aligned}&\partial_\alpha D_\alpha = 0 \quad , \quad \varepsilon_{\alpha\beta\gamma}\partial_\beta E_\gamma = -\partial_t B_\alpha \quad , \\ &\partial_\alpha B_\alpha = 0 \quad , \quad \varepsilon_{\alpha\beta\gamma}\partial_\beta H_\gamma = \partial_t D_\alpha \quad . \end{aligned} \tag{7.45}$$

Next the magnetic properties are neglected, i.e. $\boldsymbol{B}-\mu_0\boldsymbol{H}$ is set equal to zero. Then Eqs. (7.45) give rise to the following wave equation for the dielectric displacement $\boldsymbol{D}(\boldsymbol{r},t)$

$$\partial_\beta\partial_\beta D_\alpha(\boldsymbol{r},t) - \frac{1}{c_2}\partial_t^2 D_\alpha(\boldsymbol{r},t) = \partial_\beta\partial_\beta P_\alpha(\boldsymbol{r},t) - \partial_\alpha\partial_\beta P_\beta(\boldsymbol{r},t) \quad , \tag{7.46}$$

where $P_\alpha(\boldsymbol{r},t) = D_\alpha(\boldsymbol{r},t) - \varepsilon_0 E_\alpha(\boldsymbol{r},t)$ denotes the polarization of the medium. Substitution of

$$D_\alpha(\boldsymbol{r},t) = \int_{-\infty}^{+\infty} d\omega\, D_\alpha(\boldsymbol{r},\omega)\, \mathrm{e}^{-\mathrm{i}\omega t} \tag{7.47a}$$

$$P_\alpha(\boldsymbol{r},t) = \int_{-\infty}^{+\infty} d\omega\, P_\alpha(\boldsymbol{r},\omega)\, \mathrm{e}^{-\mathrm{i}\omega t} \tag{7.47b}$$

into (7.46) gives

$$(\partial_\beta\partial_\beta + k^2)D_\alpha(\boldsymbol{r},\omega) = -Z_\alpha(\boldsymbol{r},\omega) \tag{7.48}$$

where $k^2 = \omega^2/c^2$ and

$$Z_\alpha(\boldsymbol{r},\omega) = \partial_\alpha\partial_\beta P_\beta(\boldsymbol{r},\omega) - \partial_\beta\partial_\beta P_\alpha(\boldsymbol{r},\omega) \quad . \tag{7.49}$$

The formal solution of (7.48) is

$$D_\alpha(\boldsymbol{r},\omega) = D_\alpha^{(0)}(\boldsymbol{r},\omega) + \frac{1}{4\pi}\int d\boldsymbol{r}'\frac{\exp(\mathrm{i}k|\boldsymbol{r}-\boldsymbol{r}'|)}{|\boldsymbol{r}-\boldsymbol{r}'|}Z_\alpha(\boldsymbol{r}',\omega) \quad , \tag{7.50}$$

where $D_\alpha^{(0)}(\boldsymbol{r},\omega)$ is the solution of the homogeneous wave equation. The field far away from the scattering region is then given by

$$D_\alpha(\boldsymbol{r},\omega) = D_\alpha^{(0)}(\boldsymbol{r},\omega) + D_\alpha^{(\mathrm{s})}(\boldsymbol{p},\omega)\frac{\mathrm{e}^{ikr}}{r} \quad , \tag{7.51}$$

where $D_\alpha^{(\mathrm{s})}(\boldsymbol{p},\omega)$ denotes the scattering amplitude

$$D_\alpha^{(\mathrm{s})}(\boldsymbol{p},\omega) = \frac{1}{4\pi}\int d\boldsymbol{r}' \exp(-\mathrm{i}k\boldsymbol{p}\cdot\boldsymbol{r}')Z_\alpha(\boldsymbol{r}',\omega) \tag{7.52}$$

with $\boldsymbol{p}$ being the unit vector in the direction of $\boldsymbol{r}$. The integrand of (7.52) is confined to a finite region of space, namely the region occupied by the scattering medium. It is clear that the scattering amplitude is known as soon as the polarization is known. Here it suffices to approximate the polarization by

$$P_\alpha(\boldsymbol{r},t) = \varepsilon_0 \chi_{\alpha\beta}(\boldsymbol{r}) E_\beta(\boldsymbol{r},t) \quad ,$$

where $\chi_{\alpha\beta}(\boldsymbol{r})$ is the electric susceptibility tensor. Consequently the formula (7.52) for the scattering amplitude is not an explicit expression.

In the following, an approximate value of the scattering amplitude is obtained by replacing $\varepsilon_0 E_\beta(\boldsymbol{r},\omega)$ by the solution $D_\beta^{(0)}(\boldsymbol{r},\omega)$ of the homogeneous wave equation. This means that the polarization is approximated by

$$P_\alpha(\boldsymbol{r},\omega) = \chi_{\alpha\beta}(\boldsymbol{r}) D_\beta^{(0)}(\boldsymbol{r},\omega) \quad , \tag{7.53}$$

i.e. the effect of the scattered wave on the polarization is neglected. Mathematically speaking this approximation, which is called the Born approximation, corresponds to a first iteration of (7.50). The starting point of the calculation is the plane wave

$$D_\alpha^{(0)}(\boldsymbol{r},\omega) = i_\alpha D_0 \exp(\mathrm{i}k\boldsymbol{p}_0 \cdot \boldsymbol{r}) \tag{7.54}$$

which propagates in the direction $\boldsymbol{p}_0$ with amplitude D_0 and a polarization direction $\boldsymbol{i}$ such that $\boldsymbol{i} \cdot \boldsymbol{p}_0 = 0$. It is clear that this wave is a solution of the homogeneous wave equation. In order to obtain an expression for the scattering amplitude the components of the electric susceptibility tensor are expanded in a Fourier series according to

$$\chi_{\alpha\beta}(\boldsymbol{r}) = \sum_{\boldsymbol{q}} \chi_{\alpha\beta}(\boldsymbol{q}) \exp(-\mathrm{i}\boldsymbol{q} \cdot \boldsymbol{r}) \quad , \tag{7.55a}$$

where

$$\chi_{\alpha\beta}(\boldsymbol{q}) = \frac{1}{\Omega} \int d\boldsymbol{r}' \chi_{\alpha\beta}(\boldsymbol{r}') \exp(\mathrm{i}\boldsymbol{q} \cdot \boldsymbol{r}') \tag{7.55b}$$

with Ω denoting the volume of the scattering sample. Consequently the Born approximation yields

$$\begin{aligned} Z_\alpha(\boldsymbol{r},\omega) &= \partial_\alpha \partial_\beta [\chi_{\beta\mu}(\boldsymbol{r}) D_\mu^{(0)}(\boldsymbol{r},\omega)] - \partial_\beta \partial_\beta [\chi_{\alpha\mu}(\boldsymbol{r}) D_\mu^{(0)}(\boldsymbol{r},\omega)] \\ &= D_0 i_\mu \sum_{\boldsymbol{q}} [\chi_{\alpha\mu}(\boldsymbol{q})(kp_{0\beta} - q_\beta) - \chi_{\beta\mu}(\boldsymbol{q})(kp_{0\alpha} - q_\alpha)] \\ &\quad \times (kp_{0\beta} - q_\beta) \exp[\mathrm{i}(k\boldsymbol{p}_0 - \boldsymbol{q}) \cdot \boldsymbol{r}] \quad . \end{aligned} \tag{7.56}$$

Substituting (7.56) into (7.52) and using

$$\int d\boldsymbol{r}' \exp\left[\mathrm{i}(\boldsymbol{k}_1 - \boldsymbol{k}_2)\cdot\boldsymbol{r}'\right] = \Omega\delta_{\boldsymbol{k}_1\boldsymbol{k}_2} \tag{7.57}$$

results in the following expression for the scattering amplitude:

$$D_\alpha^{(\mathrm{s})}(\boldsymbol{p},\omega) = \frac{D_0\Omega\omega^2}{4\pi c^2} i_\mu p_\beta[\chi_{\alpha\mu}(\boldsymbol{q})p_\beta - \chi_{\beta\mu}(\boldsymbol{q})p_\alpha] \quad , \tag{7.58}$$

where $\boldsymbol{q} = k(\boldsymbol{p}_0 - \boldsymbol{p})$ is the difference between the incident and scattered wave vector. The projection of the scattering amplitude onto a given polarization direction $\boldsymbol{f}$, where the unit vector $\boldsymbol{f}$ is perpendicular to the direction of the outgoing wave ($p_\alpha f_\alpha = 0$), gives

$$D_\alpha^{(\mathrm{s})} f_\alpha = \frac{D_0\Omega\omega^2}{4\pi c^2} f_\alpha\chi_{\alpha\beta}(\boldsymbol{q})i_\beta \quad . \tag{7.59}$$

Now the differential scattering cross-section per unit solid angle of the outgoing beam around the direction $\boldsymbol{p}_0$ is defined as

$$\sigma = \frac{\Omega^2\omega^4}{16\pi^2 c^4}\langle|f_\alpha\chi_{\alpha\beta}(\boldsymbol{q})i_\beta|^2\rangle \quad , \tag{7.60}$$

where the brackets indicate a thermal average, which originates from the stochastic nature of the fluctuations of the components of the electric susceptibility.

The electric susceptibility tensor $\tilde{\chi}(\boldsymbol{r})$ is related to the tensor order parameter $\tilde{\boldsymbol{Q}}$ in the following way (see Sect. 5.2)

$$\chi_{\alpha\beta}(\boldsymbol{r}) = \tfrac{1}{3}(\chi_{\parallel} + 2\chi_{\perp})\delta_{\alpha\beta} + \Delta\chi_{\max} Q_{\alpha\beta}(\boldsymbol{r}) \quad , \tag{7.61}$$

where $\Delta\chi_{\max}$ denotes the anisotropy in the susceptibility in the perfectly aligned state, i.e. the phase without fluctuations ($S = 1$). The Fourier components $\chi_{\alpha\beta}(\boldsymbol{q})$ can now easily be expressed in terms of the Fourier components $Q_{\alpha\beta}(\boldsymbol{q})$ of the tensor order parameter. Because of

$$Q_{\alpha\beta}(\boldsymbol{r}) = \sum_{\boldsymbol{q}} Q_{\alpha\beta}(\boldsymbol{q})\exp(-\mathrm{i}\boldsymbol{q}\cdot\boldsymbol{r}) \tag{7.62}$$

and the identity $\Delta\chi_{\max} = \Delta\varepsilon_{\max}$ the following relation holds for $\boldsymbol{q} \neq 0$:

$$\chi_{\alpha\beta}(\boldsymbol{q}) = \Delta\varepsilon_{\max} Q_{\alpha\beta}(\boldsymbol{q}) \quad . \tag{7.63}$$

Consequently the differential scattering cross-section can be written as

$$\sigma = \frac{\Omega^2 \omega^4 \Delta\varepsilon_{\max}^2}{16\pi^2 c^4} \langle |f_\alpha Q_{\alpha\beta}(\boldsymbol{q}) i_\beta|^2 \rangle \tag{7.64}$$

and an appeal has to be made to the fluctuation theory of nematics in order to calculate this thermal average.

According to the usual theory of fluctuations $\boldsymbol{Q}_{\alpha\beta}(\boldsymbol{q})$ is approximated by $SN_{\alpha\beta}(\boldsymbol{q})$ for $\boldsymbol{q} \neq 0$. Consequently the diffrential scattering cross-section can be expressed in the following form

$$\sigma = \frac{\Omega^2 \omega^4 \Delta\varepsilon^2}{16\pi^2 c^4} \langle |f_\alpha N_{\alpha\beta}(\boldsymbol{q}) i_\beta|^2 \rangle \quad , \tag{7.65}$$

where $\Delta\varepsilon = \Delta\varepsilon_{\max} S$, or using (7.33)

$$\begin{aligned} \sigma = \frac{\Omega^2 \omega^4 \Delta\varepsilon^2}{8\pi^2 c^4} & [A_{11}^{-1}(f_x i_z + f_z i_x)^2 + A_{22}^{-1}(f_y i_z + f_z i_y)^2 \\ & + 2A_{12}^{-1}(f_x i_z + f_z i_x)(f_y i_z + f_z i_y)] \quad , \end{aligned} \tag{7.66}$$

with the optical axis of the nematic taken along the z axis. The expression (7.66) can be simplified considerably by resorting to the one-constant approximation. The scattering cross-section is then given by

$$\sigma = \frac{\Omega \omega^4 \Delta\varepsilon^2 k_{\mathrm{B}} T}{16\pi^2 c^4 (Kk^2 + \mu_0^{-1} \Delta\chi B^2)} [(f_x i_z + f_z i_x)^2 + (f_y i_z + f_z i_y)^2] \quad . \tag{7.67}$$

The formula derived for the differential scattering cross-section (7.67) allows the following conclusions, which have been confirmed experimentally (Chatelain 1951; Orsay Liquid Crystal Group 1969).

i) The scattering is only intense when the polarization vectors $\boldsymbol{i}$ and $\boldsymbol{f}$ are perpendicular to each other. Consider e.g. an incident beam with a polarization $\boldsymbol{i}$ along the optical axis of the sample (Fig. 7.5). Then the differential scattering cross section is proportional to $f_x^2 + f_y^2$. This implies directly that the scattering is maximal for $f_z = 0$, i.e. $\boldsymbol{i} \cdot \boldsymbol{f} = 0$.

ii) In the absence of a magnetic field the scattering becomes very pronounced for small scattering angles.

iii) A strong magnetic field can be expected to reduce the scattering because it suppresses the fluctuations (Martinand and Durand 1972).

The pronounced light scattering of the nematic state relative to the isotropic state is due to the fact that the elastic constants K_i $(i = 1, 2, 3)$

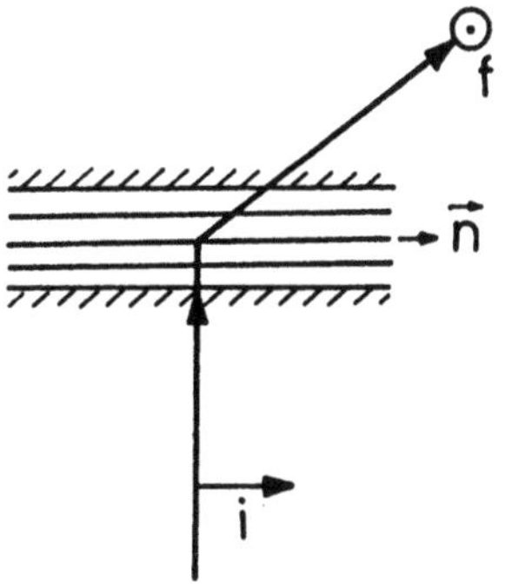

Fig. 7.5. Scattering configuration with the incident beam polarized parallel to the director

associated with the distortions of the director field are small compared to the compressibility. Accordingly, light scattering experiments have been used to determine the elastic constants of liquid crystals. Considering the approximations involved, however, the expression (7.66) for the differential scattering cross section does not seem to be reliable from a quantitative point of view, in spite of the fact that the functional dependence is observed experimentally. Hence it is not surprising that the values of the elastic constants thus obtained (Van der Meulen et al. 1984) deviate systematically from the values measured, in a much more direct way, using Frederiks transitions.

7.4 Optical Properties of Chiral Nematics

7.4.1 Introduction

Consider a planar chiral sample with pitch p. Choosing the z axis of the coordinate system along the helix axis, the helical structure is described by the director pattern

$$\boldsymbol{n} = [\cos(t_0 z),\ \sin(t_0 z),\ 0] \quad , \tag{7.68}$$

where $p = 2\pi/|t_0|$. In a right-handed coordinate system (7.68) describes a right-handed helix for $t_0>0$ and a left-handed helix for $t_0<0$. As already mentioned in Sect. 2.2, the helical structure of the medium gives rise to a number of spectacular optical phenomena occurring when a light beam of a given frequency ω incides along z. These can be summarized as follows:

(1) *Bragg Reflection.* The light beam is strongly reflected as soon as the wavelength of the beam in the medium is of the order of the pitch p, i.e. a reflection band is observed [see (2.4)].

(2) *Polarization.* For normal incidence the reflected light is circularly polarized. Decomposing the incident light beam in a right- and a left-circularly polarized light beam, the beam for which the electric vector exactly matches the helicoidal structure is completely reflected, whereas the other one is completely transmitted. The direction of rotation of the electric vector is reversed on reflection. As the direction of the light beam is also reversed, the sense of rotation of the electric vector (referred to the direction of the beam) is the same as before (see Fig. 7.6). This contrasts with normal substances, where the direction of rotation is not reversed upon reflection.

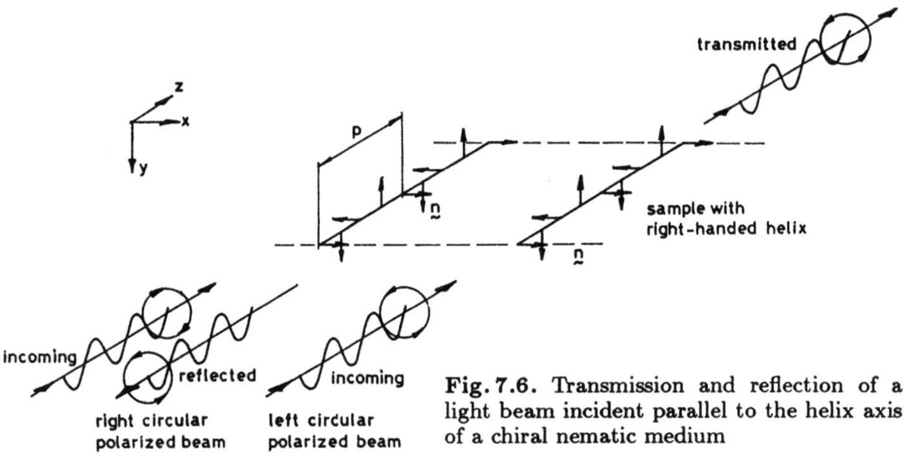

Fig. 7.6. Transmission and reflection of a light beam incident parallel to the helix axis of a chiral nematic medium

(3) *Rotary Power.* Circularly polarized waves of opposite sign traverse the medium with different speeds. Consequently a large phase difference can develop between the two waves, leading to a large rotary power. The reflection band separates two regimes of rotary power of opposite sign.

These phenomena can be completely understood in terms of the theory of De Vries (1951). This theory will be discussed in the next section, and boils down to the solving of the Maxwell equations for a chiral nematic medium. However, the observations (1) and (2) can also be directly understood in terms of the scattering properties discussed in the previous section. Consider the expression (7.60) for the differential scattering cross-section per unit solid angle. In the present case the incident wave vector $k\boldsymbol{p}_0$, the scattered wave vector $k\boldsymbol{p}$ and $\boldsymbol{q} = k(\boldsymbol{p}_0 - \boldsymbol{p})$ are all along the z axis. In order to calculate $\chi_{\alpha\beta}(\boldsymbol{q})$ use is made of the fact that a chiral nematic behaves locally as a nematic. Hence, taking into account the director field (7.68), the electric susceptibility tensor reads

$$\tilde{\chi}(\boldsymbol{r}) = \begin{pmatrix} \chi_\perp + \frac{1}{2}\Delta\chi[1+\cos(2t_0 z)] & \frac{1}{2}\Delta\chi\,\sin(2t_0 z) & 0 \\ \frac{1}{2}\Delta\chi\,\sin(2t_0 z) & \chi_\perp + \frac{1}{2}\Delta\chi[1-\cos(2t_0 z)] & 0 \\ 0 & 0 & \chi_\perp \end{pmatrix} . \tag{7.69}$$

Now it follows directly from (7.55b) that, for $\boldsymbol{q} \neq 0$, $\chi_{\alpha\beta}(\boldsymbol{q}) = 0$ except for $q = \pm 2t_0$. The physical solution corresponds to $q = 2t_0$, which is equivalent to expression (2.4) for Bragg reflection. Then the corresponding susceptibility tensor becomes

$$\tilde{\chi}(2t_0) = \tfrac{1}{4}\Delta\chi \begin{pmatrix} 1 & \mathrm{i} & 0 \\ \mathrm{i} & -1 & 0 \\ 0 & 0 & 0 \end{pmatrix} . \tag{7.70}$$

According to (7.64) the scattering depends on

$$f_\alpha \chi_{\alpha\beta}(2t_0) i_\beta = \tfrac{1}{4}\Delta\chi (f_x + \mathrm{i} f_y)(i_x + \mathrm{i} i_y) . \tag{7.71}$$

This relation implies immediately that the scattering does not occur when the incident light has the type of polarization $i_x = -\mathrm{i} i_y$. Evidently this is the situation of full transmission. The electric vector of the incident wave rotates in a direction opposite to that of the helicoidal structure. On the other hand maximal scattering occurs as soon as $i_x = \mathrm{i} i_y$, in which case the electric vector of the incident wave exactly matches the screw sense of the helix. The polarization of the reflected wave is given by $f_x = \mathrm{i} f_y$.

Before turning to the general discussion of the optical properties at arbitrary wavelength, it should be remarked that the situation with $p \gg \lambda$ is of great practical interest. Such a situation is found in the case of a twisted nematic cell, i.e. if the twist angle equals $\frac{\pi}{2}$ and the layer thickness is d the pitch is given by $p = 4d$. Such twisted nematic layers with $p \gg \lambda$ appear to act as "waveguides" in the sense that an incident light wave, which is linearly polarized parallel or perpendicular to the director at the surface of incidence, is completely transmitted, but with its polarization direction rotating with the director. This can be understood in a simple way as follows. Dividing the sample into m discrete layers, the principal optical axes of successive layers make an angle $\pi/2m$ with each other. If one assumes that the electric vector $\boldsymbol{E}$ is along $\boldsymbol{n}$ at the boundary of incidence, the component of $\boldsymbol{E}$ along $\boldsymbol{n}$ at the second layer is $E\,\cos(\pi/2m)$. The influence of the component of $\boldsymbol{E}$ perpendicular to $\boldsymbol{n}$ is neglected. This assumption can be justified for $p \gg \lambda$ as follows from the theory of De Vries (1951). After passage through m layers the component of the electric vector along the local director reads $E\,\cos^m(\pi/2m)$. Taking the limit $m \to \infty$ gives

$$E \lim_{m \to \infty} \cos^m\left(\frac{\pi}{2m}\right) = E \quad .$$

Consequently the incident light wave is fully transmitted, but its polarization direction rotates in accordance with the twist of the nematic medium.

7.4.2 General Theory

In order to understand the optical phenomena associated with the perpendicular incidence of light on a planar chiral nematic medium with the helix axis along the z axis, the Maxwell equations (7.45) must be solved. The helical structure is taken into account by substituting (7.68) for $\boldsymbol{n}(\boldsymbol{r})$ in the expressions

$$D_\alpha = \varepsilon_0 \varepsilon_\perp E_\alpha + \varepsilon_0 \Delta\varepsilon n_\alpha n_\beta E_\beta \quad , \tag{7.72a}$$

$$B_\alpha = \mu_0 (1 + \chi_\perp) H_\alpha + \mu_0 \Delta\chi n_\alpha n_\beta H_\beta \quad . \tag{7.72b}$$

Next a solution of the Maxwell equations is sought, which propagates along the z axis. Consequently the Maxwell equations simplify to

$$\partial_z E_x = -\partial_t B_y \quad , \qquad \partial_z E_y = \partial_t B_x \quad , \tag{7.73a}$$

$$\partial_z H_x = \partial_t D_y \quad , \qquad \partial_z H_y = -\partial_t D_x \quad , \tag{7.73b}$$

and $E_z = D_z = H_z = B_z = 0$. Substitution of (7.72) into (7.73) results in the following system of coupled differential equations for the quantities $E_x(z,t)$, $E_y(z,t)$, $H_x(z,t)$ and $H_y(z,t)$

$$\begin{aligned} \partial_z E_x = & - \mu_0 \partial_t [(1 + \chi_\perp + \tfrac{1}{2}\Delta\chi) H_y - \tfrac{1}{2}\Delta\chi \cos(2t_0 z) H_y \\ & + \tfrac{1}{2}\Delta\chi \sin(2t_0 z) H_x] \quad , \end{aligned} \tag{7.74a}$$

$$\begin{aligned} \partial_z E_y = & \mu_0 \partial_t [(1 + \chi_\perp + \tfrac{1}{2}\Delta\chi) H_x + \tfrac{1}{2}\Delta\chi \cos(2t_0 z) H_x \\ & + \tfrac{1}{2}\Delta\chi \sin(2t_0 z) H_y] \quad , \end{aligned} \tag{7.74b}$$

$$\begin{aligned} \partial_z H_x = & \varepsilon_0 \partial_t [(\varepsilon_\perp + \tfrac{1}{2}\Delta\varepsilon) E_y - \tfrac{1}{2}\Delta\varepsilon \cos(2t_0 z) E_y \\ & + \tfrac{1}{2}\Delta\varepsilon \sin(2t_0 z) E_x] \quad , \end{aligned} \tag{7.74c}$$

$$\begin{aligned} \partial_z H_y = & - \varepsilon_0 \partial_t [(\varepsilon_\perp + \tfrac{1}{2}\Delta\varepsilon) E_x + \tfrac{1}{2}\Delta\varepsilon \cos(2t_0 z) E_x \\ & + \tfrac{1}{2}\Delta\varepsilon \sin(2t_0 z) E_y] \quad . \end{aligned} \tag{7.74d}$$

The solution of this system has the form

$$E_x(z,t) = A_1 \cos[(k+t_0)z - \omega t + \phi_1] + A_2 \cos[(k-t_0)z - \omega t + \phi_1] \quad , \tag{7.75a}$$

$$E_y(z,t) = A_1 \sin[(k+t_0)z - \omega t + \phi_1] - A_2 \sin[(k-t_0)z - \omega t + \phi_1] \quad , \tag{7.75b}$$

$$H_x(z,t) = B_1 \sin[(k+t_0)z - \omega t + \phi_1] - B_2 \sin[(k-t_0)z - \omega t + \phi_1] \quad , \tag{7.75c}$$

$$H_y(z,t) = -B_1 \cos[(k+t_0)z - \omega t + \phi_1] - B_2 \cos[(k-t_0)z - \omega t + \phi_1] \quad , \tag{7.75d}$$

where ϕ_1 is an arbitrary phase angle that will be taken to be zero in the following. Substitution of (7.75) into (7.74) gives the dispersion relation, i.e. the connection between ω and k, and relations between the amplitudes A_1, A_2, B_1 and B_2. It is easily found that

$$(k+t_0)A_1 = -\mu_0\omega(1+\chi_\perp+\tfrac{1}{2}\Delta\chi)B_1 + \tfrac{1}{2}\mu_0\omega\Delta\chi B_2 \quad , \tag{7.76a}$$

$$(k-t_0)A_2 = \tfrac{1}{2}\mu_0\omega\Delta\chi B_1 - \mu_0\omega(1+\chi_\perp+\tfrac{1}{2}\Delta\chi)B_2 \quad , \tag{7.76b}$$

$$(k+t_0)B_1 = -\varepsilon_0\omega(\varepsilon_\perp+\tfrac{1}{2}\Delta\varepsilon)A_1 - \tfrac{1}{2}\varepsilon_0\omega\Delta\varepsilon A_2 \quad , \tag{7.76c}$$

$$(k-t_0)B_2 = -\tfrac{1}{2}\varepsilon_0\omega\Delta\varepsilon A_1 - \varepsilon_0\omega(\varepsilon_\perp+\tfrac{1}{2}\Delta\varepsilon)A_2 \quad . \tag{7.76d}$$

An analytic solution of this eigenvalue problem can be obtained without any difficulties. In practice the solution becomes even simpler, because the substances under consideration are diamagnetic so that $\chi_\|, \chi_\perp \ll 1$ and their influence may be neglected. Then the following relations between A_1 and A_2 are obtained

$$-\frac{(k+t_0)^2}{\mu_0\omega}A_1 = -\varepsilon_0\varepsilon\omega A_1 - \tfrac{1}{2}\varepsilon_0\Delta\varepsilon\omega A_2 \quad , \tag{7.77a}$$

$$-\frac{(k-t_0)^2}{\mu_0\omega}A_2 = -\tfrac{1}{2}\varepsilon_0\Delta\varepsilon\omega A_1 - \varepsilon_0\varepsilon\omega A_2 \quad , \tag{7.77b}$$

with $\varepsilon = \frac{1}{2}(\varepsilon_\| + \varepsilon_\perp)$. The dispersion relation is obtained by equating the determinant of the coefficients to zero. Using $c^2 = (\varepsilon_0\mu_0)^{-1}$ this leads to

$$(\varepsilon^2 - \frac{1}{4}\Delta\varepsilon^2)\frac{\omega^4}{c^4} - 2\varepsilon(k^2+t_0^2)\frac{\omega^2}{c^2} + (k^2-t_0^2)^2 = 0 \tag{7.78}$$

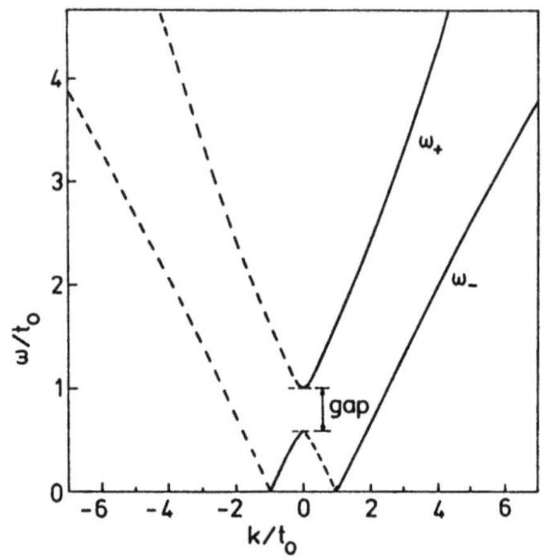

Fig. 7.7. An exaggerated representation of the general features of the dispersion relation (7.79) for $\varepsilon/c^2 = 2$, $\Delta\varepsilon/(2c^2) = 1$. The full (broken) lines correspond with waves propagating along the positive (negative) z direction

or

$$\omega_\pm^2(k) = \frac{\varepsilon c^2(k^2+t_0^2)}{\varepsilon^2 - \frac{1}{4}\Delta\varepsilon^2} \pm \frac{c^2}{\varepsilon^2 - \frac{1}{4}\Delta\varepsilon^2}\left[4\varepsilon^2 k^2 t_0^2 + \frac{1}{4}\Delta\varepsilon^2(k^2-t_0^2)^2\right]^{1/2} \quad . \tag{7.79}$$

Clearly the dispersion relation consists of two distinct branches $\omega_+(k)$ and $\omega_-(k)$. The general behaviour of this dispersion relation is shown in Fig. 7.7. The $\omega_+(k)$ branch is a monotonic increasing function of k^2, whereas the $\omega_-(k)$ branch is a monotonic decreasing function of k^2 until the value $k^2 = t_0^2$ is reached where $\omega_-(t_0^2) = 0$. As soon as $k^2 > t_0^2$ then $\omega_-(k)$ also becomes a monotonic increasing function of k^2. The full lines appearing in the drawn dispersion relation shown in Fig. 7.7 correspond to waves propagating along the positive z direction and the broken lines to waves propagating along the negative z direction. It follows that in general, four k values, arranged in pairs $(k_1, -k_1)$ and $(k_2, -k_2)$, are obtained for a given value of ω. Only one pair is obtained for frequencies ω lying in the gap

$$\omega_-(0) = \frac{c|t_0|}{n_\parallel} < \omega < \omega_+(0) = \frac{c|t_0|}{n_\perp} \quad , \tag{7.80}$$

where $n = \varepsilon^{1/2}$ denotes the index of refraction, assuming of course $n_\perp < n_\parallel$. In practice the frequency gap is relatively narrow

$$\frac{\omega_+(0) - \omega_-(0)}{\omega_+(0)} \simeq 0.05 \quad . \tag{7.81}$$

Insight into the wave behaviour in the three different frequency regions $\omega \leqq \omega_-(0)$, $\omega_-(0) < \omega < \omega_+(0)$ and $\omega_+(0) \leqq \omega$ can be obtained most easily by neglecting the influence of the optical dielectric anisotropy $\Delta\varepsilon$. Such an approximation is justified for most practical cases. Then the set of Eqs. (7.77) simplifies to

$$A_1[(k+t_0)^2 - n^2\omega^2/c^2] = 0 \quad , \tag{7.82a}$$

$$A_2[(k-t_0)^2 - n^2\omega^2/c^2] = 0 \quad . \tag{7.82b}$$

After some simple algebra a given value of ω appears to correspond to the following circularly polarized waves

$$(1) \quad k_1 = \frac{n\omega}{c} - t_0 \quad ; \quad \begin{pmatrix} E_x(z,t) \\ E_y(z,t) \end{pmatrix} = \begin{pmatrix} A_1 \cos[(k_1+t_0)z - \omega t] \\ A_1 \sin[(k_1+t_0)z - \omega t] \end{pmatrix} \quad ;$$

$$(2) \quad k_2 = -\frac{n\omega}{c} - t_0 \quad ; \begin{pmatrix} E_x(z,t) \\ E_y(z,t) \end{pmatrix} = \begin{pmatrix} A_1 \cos[(k_2+t_0)z - \omega t] \\ A_1 \sin[(k_2+t_0)z - \omega t] \end{pmatrix} \quad ;$$

$$(3) \quad k_3 = -k_2 \quad ; \quad \begin{pmatrix} E_x(z,t) \\ E_y(z,t) \end{pmatrix} = \begin{pmatrix} A_2 \cos[(-k_2-t_0)z - \omega t] \\ -A_2 \sin[(-k_2-t_0)z - \omega t] \end{pmatrix} \quad ;$$

$$(4) \quad k_4 = -k_1 \quad ; \quad \begin{pmatrix} E_x(z,t) \\ E_y(z,t) \end{pmatrix} = \begin{pmatrix} A_2 \cos[(-k_1-t_0)z - \omega t] \\ -A_2 \sin[(-k_1-t_0)z - \omega t] \end{pmatrix} \quad .$$

Now consider a plane wave with frequency ω travelling in the positive z direction and incident on a planar chiral nematic sample with the helix axis along the z direction. Without loss of generality t_0 may be taken to be positive. Then the following solutions represent waves traversing the chiral nematic medium in the positive z direction.

a) $\omega \leqq \omega_-(0)$. The solutions (1) and (3) represent circularly polarized waves of opposite sense of rotation propagating along the positive z direction. The electric field pattern of the first wave is a helix with the same sense of rotation as the helix.

b) $\omega_-(0) < \omega < \omega_+(0)$. Only the solution (3) appears. A solution of type (1) does not exist in the medium. Consequently waves, whose circular polarization has the same sense of rotation as solution (1), are reflected by the material. This is the phenomenon described earlier as Bragg reflection.

c) $\omega_+(0) \leqq \omega$. Both solution (1) and solution (3) are allowed.

As soon as the dielectric anisotropy is taken into account a frequency gap appears and the polarization is no longer circular but elliptical. For waves travelling in the negative z direction similar considerations apply to the solutions (2) and (4).

The rotary power can be easily calculated from the dispersion relation (7.78). Here, however, the influence of $\Delta\varepsilon$ may not be neglected in a first approximation, because the appearance of a dielectric anisotropy is crucial as will be shown. According to the dispersion relation, a given ω is associated with four k values except when ω is situated in the frequency gap. The four k values can be determined by rewriting the dispersion relation (7.78) in the form

$$k^4 - 2\left(t_0^2 + \frac{\varepsilon\omega^2}{c^2}\right)k^2 + \left(t_0^2 - \frac{\varepsilon\omega^2}{c^2}\right)^2 - \frac{\Delta\varepsilon^2\omega^4}{4c^4} = 0 \tag{7.83}$$

or

$$k_\pm^2 = t_0^2 + \frac{\varepsilon\omega^2}{c^2} \pm 2t_0\frac{\varepsilon^{1/2}\omega}{c}\left(1 + \frac{\Delta\varepsilon^2\omega^2}{16\varepsilon c^2 t_0^2}\right)^{1/2} \quad . \tag{7.84}$$

For small values of $\Delta\varepsilon$ the solutions of expression (7.84) may be approximated by

$$k_1 = k_- = \frac{\varepsilon^{1/2}\omega}{c} - t_0 - \frac{\Delta\varepsilon^2\omega^3}{32\varepsilon^{1/2}c^2t_0(\varepsilon^{1/2}\omega - ct_0)} \quad , \qquad k_2 = -k_3 \quad , \tag{7.85a}$$

$$k_3 = k_+ = \frac{\varepsilon^{1/2}\omega}{c} + t_0 + \frac{\Delta\varepsilon^2\omega^3}{32\varepsilon^{1/2}c^2t_0(\varepsilon^{1/2}\omega + ct_0)} \quad , \qquad k_4 = -k_1 \quad , \tag{7.85b}$$

This means that the original incident wave with frequency ω is split up into two waves which propagate in the medium with wave numbers $k_+ - t_0$ and $k_- + t_0$ and satisfy the dispersion relation

$$\omega = \frac{c}{n_+}(k_+ - t_0) \quad , \quad \omega = \frac{c}{n_-}(k_- + t_0) \quad , \tag{7.86}$$

where

$$n_\pm = \varepsilon^{1/2} + \frac{\Delta\varepsilon^2\omega^2}{32\varepsilon^{1/2}ct_0(ct_0 \pm \varepsilon^{1/2}\omega)} \quad . \tag{7.87}$$

Now the optical rotation per unit length or rotary power is

$$\frac{d\psi}{dz} = \frac{\omega}{2c}(n_- - n_+) = \frac{\Delta\varepsilon^2\omega^4}{32c^2t_0(c^2t_0^2 - \varepsilon\omega^2)} \quad . \tag{7.88}$$

Clearly the sign of the optical rotation changes on passing through $\omega = ct_0/\varepsilon^{1/2}$, which corresponds to passing through λ_{max} of the Bragg reflection. Using $\lambda = 2\pi c/\omega$ and $\lambda_{max} = 2\pi\varepsilon^{1/2}/t_0$ and introducing the reduced quantities $\lambda' = \lambda/\lambda_{max}$ and $\alpha = \frac{1}{2}\Delta\varepsilon/\varepsilon$ the expression (7.88) can be rewritten to give

$$\frac{d\psi}{dz} = -\frac{2\pi}{p}\frac{\alpha^2}{8\lambda'^2(1-\lambda'^2)}. \tag{7.89}$$

This result is shown in Fig. 7.8 together with some experimental data. The agreement is excellent. The De Vries theory has also been applied to oblique incidence, where the problem must be solved numerically. The agreement with experiment appears again to be excellent (Berreman and Scheffer 1970).

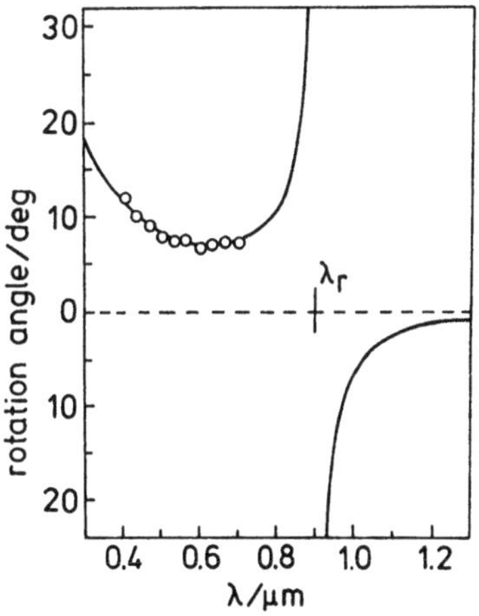

Fig. 7.8. Optical rotation as a function of wavelength for a cholesteric ester (Baessler et al. 1970). The full curves are according to (7.89)

To finish this section, the practically important limit $p \gg \lambda$, the so-called Mauguin-limit, will be considered. Then the relations (7.77) between the amplitudes A_1 and A_2 may be replaced by

$$\left(\frac{\varepsilon\omega^2}{c^2} - k^2\right)A_1 + \frac{\Delta\varepsilon\omega^2}{2c^2}A_2 = 0 \quad , \tag{7.90a}$$

$$\frac{\Delta\varepsilon\omega^2}{2c^2}A_1 + \left(\frac{\varepsilon\omega^2}{c^2} - k^2\right)A_2 = 0 \quad , \tag{7.90b}$$

and the following solutions result:

(a) $\omega = (c/n_\perp)k$, where $n_\perp = \varepsilon_\perp^{1/2}$ is the index of refraction perpendicular to the director $\boldsymbol{n}$, and $A_1 = -A_2$. The electric field is

$$\begin{pmatrix} E_x(z,t) \\ E_y(z,t) \end{pmatrix} = \begin{pmatrix} -2A_1 \sin(kz - \omega t) \sin(t_0 z) \\ 2A_1 \cos(kz - \omega t) \cos(t_0 z) \end{pmatrix} \quad .$$

It is easily verified that $E_\alpha(z,t)n_\alpha(z) = 0$, i.e. the electric field is perpendicular to $\boldsymbol{n}$.

(b) $\omega = (c/n_\parallel)k$, where $n_\parallel = \varepsilon_\parallel^{1/2}$ is the index of refraction parallel to $\boldsymbol{n}$ and $A_1 = A_2$. The electric field is now given by

$$\begin{pmatrix} E_x(z,t) \\ E_y(z,t) \end{pmatrix} = \begin{pmatrix} 2A_1 \cos(kz - \omega t) \cos(t_0 z) \\ 2A_1 \sin(kz - \omega t) \sin(t_0 z) \end{pmatrix} \quad .$$

Here the electric field $\boldsymbol{E}$ is parallel to the director $\boldsymbol{n}$.

Clearly both solutions are guided by the helix, because the amplitudes remain normal and parallel respectively, to the local director $\boldsymbol{n}(\boldsymbol{r})$. This is just the situation discussed more qualitatively in Sect. 7.4.1.

7.5 Correlations in the Smectic A Phase

In this section the correlation between smectic layers in the S_A phase is considered. This is of interest because the absence of true long-range correlation influences the physics of phase transitions involving smectic A phases rather fundamentally. This so-called Landau-Peierls instability is related to the mean-square fluctuation of the displacement of a given smectic layer. Following the continuum theory as given in Sect. 5.3 this displacement is denoted by $u_z(\boldsymbol{r})$. The relevant correlation function thus concerns $\langle u_z^2(\boldsymbol{r})\rangle$. In order to calculate the thermal averages the distortion free energy density (5.36) of the S_A continuum theory is used. The calculation proceeds analogously to the calculation of the fluctuations in the nematic phase as discussed in Sect. 7.2. Substitution of the Fourier series

$$u_z(\boldsymbol{r}) = \sum_{\boldsymbol{k}} u(\boldsymbol{k}) \exp(\mathrm{i}\boldsymbol{k}\cdot\boldsymbol{r}) \tag{7.91}$$

into (5.36) and subsequent integration over the volume Ω of the sample gives the following free energy

$$F = \frac{1}{2}\Omega \sum_{\boldsymbol{k}} [\lambda_1 k_z^2 + K_1(k_x^2 + k_y^2)^2]|u(\boldsymbol{k})|^2 \quad . \tag{7.92}$$

According to fluctuation theory the thermal average $\langle|u(\boldsymbol{k})|^2\rangle$ is given by

$$\langle|u(\boldsymbol{k})|^2\rangle = \{\beta\Omega[\lambda_1 k_z^2 + K_1(k_x^2+k_y^2)^2]\}^{-1} \quad . \tag{7.93}$$

Then the relevant correlation function reads

$$\langle u_z^2(\boldsymbol{r})\rangle = (k_\mathrm{B}T/\Omega)\sum_{\boldsymbol{k}} \frac{1}{\lambda_1 k_z^2 + K_1(k_x^2+k_y^2)^2} \quad . \tag{7.94}$$

Replacing the summation by an integration and changing to cylindrical coordinates the correlation function can be written as

$$\langle u_z^2(\boldsymbol{r})\rangle = \frac{k_\mathrm{B}T}{4\pi^2}\int dk_z \int \frac{k\,dk}{\lambda_1 k_z^2 + K_1 k^4} \quad , \tag{7.95}$$

where the length L of the sample in the z direction can be taken into account by allowing for k_z the integration range $2\pi/L \leq |k_z| \leq 2\pi/d$ with d denoting the layer periodicity. For the sake of simplicity k is allowed an integration range $0 \leq k \leq \infty$. Then it follows that

$$\begin{aligned}\langle u_z^2(\boldsymbol{r})\rangle &= \frac{k_\mathrm{B}T}{2\pi^2}\int\limits_{2\pi/L}^{2\pi/d} dk_z \int\limits_0^\infty \frac{k\,dk}{\lambda_1 k_z^2 + K_1 k^4} \\ &= \frac{k_\mathrm{B}T}{4\pi^2(\lambda_1 K_1)^{1/2}}\int\limits_a^b dl \int\limits_0^\infty \frac{dm}{l^2+m^2} \quad ,\end{aligned} \tag{7.96}$$

where $a = 2\pi\lambda_1^{1/2}/L$ and $b = 2\pi\lambda_1^{1/2}/d$. Using the result

$$\int\limits_0^\infty \frac{dm}{l^2+m^2} = \frac{\pi}{2l}$$

the required thermal average can be expressed as

$$\langle u_z^2(\boldsymbol{r})\rangle = \frac{k_\mathrm{B}T}{8\pi(\lambda_1 K_1)^{1/2}}\int\limits_a^b \frac{dl}{l} = \frac{k_\mathrm{B}T}{8\pi(\lambda_1 K_1)^{1/2}}\ln\left(\frac{L}{d}\right) \quad . \tag{7.97}$$

This means that the mean-square displacement diverges as the logarithm of the size of the sample in the z direction, i.e. as the logarithm of the number of smectic layers. It is worthwhile to mention that the coupling between the displacement field and the orientational field, which is represented by the term proportional to K_1, causes the logarithmic divergence. When this coupling is absent the mean-square displacement diverges linearly with the length L. The logarithmic divergence causes the positional order of the smectic A layering to be quasi long-range (see Table 3.1). As discussed in Sect. 3.1 the corresponding lineshape in the x-ray spectrum has been observed.

8. Fluid Dynamics of Nematics

In this chapter we deal with the basic aspects of nematodynamics which will be described from a variational point of view using Rayleigh's dissipation function as a starting point. First of all we briefly review the essentials of the fluid dynamics of simple isotropic liquids. Then the theory is generalized to nematics, i.e. anisotropic liquids. This generalization involves the so-called Leslie viscosity stress tensor, which for an incompressible fluid, contains five viscosity coefficients as follows directly from symmetry arguments. Apart from the effect of the anisotropy on the viscosity stress tensor, the orientation of the uniaxial axis will be influenced in most cases by the occurrence of flow. This effect on the orientational field is described by equations of motion, which will be derived taking into account the elastic and viscous forces. Next viscosity measurements are briefly discussed. Finally, as an example of the various types of instabilities that can occur, we consider the one-dimensional theory of electrohydrodynamic instabilities.

8.1 Fluid Dynamics of Isotropic Liquids

The macroscopic motion of simple isotropic liquids can be adequately described by regarding the liquid as a continuous medium (Landau and Lifshitz 1959b). In the Eulerian description of fluid dynamics the state of such a fluid is represented by a velocity field $\boldsymbol{v}(\boldsymbol{r},t)$, a pressure field $p(\boldsymbol{r},t)$, and a density field $\varrho(\boldsymbol{r},t)$. In general the fluids are viscous, meaning that the flow of the fluid gives rise to energy dissipation due to internal friction. This dissipation depends on the velocity gradients $\partial_\alpha v_\beta$ $(\alpha,\beta = x,y,z)$. In order to keep the discussion as simple as possible the fluids are assumed to be incompressible in the following. This means that their density field is assumed to be constant, $\varrho(\boldsymbol{r},t) = \varrho$. Now the equation of continuity

$$\partial_t \varrho + \partial_\alpha \varrho v_\alpha = 0 \quad , \tag{8.1}$$

where $\partial_t \varrho$ is a shorthand for $\partial\varrho/\partial t$, implies immediately that an incompressible fluid is described by

$$\partial_\alpha v_\alpha = 0 \quad . \tag{8.2}$$

So the divergence of the velocity field of an incompressible fluid vanishes. The limitation to incompressible fluids is not a serious one for the present discussion; the results can easily be generalized.

An ideal fluid, i.e. a fluid without viscosity, is described by an equation of motion called Euler's equation, which reads, taking also into account the incompressibility of the fluid,

$$\varrho \partial_t v_\alpha = -\partial_\beta \Pi_{\beta\alpha} \quad , \tag{8.3}$$

where $\tilde{\boldsymbol{\Pi}}$ is called the momentum flux density tensor with elements

$$\Pi_{\alpha\beta} = p\delta_{\alpha\beta} + \varrho v_\alpha v_\beta \quad . \tag{8.4}$$

The term $p\delta_{\alpha\beta}$ is due to the static pressure, whereas the term $\varrho v_\alpha v_\beta$ results from the fact that all quantities refer to the fluid at a given point $\boldsymbol{r}$. This means that the motion of the fluid particles themselves also contributes to the momentum transfer at this point.

The viscosity of the fluid is taken into account by a viscosity stress tensor $\tilde{\boldsymbol{\sigma}}'$, which originates from the irreversible viscous transfer of momentum. Clearly this irreversible transfer of momentum influences the reversible transfer of momentum (8.4) in a negative way. Consequently the momentum flux density tensor in a viscous fluid has elements of the form

$$\Pi_{\alpha\beta} = p\delta_{\alpha\beta} + \varrho v_\alpha v_\beta - \sigma'_{\alpha\beta} = -\sigma_{\alpha\beta} + \varrho v_\alpha v_\beta \quad , \tag{8.5}$$

where the tensor with elements

$$\sigma_{\alpha\beta} = -p\delta_{\alpha\beta} + \sigma'_{\alpha\beta}$$

is called the stress tensor. The equation of motion of a viscous fluid is again described by (8.3) but with a momentum flux density tensor (8.5).

The heart of the problem is to construct the appropriate expression for $\sigma'_{\alpha\beta}$. The general form of this tensor can be obtained from Rayleigh's dissipation function D, also called the dissipative function (Landau and Lifshitz 1980), by means of

$$\sigma'_{\alpha\beta} = \frac{\partial D}{\partial \partial_\alpha v_\beta} \quad . \tag{8.6}$$

The dissipation function may be conceived as a generalized potential, which depends on the derivatives $\partial_\alpha v_\beta$ of the velocity field. Frictional stresses, which are proportional to the derivatives of a velocity field, result from a dissipation function D, which is obtained by constructing all possible invariants quadratic in $\partial_\alpha v_\beta$ and adding them together. The situation is somewhat analogous to Hooke's law, which results from a potential quadratic in the strains. Consequently D reads

$$D = \tfrac{1}{2}\eta\partial_\alpha v_\beta \partial_\alpha v_\beta + \tfrac{1}{2}\xi\partial_\alpha v_\beta \partial_\beta v_\alpha + \tfrac{1}{2}\zeta\partial_\alpha v_\alpha \partial_\beta v_\beta \quad , \tag{8.7}$$

where the coefficients η, ξ and ζ are called coefficients of viscosity. Note that the last term does not contribute in the case of an incompressible fluid. The viscosity coefficients are partly dependent on each other because of the requirement that both D and $\sigma'_{\alpha\beta}$ must vanish when the whole fluid is in uniform rotation, i.e. when the velocity field is given by $\boldsymbol{v} = \boldsymbol{\omega} \times \boldsymbol{r}$ or

$$\partial_\beta v_\alpha = \varepsilon_{\alpha\mu\beta}\omega_\mu \quad . \tag{8.8}$$

Substitution of (8.8) into (8.7) gives

$$D = \tfrac{1}{2}\eta\varepsilon_{\alpha\mu\beta}\omega_\mu\varepsilon_{\alpha\gamma\beta}\omega_\gamma + \tfrac{1}{2}\xi\varepsilon_{\alpha\mu\beta}\omega_\mu\varepsilon_{\beta\gamma\alpha}\omega_\gamma = 0 \quad , \tag{8.9}$$

or

$$(\eta - \xi)\omega_\mu\omega_\mu = 0 \quad , \tag{8.10}$$

i.e. $\eta = \xi$. Consequently the dissipation function for an incompressible isotropic fluid is given by

$$D = \tfrac{1}{2}\eta(\partial_\alpha v_\beta \partial_\alpha v_\beta + \partial_\alpha v_\beta \partial_\beta v_\alpha) \quad , \tag{8.11}$$

and the viscous stress tensor is

$$\sigma'_{\alpha\beta} = \eta(\partial_\alpha v_\beta + \partial_\beta v_\alpha) \quad . \tag{8.12}$$

Substitution of (8.12) into (8.5) results in the momentum flux density tensor for an incompressible viscous fluid with the following elements

$$\Pi_{\alpha\beta} = p\delta_{\alpha\beta} + \varrho v_\alpha v_\beta - \eta(\partial_\alpha v_\beta + \partial_\beta v_\alpha) \quad . \tag{8.13}$$

According to (8.3) the equation of motion for such a fluid is

$$\varrho\partial_t v_\alpha = -\partial_\beta[p\delta_{\alpha\beta} + \varrho v_\alpha v_\beta - \eta(\partial_\alpha v_\beta + \partial_\beta v_\alpha)] \quad ,$$

or

$$\partial_t v_\alpha + v_\beta\partial_\beta v_\alpha = -\frac{1}{\varrho}\partial_\alpha p + \frac{\eta}{\varrho}\partial_\beta\partial_\beta v_\alpha \quad . \tag{8.14}$$

This equation is the well-known Navier-Stokes equation, which describes, together with the equation of state and the equation of continuity (8.2), the hydrodynamic behaviour of incompressible viscous fluids. For a further discussion of this equation the reader is referred to the literature, e.g. the book on fluid dynamics by Landau and Lifshitz (1959).

8.2 Nematodynamics

The macroscopic motion of nematic liquid crystals can, as in the case of simple liquids, be adequately described by regarding the nematic fluid as a continuous medium. Due to the anisotropy of the fluid, however, the Eulerian description must be extended with two tensor fields of the second rank, namely an inertial density tensor field $\tilde{\boldsymbol{I}}(\boldsymbol{r},t)$ and a tensor order parameter field $\tilde{\boldsymbol{Q}}(\boldsymbol{r},t)$. Because of the local uniaxiality of the moving fluid the tensor elements $Q_{\alpha\beta}(\boldsymbol{r},t)$ are

$$Q_{\alpha\beta}(\boldsymbol{r},t) = S[n_\alpha(\boldsymbol{r},t)n_\beta(\boldsymbol{r},t) - \tfrac{1}{3}\delta_{\alpha\beta}] \quad , \tag{8.15}$$

where S denotes the order parameter of the nematic and $\boldsymbol{n}(\boldsymbol{r},t)$ is the local director. All fields appearing depend, in general, on the position $\boldsymbol{r}$ and the time t.

Nematics are viscous fluids, i.e. the flow of the fluid gives rise to energy dissipation due to internal friction. This dissipation appears to depend on the velocity gradients $\partial_\alpha v_\beta$, as for isotropic fluids, and in addition on $Q_{\alpha\beta}$ as well as on the time derivatives $\dot{Q}_{\alpha\beta}$. Another way of expressing this is that the flow disturbs in general the local anisotropy of the medium and vice versa, or in molecular terms that the translational motion of the molecules is coupled to their rotational motion. The existence of a coupling between the fields $\boldsymbol{v}(\boldsymbol{r},t)$ and $\tilde{\boldsymbol{Q}}(\boldsymbol{r},t)$ is also dictated by the requirement that the energy dissipation and the associated viscosity stress tensor must vanish when the entire fluid is in uniform motion.

In order to describe the motion of anisotropic viscous fluids a generalization of the Navier-Stokes equation alone does not suffice because of the rotational motion of the orientational field. It is true that this rotational motion appears in the generalized Navier-Stokes by way of the viscosity stress tensor but additional equations of motion are needed in order to describe the time-dependent behaviour of the orientational field. Several approaches have been used. For a discussion of the Ericksen-Leslie-Parodi (ELP) approach and the Harvard approach the reader is referred to the book of De Gennes (1974). Here a slightly different formulation will be presented, which is probably more transparent.

The starting point of the present discussion is Rayleigh's dissipation function. Furthermore the nematic fluid is assumed to be incompressible for reasons of simplicity. The relevant tensors for constructing the dissipation function are $\partial_\alpha v_\beta$, $Q_{\alpha\beta}$ and $\dot{Q}_{\alpha\beta}$. However, instead of $Q_{\alpha\beta}$ and $\dot{Q}_{\alpha\beta}$, the more simple tensors $n_\alpha n_\beta$ and $\dot{n}_\alpha n_\beta$ may be taken, due to the local uniaxiality of the fluid. The dissipation function now consists of all possible invariants which can be constructed from these tensors. These invariant terms must be bilinear in the tensors $\partial_\alpha v_\beta$ and $\dot{n}_\alpha n_\beta$ because of the linear dependence of the stresses on these quantities. It will be clear that the dissipation function consists of the sum of two contributions. The first one contains only terms quadratic in $\partial_\alpha v_\beta$ and has already been given by (8.11), whereas the second one takes into account the anisotropy of the fluid. For an incompressible fluid the anisotropic contribution D_{a} is expressed in its most general form as

$$\begin{aligned} D_{\mathrm{a}} = {} & \tfrac{1}{2}\xi_1 n_\alpha n_\beta \partial_\alpha v_\gamma \partial_\beta v_\gamma + \tfrac{1}{2}\xi_2 n_\alpha n_\beta \partial_\gamma v_\alpha \partial_\gamma v_\beta + \xi_3 n_\alpha n_\beta \partial_\gamma v_\alpha \partial_\beta v_\gamma \\ & + \tfrac{1}{2}\xi_4 n_\alpha n_\beta n_\gamma n_\delta \partial_\alpha v_\beta \partial_\gamma v_\delta + \xi_5 \dot{n}_\alpha n_\beta \partial_\beta v_\alpha \\ & + \xi_6 \dot{n}_\alpha n_\beta \partial_\alpha v_\beta + \tfrac{1}{2}\xi_7 \dot{n}_\alpha \dot{n}_\alpha \quad . \end{aligned} \tag{8.16}$$

Using (8.6) the anisotropic contribution to the viscosity stress tensor $\sigma'_{\alpha\beta}$ is now given by

$$\begin{aligned} \sigma'_{\alpha\beta} = {} & \xi_1 n_\alpha n_\gamma \partial_\gamma v_\beta + \xi_2 n_\beta n_\gamma \partial_\alpha v_\gamma + \xi_3 n_\beta n_\gamma \partial_\gamma v_\alpha + \xi_3 n_\alpha n_\gamma \partial_\beta v_\gamma \\ & + \xi_4 n_\alpha n_\beta n_\gamma n_\delta \partial_\gamma v_\delta + \xi_5 \dot{n}_\beta n_\alpha + \xi_6 \dot{n}_\alpha n_\beta \quad . \end{aligned} \tag{8.17}$$

The appearing coefficients ξ are not all independent, because both D_{a} and $\sigma'_{\alpha\beta}$ must vanish when the whole fluid is in uniform rotation or, stated in mathematical terms, when

$$\partial_\beta v_\alpha = \varepsilon_{\alpha\mu\beta}\omega_\mu \quad , \tag{8.18a}$$

$$\dot{n}_\alpha = \varepsilon_{\alpha\mu\beta}\omega_\mu n_\beta \quad . \tag{8.18b}$$

Substitution of (8.18) into (8.16) and (8.17) now gives

$$(\tfrac{1}{2}\xi_1 + \tfrac{1}{2}\xi_2 - \xi_3 + \xi_5 - \xi_6 + \tfrac{1}{2}\xi_7)(\omega_\alpha\omega_\alpha - n_\alpha n_\beta \omega_\alpha \omega_\beta) = 0 \quad , \tag{8.19a}$$

$$(\xi_1 - \xi_3 + \xi_5)\varepsilon_{\beta\mu\gamma} n_\alpha n_\gamma \omega_\mu + (\xi_2 - \xi_3 - \xi_6)\varepsilon_{\gamma\mu\alpha} n_\beta n_\gamma \omega_\mu = 0 \quad . \tag{8.19b}$$

Equations (8.19) immediately imply

$$\xi_5 = -\xi_1 + \xi_3 \quad , \tag{8.20a}$$

$$\xi_6 = \xi_2 - \xi_3 \quad , \tag{8.20b}$$

$$\xi_7 = \xi_1 + \xi_2 - 2\xi_3 \quad , \tag{8.20c}$$

i.e. only four independent coefficients of viscosity appear in the anisotropic contribution to the dissipation function and in the anisotropic contribution to the viscosity stress tensor. Consequently Rayleigh's dissipation function for an incompressible anisotropic fluid is

$$\begin{aligned} D = {} & \tfrac{1}{2}\eta(\partial_\alpha v_\beta \partial_\alpha v_\beta + \partial_\alpha v_\beta \partial_\beta v_\alpha) + \tfrac{1}{2}\xi_1(n_\alpha \partial_\alpha v_\gamma - \dot{n}_\gamma)(n_\beta \partial_\beta v_\gamma - \dot{n}_\gamma) \\ & + \tfrac{1}{2}\xi_2(n_\alpha \partial_\gamma v_\alpha + \dot{n}_\gamma)(n_\beta \partial_\gamma v_\beta + \dot{n}_\gamma) \\ & + \xi_3(n_\alpha \partial_\alpha v_\gamma - \dot{n}_\gamma)(n_\beta \partial_\gamma v_\beta + \dot{n}_\gamma) + \tfrac{1}{2}\xi_4 n_\alpha n_\beta n_\gamma n_\delta \partial_\alpha v_\beta \partial_\gamma v_\delta \quad , \end{aligned} \tag{8.21}$$

whereas the associated viscosity stress tensor is given by

$$\begin{aligned} \sigma'_{\alpha\beta} = {} & \eta(\partial_\alpha v_\beta + \partial_\beta v_\alpha) + \xi_1 n_\alpha(n_\gamma \partial_\gamma v_\beta - \dot{n}_\beta) + \xi_2 n_\beta(n_\gamma \partial_\alpha v_\gamma + \dot{n}_\alpha) \\ & + \xi_3 n_\beta(n_\gamma \partial_\gamma v_\alpha - \dot{n}_\alpha) + \xi_3 n_\alpha(n_\gamma \partial_\beta v_\gamma + \dot{n}_\beta) \\ & + \xi_4 n_\alpha n_\beta n_\gamma n_\delta \partial_\gamma v_\delta \quad . \end{aligned} \tag{8.22}$$

The expressions (8.21) and (8.22) can also be written down immediately by realizing that the requirement that D and $\sigma'_{\alpha\beta}$ must both vanish in the case of uniform rotation, is equivalent to the requirement that D and $\sigma'_{\alpha\beta}$ must be composed of tensors satisfying that requirement. The only relevant tensors are clearly just $n_\alpha n_\gamma \partial_\gamma v_\beta - n_\alpha \dot{n}_\beta$, $n_\alpha n_\gamma \partial_\beta v_\gamma + n_\alpha \dot{n}_\beta$ and $n_\alpha n_\beta n_\gamma n_\delta \partial_\gamma v_\delta$.

The coefficienst ξ can be expressed in terms of the Leslie coefficients α, which are widely used in the literature. The Leslie viscosity stress tensor reads (Leslie 1979)

$$\begin{aligned} \sigma'_{\alpha\beta} = {} & \tfrac{1}{2}\alpha_4(\partial_\alpha v_\beta + \partial_\beta v_\alpha) + \tfrac{1}{2}(\alpha_5 - \alpha_2) n_\alpha n_\gamma \partial_\gamma v_\beta \\ & + \tfrac{1}{2}(\alpha_6 + \alpha_3) n_\beta n_\gamma \partial_\alpha v_\gamma + \tfrac{1}{2}(\alpha_6 - \alpha_3) n_\beta n_\gamma \partial_\gamma v_\alpha \\ & + \tfrac{1}{2}(\alpha_5 + \alpha_2) n_\alpha n_\gamma \partial_\beta v_\gamma + \alpha_2 n_\alpha \dot{n}_\beta + \alpha_3 n_\beta \dot{n}_\alpha \\ & + \alpha_1 n_\alpha n_\beta n_\gamma n_\delta \partial_\gamma v_\delta \quad . \end{aligned} \tag{8.23}$$

A comparison between (8.22) and (8.23) results in

$$\begin{aligned} \xi_1 &= \tfrac{1}{2}(\alpha_5 - \alpha_2) \quad , \\ \xi_2 &= \tfrac{1}{2}(\alpha_6 + \alpha_3) \quad , \\ \xi_3 &= \tfrac{1}{2}(\alpha_6 - \alpha_3) = \tfrac{1}{2}(\alpha_5 + \alpha_2) \quad , \\ \xi_4 &= \alpha_1 \quad , \\ \eta &= \tfrac{1}{2}\alpha_4 \quad . \end{aligned} \tag{8.24}$$

It follows directly that Leslie's original assumption of six independent coefficients does not hold since

$$\alpha_6 - \alpha_3 = \alpha_5 + \alpha_2 \quad . \tag{8.25}$$

This relation has also been derived by Parodi (1970) starting from the Onsager reciprocal relations.

Substitution of (8.22) into (8.5) gives the momentum flux density tensor of the incompressible anisotropic fluid and the required equation of motion follows then directly from (8.3). Clearly this equation does not suffice. Additional equations of motion are required for the description of the time-dependent behaviour of the director, i.e. $\dot{n}_\alpha n_\beta$ must be calculated. The time-dependent behaviour of the orientational field $n_\alpha n_\beta$ is described in terms of the equations of motions for the functions $\phi(\boldsymbol{r},t)$ and $\theta(\boldsymbol{r},t)$, which specify the director with respect to the laboratory system which is denoted by the unit vectors $\boldsymbol{e}_1$, $\boldsymbol{e}_2$ and $\boldsymbol{e}_3$. According to Lagrange these functions satisfy the following two equations of motion:

$$\frac{d}{dt}\left(\frac{\partial T}{\partial \dot{\theta}}\right) - \frac{\partial T}{\partial \theta} = -\frac{\partial D}{\partial \dot{\theta}} - F_{\mathrm{el},\theta} \quad , \tag{8.26a}$$

$$\frac{d}{dt}\left(\frac{\partial T}{\partial \dot{\phi}}\right) - \frac{\partial T}{\partial \phi} = -\frac{\partial D}{\partial \dot{\phi}} - F_{\mathrm{el},\phi} \quad , \tag{8.26b}$$

where T denotes the rotational kinetic energy density, $\partial D/\partial\dot{\theta}$ and $\partial D/\partial\dot{\phi}$ are generalized frictional force densities, and $F_{\mathrm{el},\theta}$ and $F_{\mathrm{el},\phi}$ are generalized elastic force densities, which arise because of the elastic distortion of the medium. For all these quantities expressions still have to be given.

The rotational kinetic energy density T is obtained from the inertia density tensor field $\tilde{\boldsymbol{I}}(\boldsymbol{r},t)$. Because of the uniaxiality of the system this tensor is given by

$$I_{\alpha\beta} = I_\perp \delta_{\alpha\beta} + (I_\parallel - I_\perp) n_\alpha n_\beta \quad , \tag{8.27}$$

where $I_\parallel$ and $I_\perp$ are respectively the components of the inertia tensor along and perpendicular to the director. This means in the terminology of Sect. 5.1 that $\tilde{\boldsymbol{I}}$ is diagonal with respect to the rotated coordinate system denoted by the unit vectors $\overline{\boldsymbol{e}}_1$, $\overline{\boldsymbol{e}}_2$ and $\overline{\boldsymbol{e}}_3$. The direction of $\boldsymbol{n}$ coincides with the basis vector $\overline{\boldsymbol{e}}_3$ or with $-\overline{\boldsymbol{e}}_3$, where

$$\overline{\boldsymbol{e}}_3 = \sin\theta\,\cos\phi\,\boldsymbol{e}_1 + \sin\theta\,\sin\phi\,\boldsymbol{e}_2 + \cos\theta\,\boldsymbol{e}_3 \quad , \tag{8.28}$$

with $\boldsymbol{e}_1$, $\boldsymbol{e}_2$ and $\boldsymbol{e}_3$ indicating the basis vectors along the x, y and z axes respectively of the original non-rotated coordinate system (laboratory system). The kinetic energy density which is associated with this rotational motion, is given by

$$T = \tfrac{1}{2} I_{\alpha\beta}\omega_\alpha\omega_\beta \quad , \tag{8.29}$$

where $\boldsymbol{\omega}$ denotes the angular velocity. Using the representation in Eulerian angles as mentioned in Sect. 5.1 the angular velocity $\boldsymbol{\omega}$ equals

$$\begin{aligned}\boldsymbol{\omega} = {} & (\dot{\theta} \sin \psi - \dot{\phi} \sin \theta \cos \psi)\overline{\boldsymbol{e}}_1 + (\dot{\phi} \sin \theta \sin \psi + \dot{\theta} \cos \psi)\overline{\boldsymbol{e}}_2 \\ & + (\dot{\phi} \cos \theta + \dot{\psi})\overline{\boldsymbol{e}}_3 \quad .\end{aligned} \tag{8.30}$$

Substitution of (8.27) and (8.30) into (8.29) and taking $\dot{\psi} = 0$, i.e. neglecting any possible "spin" of the orientational field, results in

$$T = \tfrac{1}{2} I_\perp(\dot{\phi}^2 \sin^2 \theta + \dot{\theta}^2) + \tfrac{1}{2} I_{||}\dot{\phi}^2 \cos^2 \theta \quad . \tag{8.31}$$

The generalized frictional forces are given by

$$\begin{aligned}\frac{\partial D}{\partial \dot{\theta}} = {} & [(\xi_1 + \xi_2 - 2\xi_3)\dot{n}_\gamma - (\xi_1 - \xi_3)n_\alpha\partial_\alpha v_\gamma + (\xi_2 - \xi_3)n_\alpha\partial_\gamma v_\alpha] \\ & \times \frac{\partial \dot{n}_\gamma}{\partial \dot{\theta}} \quad ,\end{aligned} \tag{8.32a}$$

$$\begin{aligned}\frac{\partial D}{\partial \dot{\phi}} = {} & [(\xi_1 + \xi_2 - 2\xi_3)\dot{n}_\gamma - (\xi_1 - \xi_3)n_\alpha\partial_\alpha v_\gamma + (\xi_2 - \xi_3)n_\alpha\partial_\gamma v_\alpha] \\ & \times \frac{\partial \dot{n}_\gamma}{\partial \dot{\phi}} \quad .\end{aligned} \tag{8.32b}$$

Regarding the calculation of the generalized frictional forces (8.32) it is worthwhile to observe here that the usual choice $\boldsymbol{n} = \overline{\boldsymbol{e}}_3$ gives rise to

$$\begin{aligned}\dot{\boldsymbol{n}} = {} & (\dot{\theta} \cos \theta \cos \phi - \dot{\phi} \sin \theta \sin \phi)\boldsymbol{e}_1 \\ & + (\dot{\theta} \cos \theta \sin \phi + \dot{\phi} \sin \theta \cos \phi)\boldsymbol{e}_2 \\ & - \dot{\theta} \sin \theta\, \boldsymbol{e}_3 \quad ,\end{aligned} \tag{8.33a}$$

i.e.

$$\frac{\partial \dot{\boldsymbol{n}}}{\partial \dot{\theta}} = \cos \theta \cos \phi\, \boldsymbol{e}_1 + \cos \theta \sin \phi\, \boldsymbol{e}_2 - \sin \theta\, \boldsymbol{e}_3 \quad , \tag{8.33b}$$

$$\frac{\partial \dot{\boldsymbol{n}}}{\partial \dot{\phi}} = - \sin \theta \sin \phi\, \boldsymbol{e}_1 + \sin \theta \cos \phi\, \boldsymbol{e}_2 \quad . \tag{8.33c}$$

Expressions for the components $\dot{n}_\gamma$, $\partial\dot{n}_\gamma/\partial\dot{\phi}$ and $\partial\dot{n}_\gamma/\partial\dot{\theta}$ can easily be deduced from the relations (8.33). The combination of ξ coefficients associated with $\dot{n}_\gamma$ in the expressions (8.32) (pure rotation of the director) is usually written as

$$\gamma_1 = \xi_1 + \xi_2 - 2\xi_3 \quad . \tag{8.34}$$

Using (8.24) the coefficient γ_1 can conveniently be expressed in terms of Leslie coefficients as $\gamma_1 = \alpha_3 - \alpha_2$.

As for the generalized elastic forces $F_{\mathrm{el},\theta}$ and $F_{\mathrm{el},\phi}$, they follow directly from the expression (5.24) for the distortion free energy density. In order to rewrite this expression in terms of the functions $\theta(\boldsymbol{r},t)$ and $\phi(\boldsymbol{r},t)$ it appears to be advantageous to introduce the unit vectors

$$\begin{aligned} \boldsymbol{a} &= \cos\phi\,\boldsymbol{e}_1 + \sin\phi\,\boldsymbol{e}_2 \quad ; \\ \boldsymbol{b} &= -\sin\phi\,\boldsymbol{e}_1 + \cos\phi\,\boldsymbol{e}_2 \quad ; \\ \boldsymbol{c} &= \boldsymbol{a} \times \boldsymbol{b} (= \boldsymbol{e}_3) \quad . \end{aligned} \tag{8.35}$$

Then it holds that

$$\begin{aligned} \operatorname{div}\boldsymbol{n} &= \partial_\alpha(\sin\theta\, a_\alpha + \cos\theta\, c_\alpha) \\ &= \sin\theta\, b_\alpha\partial_\alpha\phi + \cos\theta\, a_\alpha\partial_\alpha\theta - \sin\theta\, c_\alpha\partial_\alpha\theta \quad , \end{aligned} \tag{8.36a}$$

$$\begin{aligned} \operatorname{curl}\boldsymbol{n} = &-\boldsymbol{a}\sin\theta(b_\alpha\partial_\alpha\theta + c_\alpha\partial_\alpha\phi) + \boldsymbol{b}(\sin\theta\, a_\alpha\partial_\alpha\theta + \cos\theta\, c_\alpha\partial_\alpha\theta) \\ &+ \boldsymbol{c}(\sin\theta\, a_\alpha\partial_\alpha\phi - \cos\theta\, b_\alpha\partial_\alpha\theta) \quad , \end{aligned} \tag{8.36b}$$

where use has been made of $\partial_\alpha a_\alpha = b_\alpha\partial_\alpha\phi$ and $a_\alpha = e_{\alpha\beta\gamma}b_\beta c_\gamma$ (and cyclic permutations). Substitution of (8.36) into (5.24) gives

$$\begin{aligned} f_\mathrm{d} = &\tfrac{1}{2}K_1(\sin\theta\, b_\alpha\partial_\alpha\phi + \cos\theta\, a_\alpha\partial_\alpha\theta - \sin\theta\, c_\alpha\partial_\alpha\theta)^2 \\ &+ \tfrac{1}{2}(K_2 - K_3)[\sin^2\theta(b_\alpha\partial_\alpha\theta + c_\alpha\partial_\alpha\phi) \\ &+ \cos\theta(\cos\theta\, b_\alpha\partial_\alpha\theta - \sin\theta\, a_\alpha\partial_\alpha\phi)]^2 \\ &+ \tfrac{1}{2}K_3[\sin^2\theta(b_\alpha\partial_\alpha\theta + c_\alpha\partial_\alpha\phi)^2 + (\sin\theta\, a_\alpha\partial_\alpha\theta + \cos\theta\, c_\alpha\partial_\alpha\theta)^2 \\ &+ (\sin\theta\, a_\alpha\partial_\alpha\phi - \cos\theta\, b_\alpha\partial_\alpha\theta)^2] \quad . \end{aligned} \tag{8.37}$$

This distortion free energy density gives rise to the generalized forces

$$F_{\mathrm{el},\theta} = \frac{\partial f_\mathrm{d}}{\partial\theta} - \partial_\alpha\frac{\partial f_\mathrm{d}}{\partial\partial_\alpha\theta} \quad , \tag{8.38a}$$

$$F_{\mathrm{el},\phi} = \frac{\partial f_\mathrm{d}}{\partial\phi} - \partial_\alpha\frac{\partial f_\mathrm{d}}{\partial\partial_\alpha\phi} \quad . \tag{8.38b}$$

It should be noted here that the presence of a magnetic field gives the additional term

$$f_{\mathrm{M}} = -\tfrac{1}{2}\mu_0^{-1}\Delta\chi(\boldsymbol{n}\cdot\boldsymbol{B})^2 \tag{8.39}$$

to the distortion free energy density. The resultant generalized forces have the same form as (8.38), but f_{d} is replaced by $f_{\mathrm{d}} + f_{\mathrm{M}}$.

In principle the effect of the elastic distortion must be incorporated into the Navier-Stokes equation for a nematic. The resultant elastic force $f_{\mathrm{el},\alpha}$ is taken into account by adding this term to the right-hand side of (8.3) where

$$f_{\mathrm{el},\alpha} = -\frac{\partial F_{\mathrm{el}}}{\partial\theta}\partial_\alpha\theta - \frac{\partial F_{\mathrm{el}}}{\partial\phi}\partial_\alpha\phi - \frac{\partial F_{\mathrm{el}}}{\partial\partial_\beta\theta}\partial_\alpha\partial_\beta\theta - \frac{\partial F_{\mathrm{el}}}{\partial\partial_\beta\phi}\partial_\alpha\partial_\beta\phi \quad . \tag{8.40}$$

For small deformations, however, this force is quite small and will be henceforth neglected.

Finally two remarks should be made in view of the existing alternative approaches. The first one concerns the present formulation of the equations of motion for the orientational field, i.e. a second rank tensor field, in terms of Eulerian angles. In the existing literature usually the equations of motion for the director $\boldsymbol{n}$ are postulated, where the fact that $\boldsymbol{n}$ is a unit vector is taken into account by means of a constraint, the so-called "molecular field". The second remark concerns the connection of the present approach with the ELP approach and the Harvard approach (De Gennes, 1974). For that purpose the dissipative function (8.21) and the viscosity stress tensor (8.22) can be rewritten in terms of the quantities

$$A_{\alpha\beta} = \tfrac{1}{2}(\partial_\alpha v_\beta + \partial_\beta v_\alpha) \quad , \tag{8.41}$$

$$N_\alpha = \dot{n}_\alpha + n_\beta W_{\alpha\beta} \quad , \quad \text{where} \tag{8.42}$$

$$W_{\alpha\beta} = \tfrac{1}{2}(\partial_\alpha v_\beta - \partial_\beta v_\alpha) \quad . \tag{8.43}$$

It follows easily that

$$D = \eta A_{\alpha\beta}A_{\alpha\beta} + \frac{1}{2}\left[\xi_1 + \xi_2 + 2\xi_3 - \frac{(\xi_2-\xi_1)^2}{(\xi_1+\xi_2-2\xi_3)}\right] n_\alpha n_\beta A_{\alpha\gamma}A_{\beta\gamma}$$
$$+ \frac{1}{2}\xi_4 n_\alpha n_\beta n_\gamma n_\delta A_{\alpha\beta}A_{\gamma\delta} + \frac{1}{2}(\xi_1+\xi_2-2\xi_3)^{-1} h_\alpha h_\alpha \quad , \tag{8.44}$$

where

$$h_\alpha = (\xi_1+\xi_2-2\xi_3)N_\alpha + (\xi_2-\xi_1)n_\alpha A_{\alpha\beta} \quad . \tag{8.45}$$

Clearly the dissipation function can be formulated in terms of four properly chosen viscosity coefficients. Furthermore a reactive parameter λ appears, given by

$$\lambda = \frac{\xi_1 - \xi_2}{\xi_1 + \xi_2 - 2\xi_3} = (\xi_1 - \xi_2)/\gamma_1 \quad . \tag{8.46}$$

The viscosity stress tensor is usually written as the sum of a symmetric tensor $\tilde{\boldsymbol{\sigma}}'_s$ and an antisymmetric tensor $\tilde{\boldsymbol{\sigma}}'_a$. The elements of the symmetric tensor are

$$\begin{aligned}\sigma'_{\alpha\beta,s} &= 2\eta A_{\alpha\beta} + \tfrac{1}{2}[\xi_1 + \xi_2 + 2\xi_3 + \lambda(\xi_2 - \xi_1)](n_\alpha n_\gamma A_{\gamma\beta} + n_\beta n_\gamma A_{\gamma\alpha}) \\ &\quad + \xi_4 n_\alpha n_\beta n_\gamma n_\delta A_{\gamma\delta} - \tfrac{1}{2}\lambda(n_\alpha h_\beta + n_\beta h_\alpha) \quad ,\end{aligned} \tag{8.47}$$

whereas the elements of the antisymmetric tensor are given by

$$\sigma'_{\alpha\beta,a} = -\tfrac{1}{2}(n_\alpha h_\beta - n_\beta h_\alpha). \tag{8.48}$$

The expressions (8.47) and (8.48) provide the direct connection with the other formulations. The dissipation function is just half the loss of the total free energy density, i.e. integration of this function over the volume of the system gives half the energy dissipation in an isothermal process.

The present discussion can readily be generalized to the case of compressible nematics, where two additional coefficients of viscosity are required for a complete description. Their appearance originates from terms being composed of $A_{\gamma\gamma}$, a quantity that is equal to zero in case of incompressible fluids.

8.3 Viscosity Measurements

The five coefficients of viscosity can be measured directly. The coefficients η, ξ_1, ξ_2 and ξ_4 are determined by measuring the effective viscosity in a shear flow experiment, where the director is fixed by an external field. The coefficient γ_1, which depends on the coefficient ξ_3, can be determined from the motion of the director in certain geometries.

Consider a nematic cell consisting of two parallel glass plates. The fluid flows along the z axis, which is defined to be parallel to the plates, whereas a velocity gradient exists along the x axis, which is normal to these plates (Fig. 8.1). Next a magnetic field is applied in order to define the director. This field must be sufficiently strong to minimize the wall effects and the disturbances of the alignment by the flow, i.e. the rotational motion of the director is suppressed. Substitution of the velocity field $\boldsymbol{v} = v_z(x)\boldsymbol{e}_3$ into (8.22) and using the expression (8.28) for $\boldsymbol{n}$ results in

$$\sigma_{xz} = \eta(\theta, \phi)\partial_x v_z \quad , \tag{8.49}$$

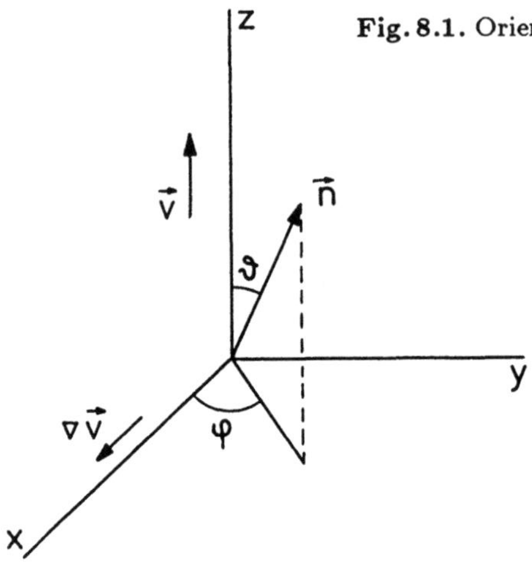

Fig. 8.1. Orientation of $\boldsymbol{n}$ with respect to the shear plane

where the effective viscosity $\eta(\theta,\phi)$ is given by

$$\eta(\theta,\phi) = \eta + \xi_1 \sin^2\theta\cos^2\phi + \xi_2\cos^2\theta + \xi_4\sin^2\theta\cos^2\theta\cos^2\phi \quad . \tag{8.50}$$

The various coefficients of viscosity are obtained most easily by choosing special orientations of the magnetic field. The three principal viscosity coefficients η_1, η_2 and η_3, the Miesowicz coefficients in the notation of Helfrich (1970), are obtained by considering the following three geometries:

i) $\theta = \frac{\pi}{2}$, $\phi = 0$, i.e. the director is perpendicular to the flow and parallel to the velocity gradient, resulting in an effective viscosity

$$\eta_1 = \eta + \xi_1 \quad ; \tag{8.51}$$

ii) $\theta = 0$, i.e. the director is parallel to the flow, resulting into an effective viscosity

$$\eta_2 = \eta + \xi_2 \quad ; \tag{8.52}$$

iii) $\theta = \frac{\pi}{2}$, $\phi = \frac{\pi}{2}$, i.e. the director is perpendicular to both the flow and the velocity gradient, resulting into an effective viscosity

$$\eta_3 = \eta \quad . \tag{8.53}$$

The remaining coefficient ξ_4, which has also been denoted by η_{12}, can be obtained by considering a geometry, in which the last term in (8.50) con-

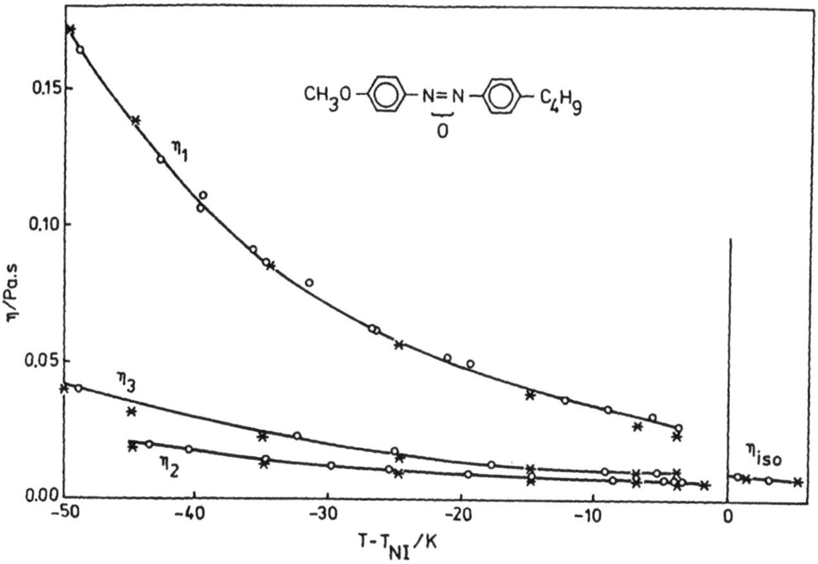

Fig. 8.2. Miesowicz viscosities of the *p*-methoxy-*p*′-butylazoxybenzene (N4); (o) Beens and De Jeu (1983); (*) Kneppe et al. (1981)

tributes maximally. This means $\theta = \frac{\pi}{4}$, $\phi = 0$, i.e. the field and thus the director is in the shear plane (xz plane) at an angle $\frac{\pi}{4}$ with the direction of both the flow and velocity gradient. In that situation the effective viscosity is given by

$$\eta(\tfrac{1}{4}\pi, 0) = \eta + \tfrac{1}{2}(\xi_1 + \xi_2) + \tfrac{1}{4}\xi_4 \quad . \tag{8.54}$$

Though measurements of shear flow using a strong magnetic field to control the orientational field are in principle already quite old (Miesowicz 1936), only recently more extensive data have become available (Kneppe and Schneider 1981; Beens and De Jeu 1983). In Fig. 8.2 some typical results are given. The temperature dependence of the Miesowicz viscosities is mainly determined by that of η. Like in isotropic liquids η varies approximately as

$$\eta = \eta_0 \exp(E/k_\mathrm{B}T) \quad , \tag{8.55}$$

where $E>0$ is an activation energy for diffusion. In the nematic phase an additional temperature dependence will come into play due to the dependence of ξ_i ($i = 1,2,3,4$) on the orientational order. The temperature dependence of the orientational order, however, is weak compared with that given in (8.55) and therefore difficult to measure. In view of this dependence, plots of $\xi_1 = \eta_1 - \eta_3$ and $\xi_2 = \eta_2 - \eta_3$ versus temperature would

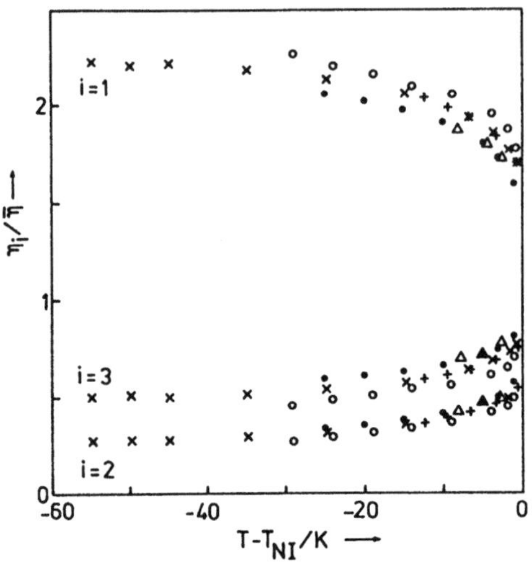

Fig. 8.3. Reduced viscosities $\eta_i/\overline{\eta}$ for various compounds (after Kneppe et al. 1981)

be expected to be rather universal. Experimentally this relation has been investigated by Kneppe et al. (1981) for several compounds. Indeed the temperature dependence of $\eta_2 - \eta_3$ shows a fairly universal behaviour, but this is not the case for $\eta_1 - \eta_3$. In fact the reduced viscosity $\eta_i/\overline{\eta}$, where $\overline{\eta} = \frac{1}{3}(\eta_1+\eta_2+\eta_3)$, shows a quite similar temperature-dependent behaviour for several substances (Fig. 8.3).

The coefficient γ_1 can be determined by considering a motion of the director, which does not involve a flow of the fluid. This can be accomplished by considering a sample in a rotating external field (Gasparoux and Prost 1971; Kneppe and Schneider 1983). Another way is by considering the dynamics of the Frederiks transition in the twist geometry (Sect. 6.2). An applied magnetic field, which is stronger than the threshold value, is suddenly removed, say at time $t = 0$. The resulting director field is described by

$$\boldsymbol{n} = [\cos\phi(z,t),\ \sin\phi(z,t),\ 0] \quad , \tag{8.56}$$

and satisfies the boundary conditions $\phi(0,t) = \phi(d,t) = 0$, where d denotes the distance between the boundaries. Switching off the field does not give rise to flow because the terms $\partial_\alpha(n_\alpha \dot{n}_\beta)$ and $\partial_\alpha(n_\beta \dot{n}_\alpha)$ do not contribute here. Neglecting the inertial term, the function $\phi(z,t)$ satisfies the differential equation

$$\frac{\partial D}{\partial \dot{\phi}} + F_{\mathrm{el},\phi} = 0 \quad , \quad \text{or} \tag{8.57}$$

$$\gamma_1 \frac{\partial \phi}{\partial t} = K_2 \frac{\partial^2 \phi}{\partial z^2} \quad . \tag{8.58}$$

The solution of this partial differential equation is

$$\phi(z,t) = \sum_{n=1}^{\infty} A_n \sin(n\pi z/d) \exp(-t/\tau_n) \quad , \tag{8.59}$$

where the relaxation times τ_n are given by (Pieranski et al. 1973)

$$\tau_n = \frac{\gamma_1}{K_2 n^2}\left(\frac{d}{\pi}\right)^2 = \frac{\mu_0\gamma_1}{\Delta\chi B_c^2 n^2} \quad , \tag{8.60}$$

with B_c denoting the threshold field. The coefficients A_n are determined by the condition that $\phi(z,0)$ is the solution of the differential equation:

$$K_2\phi_{zz} + \tfrac{1}{2}\mu_0^{-1}\Delta\chi B^2 \sin(2\phi) = 0 \quad , \tag{8.61}$$

which satisfies the proper boundary conditions. In the case where the applied field B is not much stronger than B_c the solution may be replaced by

$$\phi(z,t) = A_1 \sin(\pi z/d) \exp\left(-\frac{\mu_0\gamma_1 t}{\Delta\chi B_c^2}\right) \quad , \tag{8.62}$$

where A_1 is the distortion in the middle of the sample at $t = 0$. Provided $\Delta\chi$ and B_c are known, the coefficient of viscosity γ_1 can now be directly determined by measuring the relaxation time τ_1 appearing in expression (8.62). This can conveniently be done optically using a conoscope (Van Dijk et al. 1983).

In a number of cases the viscosity coefficient ξ_3, or γ_1, can also be obtained by considering the coupling between the flow and the rotational motion of the director in the absence of a magnetic field. Neglecting the elastic forces $F_{\mathrm{el},\theta}$ and $F_{\mathrm{el},\phi}$ the rotational motion is governed by the generalized viscous forces described in (8.32). Clearly there is no coupling, i.e. the flow does not disturb the alignment and $\dot{\boldsymbol{n}} = 0$, provided that these generalized viscous forces disappear, i.e.

$$-(\xi_1 - \xi_3)n_\alpha \frac{\partial \dot{n}_\gamma}{\partial \dot{\theta}}\partial_\alpha v_\gamma + (\xi_2 - \xi_3)n_\alpha \frac{\partial \dot{n}_\gamma}{\partial \dot{\theta}}\partial_\gamma v_\alpha = 0 \quad , \tag{8.63a}$$

$$-(\xi_1 - \xi_3)n_\alpha \frac{\partial \dot{n}_\gamma}{\partial \dot{\phi}}\partial_\alpha v_\gamma + (\xi_2 - \xi_3)n_\alpha \frac{\partial \dot{n}_\gamma}{\partial \dot{\phi}}\partial_\gamma v_\alpha = 0 \quad . \tag{8.63b}$$

For the velocity field $\boldsymbol{v} = v_z(x)\boldsymbol{e}_3$ this means

$$(\xi_1 - \xi_3)\sin^2\theta\cos\phi + (\xi_2 - \xi_3)\cos^2\theta\cos\phi = 0 \quad , \tag{8.64a}$$

$$-(\xi_2 - \xi_3) \cos\theta \sin\theta \sin\phi = 0 \quad , \tag{8.64b}$$

where use has been made of (8.33). Clearly the flow and anisotropy are decoupled provided that (a) the director lies in the shear plane, $\phi = 0$, and (b) there exists an angle θ_0, the so-called flow alignment angle, between the flow direction and the director given by

$$\tan^2\theta_0 = -\frac{(\xi_2 - \xi_3)}{(\xi_1 - \xi_3)} \quad . \tag{8.65}$$

The flow alignment angle, which is independent of $\partial_x v_z$, exists only if $(\xi_2 - \xi_3)/(\xi_1 - \xi_3) < 0$. This condition imposes a limit on the measurability of ξ_3 by means of this method as seen in the results shown in Fig. 8.4.

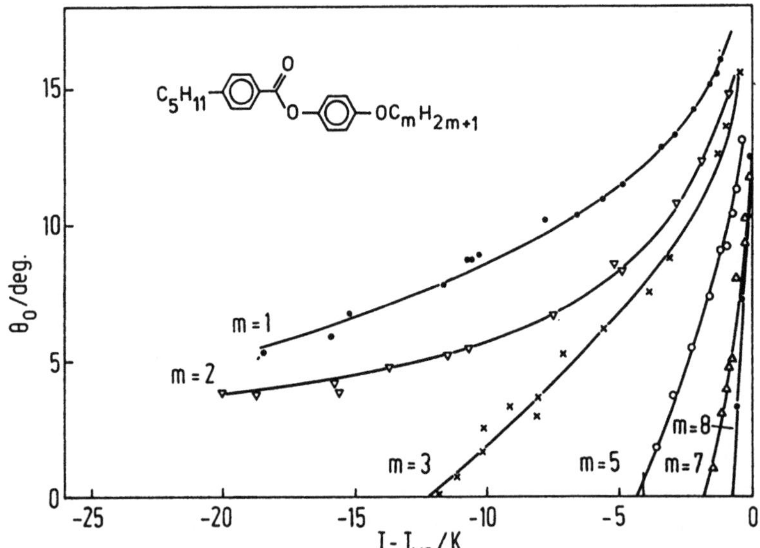

Fig. 8.4. Flow-alignment of a homologous series of compounds (Beens and De Jeu 1985)

Table 8.1. Viscosities and flow-alignment angle of p,p'-dibutylazoxybenzene (4AB, T_{NI} = 32° C) and p,p'-pentylcyanobiphenyl (5CB, T_{NI} = 35° C) around room temperature (23° C)

	4AB [a]	5CB
$\eta_1/10^{-3}$ Pa·s	76.0	129.6 [b]
$\eta_2/10^{-3}$ Pa·s	15.3	22.9 [b]
$\eta_3/10^{-3}$ Pa·s	26.5	37.4 [b]
$\eta_{12}/10^{-3}$ Pa·s		−11.1 [b]
$\gamma_1/10^{-3}$ Pa·s	61.0	81.0 [a]
θ_0/degrees	6.2	9.6 [a]

[a] Beens and De Jeu (1983);
[b] Kneppe et al. (1981)

Summarizing, the coefficients of viscosity $\eta_1 = \eta + \xi_1$, $\eta_2 = \eta + \xi_2$, $\eta_3 = \eta$, $\eta(\frac{1}{4}\pi, 0) = \frac{1}{2}(\eta_1 + \eta_2) + \frac{1}{4}\xi_4$ and $\gamma_1 = \xi_1 + \xi_2 - 2\xi_3$ are found to be directly observable experimentally. Some typical results for two nematics are given in Table 8.1. If desired, the Leslie coefficients α can be determined from these observables.

8.4 Electrohydrodynamic Instabilities

One of the interesting aspects of the fluid dynamics of nematics is the coupling between the flow and the orientational field. In fact a wide variety of hydrodynamic phenomena originate from this mechanism. Consider, for example, the classical Rayleigh-Bénard instability. This phenomenon refers to the convection of a liquid between two plates at different temperature under influence of a density gradient. In a nematic liquid the convective cells arising from this effect are directly visible due to the associated variations in birefringence, whereas convection can also be obtained under certain conditions if the sample is heated from above. For a review of this instability as well as other types of interesting instabilities the reader is referred to a review article by Dubois-Violette et al. (1978). Here the discussion will be restricted to one particular type of hydrodynamic motion, namely the motion under the influence of an electric field. These electrohydrodynamic instabilities have been known since the beginning of this century, being forgotten and rediscovered periodically. The interpretation of these instabilities has long been a matter of controversy, mainly due to the fact that it was not appreciated that steady cellular flow of a nematic can manifest itself as a static distorted director pattern. This manifestation can be fully substantiated mathematically by taking into account the elastic continuum theory. In fact the present treatment of the electrohydrodynamic instabilities can be considered as a nice, though somewhat complicated, application of the continuum theory. The complications arise because of the interplay of various effects.

The experimental situation to be discussed is as follows. Consider a nematic layer of thickness d with uniform planar boundary conditions and apply an electric field perpendicular to the boundaries, which are defined by the planes $x = 0$ and $x = d$. The unperturbed state of the nematic, i.e. the configuration in the absence of the field, is described by a director parallel to the z axis. The nematic compound in question has a *negative* dielectric anisotropy ($\Delta\varepsilon = \varepsilon_{\parallel} - \varepsilon_{\perp} < 0$). In addition the conductivity is assumed to have a *positive* anisotropy, which is usually the case, with $\sigma_{\parallel}/\sigma_{\perp} \approx 1.5$ (see Sect. 10.3). Now, according to Sect. 6.2, no Frederiks transition is expected when an electric field is applied, for the nematic slab is already in the state

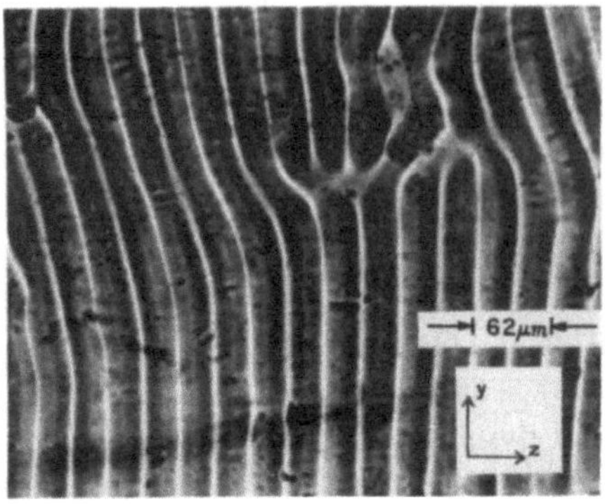

Fig. 8.5. Williams-domains just above the threshold voltage as observed with plane polarized light along z direction (Penz 1970)

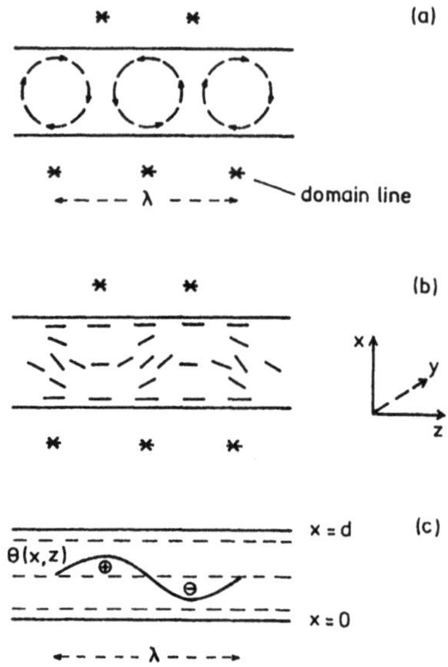

Fig. 8.6. **(a)** Cellular flow-pattern associated with the Williams domains. **(b)** Director pattern in the situation of **(a)**. **(c)** Separation of ions leading to space charge in the case of a director pattern as in **(b)**

of minimum dielectric energy. It appears, however, that a periodic distortion of the director field is observed, when the voltage reaches a threshold value of a few volts. This distortion is often a one-dimensional pattern (William's domains) as pictured in Fig. 8.5. The pattern consists of a set of regularly spaced lines parallel to the y axis with period d. The distortion can be easily detected with the aid of light propagating along the x direction and polarized along the z direction. Because of the hydrodynamic motion of the fluid, which appears as cellular flow in the form of long rolls (see Fig. 8.6a), this light experiences a periodically varying refractive index (between $n_{\|}$ and $n_{\text{eff}} < n_{\|}$). Hence the slab behaves as a periodic array of cylindrical lenses and the incident plane wave is then focused at a series of lines as shown in Fig. 8.5. In agreement with this picture two series of lines are observed, one when the focus is above, the other when the focus is below the sample. This effect disappears completely if the incident light is polarized along the y direction instead of the z direction. Consequently the hydrodynamic motion here only takes place in the xz plane, see Fig. 8.6b. The cellular flow can be displayed by following the motion of dust particles. The Williams' domains exist only over a small voltage range. With increasing voltage this type of distortion turns into a complicated texture due to turbulence (dynamic scattering). In order to simplify the discussion of the instability as far as possible only the one-dimensional theory of electrodynamic instabilities will be considered, in much the same spirit as Helfrich (1969) and Dubois-Violette et al. (1971). For a discussion of the two-dimensional theory the reader is referred to Meyerhofer (1975).

Because of symmetry the perturbed state is described by a two-dimensional director field

$$\boldsymbol{n} = [\sin\theta(x,z),\, 0,\, \cos\theta(x,z)] \quad , \tag{8.66}$$

where $\theta(x,z)$ satisfies the boundary conditions

$$\theta(0,z) = \theta(d,z) = 0 \quad . \tag{8.67}$$

In order to determine the critical field corresponding with the onset of the instability, i.e. the appearance of Williams' domains, it suffices to linearize all relevant equations around $\theta(x,z) = 0$, which is the solution below threshold. Consequently the director field may be replaced by

$$\boldsymbol{n} = (\theta,\, 0,\, 1) \quad . \tag{8.68}$$

Basic to the theory of electrohydrodynamic instabilities is that in the case of an anisotropic conductivity a distortion of the orientational field leads to a space charge. This can be understood as follows. Assume that $\sigma_{\|} > \sigma_{\perp}$ and that the electric field is perpendicular to $\boldsymbol{n}$. A distortion of the orientational field at a certain position now causes a certain distortion-dependent contribution of $\sigma_{\|}$ to be added to $\sigma_{\perp}$ at that position. This means that more charge departs then arrives. For the director pattern of Fig. 8.6b this means a separation of positive and negative charges as shown in Fig. 8.6c. Consequently transverse electric fields will also arise. Just above threshold the relation between this charge density and the distortion is rather simple and is derived as follows. Neglecting magnetic effects the charge density ϱ must follow from the following equations

$$\partial_\alpha D_\alpha = \varrho \quad , \tag{8.69}$$

$$\partial_\alpha j_\alpha = -\dot{\varrho} \quad , \quad \text{(equation of continuity)} \tag{8.70}$$

$$D_\alpha = \varepsilon_0 \varepsilon_{\alpha\beta} E_\beta = \varepsilon_0(\varepsilon_\perp \delta_{\alpha\beta} + \Delta\varepsilon n_\alpha n_\beta) E_\beta \quad , \tag{8.71}$$

$$j_\alpha = \sigma_{\alpha\beta} E_\beta = (\sigma_\perp \delta_{\alpha\beta} + \Delta\sigma n_\alpha n_\beta) E_\beta \quad , \quad \text{(Ohm's law)} \tag{8.72}$$

$$\varepsilon_{\alpha\beta\gamma} \partial_\beta E_\gamma = 0 \quad , \tag{8.73}$$

where $\boldsymbol{D}$ denotes the electric displacement. Because of (8.73) the electric field can be written as

$$E_x = E_0 + \partial_x V'(x,z,t) \quad , \quad E_z = \partial_z V'(x,z,t) \quad , \tag{8.74}$$

where $E_0 = V/d$ with V denoting the externally applied potential and $V'(x,z,t)$ a potential induced by the distortion. Clearly $V'(x,z,t)$ is of the order of the distortion just above threshold. Using this fact and (8.68) the expressions (8.71) and (8.72), up to order θ, become

$$\begin{aligned} D_x &= \varepsilon_0\varepsilon_\perp E_0 + \varepsilon_0\varepsilon_\perp \partial_x V' \quad ; \quad D_y = 0 \quad ; \\ D_z &= \varepsilon_0\varepsilon_\parallel \partial_z V' + \varepsilon_0 \Delta\varepsilon E_0 \theta \quad , \end{aligned} \tag{8.75}$$

$$\begin{aligned} j_x &= \sigma_\perp E_0 + \sigma_\perp \partial_x V' \quad ; \quad j_y = 0 \quad ; \\ j_z &= \sigma_\parallel \partial_z V' + \Delta\sigma E_0 \theta \quad . \end{aligned} \tag{8.76}$$

Substitution of (8.75) and (8.76) into (8.69) and (8.70) respectively gives the following equations for the charge density

$$\varepsilon_0\varepsilon_\perp \partial_x^2 V' + \varepsilon_0\varepsilon_\parallel \partial_z^2 V' + \varepsilon_0 \Delta\varepsilon E_0 \partial_z \theta = \varrho \quad , \tag{8.77}$$

$$\sigma_\perp \partial_x^2 V' + \sigma_\parallel \partial_z^2 V' + \Delta\sigma E_0 \partial_z \theta = -\dot{\varrho} \quad . \tag{8.78}$$

Next the influence of the electric field on the equations of motion, (8.3) and (8.26), of the fluid is considered. In a first approximation the inertial terms which appear in these equations may be neglected, because they are small in comparison with the remaining terms. The equations of motion now become

$$-\partial_\beta (p\delta_{\beta\alpha} - \sigma'_{\beta\alpha}) + \varrho E_\alpha = 0 \quad , \tag{8.79}$$

$$\frac{\partial D}{\partial \dot{\theta}} + F_{\mathrm{el},\theta} = 0 \quad , \tag{8.80}$$

where the contribution of the electric field to the distortion free energy density is given by

$$-\tfrac{1}{2}\varepsilon_0 \Delta\varepsilon (\sin\theta\, E_x + \cos\theta\, E_z)^2 . \tag{8.81}$$

Linearization of these equations results in

$$\begin{aligned} &-\partial_x p + 2\eta \partial_x^2 v_x + \eta_1 \partial_z^2 v_x + (\eta + \xi_3)\partial_z \partial_x v_z \\ &\quad -(\xi_1 - \xi_3)\partial_z \dot{\theta} + \varrho E_0 = 0 \quad , \end{aligned} \tag{8.82}$$

$$\begin{aligned} &-\partial_z p + \eta_2 \partial_x^2 v_z + (\eta + \xi_3)\partial_x \partial_z v_x + (\xi_2 - \xi_3)\partial_x \dot{\theta} \\ &\quad +(\eta_1 + \eta_2 + 2\xi_3 + \xi_4)\partial_z^2 v_z = 0 \quad , \end{aligned} \tag{8.83}$$

$$\gamma_1\dot{\theta} - (\xi_1 - \xi_3)\partial_z v_x + (\xi_2 - \xi_3)\partial_x v_z - K_1\partial_x^2\theta - K_3\partial_z^2\theta$$
$$-\varepsilon_0\Delta\varepsilon E_0^2\theta - \varepsilon_0\Delta\varepsilon E_0\partial_z V' = 0 \quad . \tag{8.84}$$

Because of the assumed incompressibility of the fluid the velocity $\boldsymbol{v}$ can be derived from a vector potential $\boldsymbol{a}$, i.e. $v_\alpha = \varepsilon_{\alpha\beta\gamma}\partial_\beta a_\gamma$. The vector potential has the special form $\boldsymbol{a} = a(x, z, t)\boldsymbol{e}_2$. This means

$$v_x = -\partial_z a \quad , \quad v_y = 0 \quad , \quad v_z = \partial_x a \quad . \tag{8.85}$$

Consequently Eqs. (8.82–84) can be written as

$$\partial_x p + [\eta_1\partial_z^3 + (\eta + \xi_3)\partial_z\partial_x^2]a + (\xi_1 - \xi_3)\partial_z\dot{\theta} - \varrho E_0 = 0 \quad , \tag{8.86}$$

$$-\partial_z p + [\eta_2\partial_x^3 + (\eta_1 + \xi_1 + \xi_2 + \xi_3 + \xi_4)\partial_x\partial_z^2]a$$
$$+(\xi_2 - \xi_3)\partial_x\dot{\theta} = 0 \quad , \tag{8.87}$$

$$\gamma_1\dot{\theta} + [(\xi_1 - \xi_3)\partial_z^2 + (\xi_2 - \xi_3)\partial_x^2]a - (K_1\partial_x^2 + K_3\partial_z^2 + \varepsilon_0\Delta\varepsilon E_0^2)\theta$$
$$-\varepsilon_0\Delta\varepsilon E_0\partial_z V' = 0 \quad . \tag{8.88}$$

The pressure terms $\partial_x p$ and $\partial_z p$ can easily be eliminated from (8.86) and (8.87) resulting in

$$[(\xi_1 - \xi_3)\partial_z^2 + (\xi_2 - \xi_3)\partial_x^2]\dot{\theta} + [\eta_2\partial_x^4 + \eta_1\partial_z^4 + (\eta_1 + \eta_2 + \xi_4)\partial_x^2\partial_z^2]a$$
$$-E_0\partial_z\varrho = 0 \quad . \tag{8.89}$$

Equations (8.77, 78, 88, 89) are the fundamental equations of the linearized theory of the electrohydrodynamic instabilities in nematic liquid crystals with a uniform planar texture. Together with the boundary conditions that $\partial_z V'$, θ, v_x and v_z must all be zero at $x = 0$ and $x = d$, these equations determine the threshold voltage of the appearance of instabilities. An analytical solution of this problem does not exist at the moment. Only numerical solutions are available (Penz 1974). However, the essential physics underlying the electrohydrodynamic instabilities can be quite well understood in analytical terms by simplifying the problem to a one-dimensional problem.

The one-dimensional theory of electrohydrodynamic instabilities is obtained by neglecting the boundary conditions. Retaining in (8.77, 78, 88, 89) only the terms that depend on the coordinate z, gives the following set of equations

$$\varepsilon_0\varepsilon_{\|}\partial_z^2 V' + \varepsilon_0\Delta\varepsilon E_0\partial_z\theta = \varrho \quad , \tag{8.90}$$

$$\sigma_{\|}\partial_z^2 V' + \Delta\sigma E_0\partial_z\theta = -\dot{\varrho} \quad , \tag{8.91}$$

$$\gamma_1\dot{\theta} + (\xi_1 - \xi_3)\partial_z^2 a - K_3\partial_z^2\theta - \varepsilon_0\Delta\varepsilon E_0^2\theta - \varepsilon_0\Delta\varepsilon E_0\partial_z V' = 0 \quad , \tag{8.92}$$

$$(\xi_1 - \xi_3)\partial_z^2\dot{\theta} + \eta_1\partial_z^4 a - E_0\partial_z\varrho = 0 \quad . \tag{8.93}$$

In order to describe the onset of Williams' domains, solutions must be found, that are periodic in space. Therefore the following trial functions are considered:

$$V' = V_0 \sin(\lambda z) \quad , \tag{8.94a}$$

$$\theta = \theta_0 \cos(\lambda z) \quad , \tag{8.94b}$$

$$\varrho = \varrho_0 \sin(\lambda z) \quad , \tag{8.94c}$$

$$a = a_0 \cos(\lambda z) \quad . \tag{8.94d}$$

The parameter λ must still be determined, just like the coefficients V_0, θ_0, ϱ_0 and a_0, which may be functions of the time t. Substitution of (8.94) into (8.90–93) gives

$$\varepsilon_0\varepsilon_{||}\lambda^2 V_0 + \varepsilon_0\Delta\varepsilon E_0\lambda\theta_0 = -\varrho_0 \quad , \tag{8.95}$$

$$\sigma_{||}\lambda^2 V_0 + \Delta\sigma E_0\lambda\theta_0 = \dot{\varrho}_0 \quad , \tag{8.96}$$

$$\gamma_1\dot{\theta}_0 - (\xi_1 - \xi_3)\lambda^2 a_0 + K_3\lambda^2\theta_0 - \varepsilon_0\Delta\varepsilon E_0^2\theta_0 - \varepsilon_0\Delta\varepsilon E_0\lambda V_0 = 0 \quad , \tag{8.97}$$

$$(\xi_1 - \xi_3)\lambda^2\dot{\theta}_0 - \eta_1\lambda^4 a_0 + E_0\lambda\varrho_0 = 0 \quad . \tag{8.98}$$

Next a_0 and V_0 are eliminated by making use of (8.95) and (8.97). Thus the following two coupled differential equations result for the distortion θ_0 and the space charge density $\varrho_0(t)$

$$\dot{\varrho}_0 + \varrho_0/\tau - \sigma_H E_0\lambda\theta_0 = 0 \quad , \tag{8.99}$$

$$\eta_H\dot{\theta}_0 + \left(K_3\lambda^2 - \frac{\varepsilon_0\Delta\varepsilon\varepsilon_\perp}{\varepsilon_{||}}E_0^2\right)\theta_0 + \frac{\varepsilon_0\Delta\varepsilon\varepsilon_\perp}{\varepsilon_{||}\tau\sigma_H\lambda}\zeta^2 E_0\varrho_0 = 0 \quad , \tag{8.100}$$

where

$$\tau = \varepsilon_0\varepsilon_{||}/\sigma_{||} \quad , \tag{8.101}$$

$$\sigma_H = \Delta\sigma - \varepsilon_0\Delta\varepsilon/\tau \quad , \tag{8.102}$$

$$\eta_H = \gamma_1 - (\xi_1 - \xi_3)^2/\eta_1 \quad , \tag{8.103}$$

$$\zeta^2 = \frac{\tau\sigma_H\varepsilon_\parallel}{\varepsilon_0 \Delta\varepsilon\varepsilon_\perp}\left[\frac{\Delta\varepsilon}{\varepsilon_\parallel} - \frac{(\xi_1 - \xi_3)}{\eta_1}\right] \quad . \tag{8.104}$$

These two coupled differential equations contain the information concerning the stability of a uniform planar texture against electric fields. Before considering, however, the instabilities caused by the application of dc and ac electric fields, the equations are solved in the case $E_0 = 0$. Next the excitations with dc and ac electric fields will be discussed.

(1) $E_0 = 0$. Here the equations are decoupled. Their solution reads

$$\varrho_0(t) = \varrho_0(0)\exp(-t/\tau) \quad , \tag{8.105}$$

$$\theta_0(t) = \theta_0(0)\exp\left(-\frac{K_3\lambda^2}{\eta_H}t\right) \quad . \tag{8.106}$$

Clearly the system is stable provided that $\eta_H > 0$. Fluctuations of the space charge as well as the orientation of the director decay to zero.

(2) DC *Electric Field.* The solution of the set of coupled equations is obtained by substitution of $\varrho_0(t) = A\exp(-\Gamma t)$, $\theta_0(t) = B\exp(-\Gamma t)$. This results in

$$\left(\frac{1}{\tau} - \Gamma\right)A - \sigma_H E_0 \lambda B = 0 \quad , \tag{8.107}$$

$$\left[\left(K_3\lambda^2 - \frac{\varepsilon_0\Delta\varepsilon\varepsilon_\perp}{\varepsilon_\parallel}E_0^2\right) - \eta_H\Gamma\right]B + \frac{\varepsilon_0\Delta\varepsilon\varepsilon_\perp}{\varepsilon_\parallel\tau\sigma_H\lambda}\zeta^2 E_0 A = 0 \quad . \tag{8.108}$$

A non-trivial solution exists only if

$$\left(\frac{1}{\tau} - \Gamma\right)\left[\left(K_3\lambda^2 - \frac{\varepsilon_0\Delta\varepsilon\varepsilon_\perp}{\varepsilon_\parallel}E_0^2\right) - \eta_H\Gamma\right] + \frac{\varepsilon_0\Delta\varepsilon\varepsilon_\perp}{\varepsilon_\parallel\tau}\zeta^2 E_0^2 = 0 \quad , \tag{8.109}$$

or

$$\Gamma = \frac{\alpha \pm (\alpha^2 - \beta^2)^{1/2}}{2\eta_H} \quad , \quad \text{where} \tag{8.110}$$

$$\alpha = K_3\lambda^2 - \frac{\varepsilon_0\Delta\varepsilon\varepsilon_\perp}{\varepsilon_\parallel}E_0^2 + \frac{\eta_H}{\tau} \quad ,$$

$$\beta = \frac{4\eta_H}{\tau}\left[K_3\lambda^2 - \frac{\varepsilon_0\Delta\varepsilon\varepsilon_\perp}{\varepsilon_\parallel}E_0^2(1 - \zeta^2)\right] \quad .$$

Now attention is confined to the interesting case $\Delta\varepsilon<0$. The coefficient α is always positive, but the coefficient β may be positive or negative provided that $\zeta^2>1$. The nature of the solution depends critically on the sign of β. The positive sign corresponds to a solution that decays to zero with increasing time. The system is then stable. On the other hand a negative sign gives rise to a non-decaying solution. The system is now unstable against the excitation with a dc electric field. Clearly the onset of the instability is characterized by $\beta = 0$ or

$$E_{0c}^2 = \frac{\varepsilon_{\|} K_3 \lambda^2}{\varepsilon_0 \Delta\varepsilon \varepsilon_{\perp}(1-\zeta^2)} \quad , \tag{8.111}$$

where $\Delta\varepsilon<0$ and $\zeta^2>1$.

It should be remarked here that according to (8.111) the system is unstable against the application of any dc electric field, because the parameter λ, i.e. the period of the distortion, can be chosen freely. However, this appears to be an artefact of the one-dimensional theory. As soon as the boundary conditions are taken into account it is found that the minimum allowable value of λ is inversely proportional to the thickness d of the sample. This fact implies immediately that a finite threshold voltage $V_c = E_{0c}d$ exists independent of the thickness, which is in accordance with experiment. The threshold voltage is of the order of a few volts.

(3) AC *Electric Field.* Consider an ac electric field $E_0 = F_0 \cos(\omega_0 t)$. In order to investigate the instability of the uniform planar structure against excitations of this field the space charge and distortion are written in terms of the Fourier integrals

$$\varrho_0(t) = \int_{-\infty}^{\infty} d\omega\, \varrho(\omega) \exp(i\omega t) \quad , \tag{8.112a}$$

$$\theta_0(t) = \int_{-\infty}^{\infty} d\omega\, \theta(\omega) \exp(i\omega t) \quad . \tag{8.112b}$$

Substitution of the oscillating field and (8.112) into the coupled equations (8.99) and (8.100) gives rise to the following infinite set of coupled linear equations

$$\left(i\omega + \frac{1}{\tau}\right)\varrho(\omega) - \frac{1}{2}\sigma_H \lambda F_0[\theta(\omega-\omega_0) + \theta(\omega+\omega_0)] = 0 \quad , \tag{8.113a}$$

$$\left(\mathrm{i}\omega\eta_H + K_3\lambda^2 - \frac{\varepsilon_0 \Delta\varepsilon\varepsilon_\perp}{2\varepsilon_\parallel} F_0^2\right)\theta(\omega)$$
$$-\frac{\varepsilon_0 \Delta\varepsilon\varepsilon_\perp}{4\varepsilon_\parallel} F_0^2[\theta(\omega - 2\omega_0) + \theta(\omega + 2\omega_0)]$$
$$+\frac{\varepsilon_0 \Delta\varepsilon\varepsilon_\perp}{2\varepsilon_\parallel \tau\lambda\sigma_H}\zeta^2 F_0[\varrho(\omega - \omega_0) + \varrho(\omega + \omega_0)] = 0 \quad . \tag{8.113b}$$

An approximate solution can be obtained by truncating this infinite set of coupled equations. The simplest approximation is obtained by putting all amplitudes $\theta(\omega)$ and $\varrho(\omega)$ equal to zero except $\theta(0)$, $\varrho(\omega_0)$ and $\varrho(-\omega_0)$. Such a procedure results in

$$\left(K_3\lambda^2 - \frac{\varepsilon_0 \Delta\varepsilon\varepsilon_\perp}{2\varepsilon_\parallel} F_0^2\right)\theta(0) + \frac{\varepsilon_0 \Delta\varepsilon\varepsilon_\perp}{2\varepsilon_\parallel \tau\lambda\sigma_H}\zeta^2 F_0[\varrho(-\omega_0) + \varrho(\omega_0)] = 0 \quad ,$$

$$\left(\mathrm{i}\omega_0 + \frac{1}{\tau}\right)\varrho(\omega_0) - \frac{1}{2}\sigma_H \lambda F_0 \theta(0) = 0 \quad ,$$

$$\left(-\mathrm{i}\omega_0 + \frac{1}{\tau}\right)\varrho(-\omega_0) - \frac{1}{2}\sigma_H \lambda F_0 \theta(0) = 0 \quad ,$$

which gives rise to

$$\left[K_3\lambda^2 - \frac{\varepsilon_0 \Delta\varepsilon\varepsilon_\perp}{2\varepsilon_\parallel} F_0^2\left(1 - \frac{\zeta^2}{1 + \omega_0^2\tau^2}\right)\right]\theta(0) = 0 \quad .$$

This means that the threshold value is determined by the requirement that the coefficient in front of $\theta(0)$ must be zero, i.e. the amplitude of the threshold field is

$$F_{0c}^2 = \frac{2\varepsilon_\parallel K_3\lambda^2(1 + \omega_0^2\tau^2)}{\varepsilon_0 \Delta\varepsilon\varepsilon_\perp(1 + \omega_0^2\tau^2 - \zeta^2)} \quad . \tag{8.114}$$

This implies immediately that the threshold value is given by

$$\langle E_{0c}^2(t)\rangle = \frac{1}{2}F_{0c}^2 = \frac{\varepsilon_\parallel K_3\lambda^2(1 + \omega_0^2\tau^2)}{\varepsilon_0 \Delta\varepsilon\varepsilon_\perp(1 + \omega_0^2\tau^2 - \zeta^2)} \quad . \tag{8.115}$$

Note that the expression (8.115) gives the exact dc threshold (8.111) in the limit $\omega_0 \to 0$. Just as in the dc case the one-dimensional theory introduces the artefact that the system is always unstable due to the free choice of λ. The boundary conditions, however, impose a lower bound of the order d on the period. Consequently a finite threshold voltage is obtained, which is independent of the thickness of the sample.

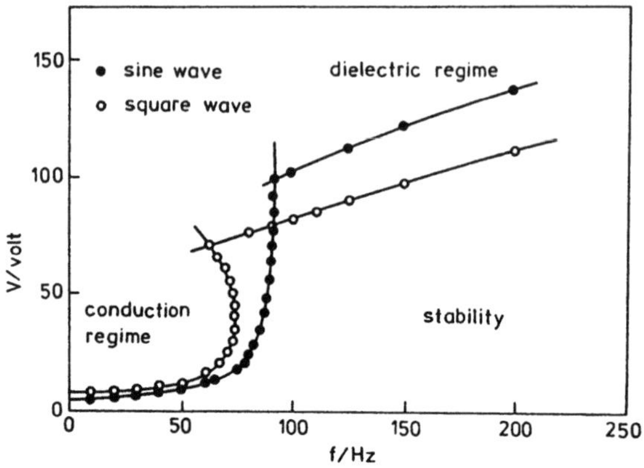

Fig. 8.7. Threshold voltage versus frequency for a uniform planar layer of nematic MBBA (Orsay Liquid Crystal Group 1972)

Clearly the calculation above breaks down when ω_0 reaches the cut-off frequency

$$\omega_c = \frac{1}{\tau}(\zeta^2 - 1)^{1/2} = \frac{\sigma_{\|}}{\varepsilon_0 \varepsilon_{\|}}(\zeta^2 - 1)^{1/2} \quad . \tag{8.116}$$

This behaviour is shown in Fig. 8.7. The frequency region $\omega<\omega_c$ is known as the conducting regime. Here the distortion is static, $\theta(0)\neq 0$, whereas the space charge oscillates with the frequency of the applied field. The high frequency regime, $\omega\gg\omega_c$, is called the dielectric regime. Here the reverse situation appears, namely the space charge is static and the distortion oscillates. Furthermore the threshold V_c now appears to be proportional to the thickness d of the sample, whereas the threshold field $E_{0c} = V_c/d$ varies as $\omega^{1/2}$. Clearly the cut-off frequency ω_c is directly proportional to the conductivity and can thus be varied by doping the nematic with ionic impurities.

Many interesting aspects of the electrohydrodynamic instabilities have been studied. Some of these will just be mentioned here.

i) Variation of the material parameters can influence the behaviour around ω_c. In particular low ζ^2 values can be reached for large negative $\Delta\varepsilon$ values, in which case the threshold curve is S-shaped, leading to stability again at higher voltages. This is illustrated in Fig. 8.7 for square-waves, where the effect is more pronounced.

ii) Extensions can be made to other signs of the dielectric anisotropy and conductivity anisotropy, and to other geometries. Some combinations again lead to instabilities.

iii) In the case of dc excitation, instabilities can be due to injected space charge. This is evident from the occurrence of similar instabilities in the isotropic phase, where the mechanism of charge separation due to an anisotropic conductivity cannot work.

For information concerning these various aspects, which are not treated here, the reader is referred to reviews by Dubois-Violette et al. (1978) and Blinov (1979). The problem of the transition to turbulence has been discussed by Manneville (1981).

Part III

Orientational Order and Anisotropic Properties

9. Orientational Order

In this chapter we consider the orientational order from the microscopic point of view. A central role is played here by the orientational distribution function, which in general depends on all three Eulerian angles. In the case of a uniaxial medium like the nematic or smectic A phase, however, the dependence on one of these angles can be disregarded. Experimentally it is very difficult to determine the full orientational distribution function. For this reason one often resorts to an expansion of this function. The expansion coefficients form a set of microscopic order parameters, of which only a few can be determined experimentally. Several of the appropriate methods will be discussed in some detail. If one is prepared to accept certain models for the molecules, the microscopic order parameters can be related to the macroscopic tensor order parameter introduced in Sect. 5.2. In the last section scattering methods are discussed that, at least in principle, give information on the full orientational distribution function.

9.1 Microscopic Order Parameters

In order to give a microscopic description of the orientational order, a nematic phase will be considered with the director $\boldsymbol{n}$ along the z axis of a laboratory-fixed coordinate system x, y, z. A second coordinate system ξ, η, ζ is fixed to a molecule, where the ζ axis is the unique molecular axis. The basis vectors of the molecular ξ, η, ζ coordinate system are the unit vectors $\boldsymbol{\xi}$, $\boldsymbol{\eta}$ and $\boldsymbol{\zeta}$, respectively. A molecule can be specified by the position of its centre of mass and by the orientation of the ξ, η, ζ system. The orientation can be described by the three Eulerian angles introduced in Sect. 5.1. This is illustrated once more in Fig. 9.1:

- ϕ is the angle between the y axis and the normal to the z, ζ plane, and describes a rotation around the z axis;
- θ is the angle between the z axis and the ζ axis and thus determines the deviation of the long molecular axis from its average direction;
- ψ is the angle between the η axis and the normal to the z, ζ plane, and describes a rotation around the ζ axis.

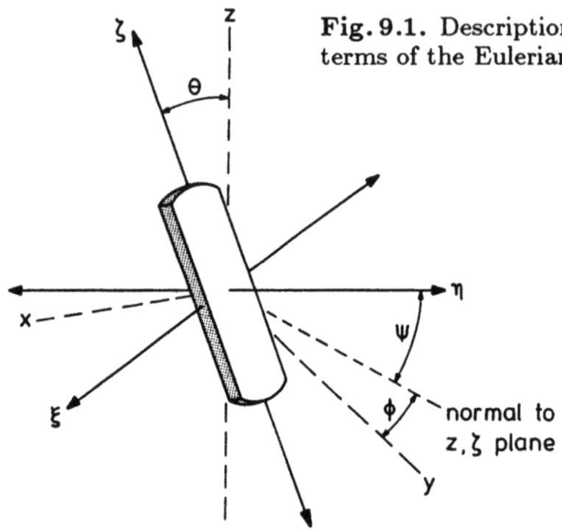

Fig. 9.1. Description of a body with two mirror planes in terms of the Eulerian angles

Now the orientational order of an ensemble of molecules is specified by the distribution function $P(\boldsymbol{r}, \Omega)$, which gives the probability of finding a molecule at a particular position $\boldsymbol{r}$ and with a particular orientation Ω, where Ω denotes the set of three Eulerian angles ϕ, θ, ψ. For the nematic phase it holds that

$$P(\boldsymbol{r}, \Omega) = \varrho f(\Omega) \quad , \tag{9.1}$$

where ϱ is the (uniform) mass density and $f(\Omega)$ is the orientational distribution function. The fraction of molecules with their orientation between ϕ and $\phi + d\phi$, θ and $\theta + d\theta$, and ψ and $\psi + d\psi$ is given by

$$N(\Omega) = f(\Omega)d\Omega = f(\phi, \theta, \psi) \sin\theta \, d\phi \, d\theta \, d\psi \quad . \tag{9.2}$$

The orientational distribution function is normalized, i.e.

$$\int d\Omega \, f(\Omega) = \int_0^{2\pi} d\phi \int_0^{\pi} \sin\theta \, d\theta \int_0^{2\pi} d\psi \, f(\phi, \theta, \psi) = 1 \quad . \tag{9.3}$$

Thus for the isotropic phase one finds

$$f(\Omega) = 1/(8\pi^2) \quad . \tag{9.4}$$

In the case of real molecules the first problem concerns the definition of the molecular frame. Usually the relatively rigid aromatic core is considered to be the most relevant part for this choice. For molecules with a *trans* configuration (Table 1.1, examples 1–2), the axis through the two outer *para* carbon atoms of the aromatic part is often taken to be the ζ axis. If the various benzene rings are coplanar (as for example with azobenzenes) this plane is identified with the ξ,ζ plane. Where such a coplanarity is not present, the ξ,ζ plane can be fixed by requiring that the angles with respect to the planes of the benzene rings are equal. In practice, the differences between the various possible choices are probably not very important. The second problem arises because mesogenic molecules usually contain flexible groups such as, for example, alkyl chains. In principle order parameters should be defined for each CC bond; these parameters will in general have different values due to differences in flexibility. This effect can indeed be observed using techniques such as nuclear magnetic resonance (NMR, see Sect. 9.3). Fortunately it turns out that the orientational order of the central core of a mesogenic molecule hardly differs from the average orientational order, if the various chain conformations are taken into account (Emsley et al. 1982). For this reason the description of a molecule (in its average conformation) in terms of a rigid body seems to be a reasonable approximation.

Experimental knowledge of $f(\Omega)$ is of primary importance in testing theoretical results for the transition from the isotropic liquid to the nematic phase, which necessarily involves a change from a constant orientational distribution function to an anisotropic one. Only a few experimental methods allow, at least in principle, the measurement of $f(\Omega)$. These will be discussed in Sect. 9.5. The remaining methods, e.g. NMR, are analyzed in terms of an expansion of $f(\Omega)$, where the expansion coefficients are microscopic order parameters. Three levels of approximation can be considered for performing such an expansion. They are described in the following.

(1) Uniaxial Phase of Cylindrical Rods

Rigid rods, for example prolate spheroids, cylinders, or spherocylinders (cylinders capped with hemispheres), are the simplest types of object which allow nematic behaviour. They are cylindrically symmetric around the ζ axis. Evidently this type of symmetry is not present in real mesogenic molecules. Nevertheless, the model is useful if properties are considered that are averaged with respect to a rotation around the ζ axis. In this case the orientational distribution is independent of ψ. Due to the uniaxiality of the phase the distribution is also independent of ϕ, i.e.

$$f(\Omega) = f(\theta)/(4\pi^2) \quad . \tag{9.5}$$

The quantity $f(\theta)\sin\theta\, d\theta$ is the fraction of molecules that have their long axes at an angle between θ and $\theta + d\theta$ with respect to the director. An expansion of $f(\theta)$ in terms of the complete set of Legendre polynomials gives

$$f(\theta) = \sum_{L=0}^{\infty} \frac{1}{2}(2L+1) S_L P_L(\cos\theta). \tag{9.6}$$

Because of the head-tail symmetry of the nematic phase it must hold that $f(\pi - \theta) = f(\theta)$. The relation $\cos(\pi - \theta) = -\cos\theta$ immediately implies that only even terms can occur in (9.6), i.e.

$$f(\theta) = \sum_{l=0}^{\infty} \frac{1}{2}(4l+1) S_{2l} P_{2l}(\cos\theta) \quad . \tag{9.7}$$

The expansion coefficients S_{2l} are found, using the orthogonality relations of the Legendre polynomials, by

$$S_{2l} = \langle P_{2l}(\cos\theta)\rangle = \int_{-1}^{1} P_{2l}(\cos\theta) f(\theta) d(\cos\theta) \quad . \tag{9.8}$$

Explicitly, the first three expansion coefficients are:

$$\begin{aligned} S_0 &\equiv \langle P_0(\cos\theta)\rangle = 1 \quad , \\ S_2 &\equiv \langle P_2(\cos\theta)\rangle = \tfrac{1}{2}\langle 3\cos^2\theta - 1\rangle \quad , \\ S_4 &\equiv \langle P_4(\cos\theta)\rangle = \tfrac{1}{8}\langle 35\cos^4\theta - 30\cos^2\theta + 3\rangle \quad . \end{aligned} \tag{9.9}$$

The coefficient $S_2 = \langle P_2(\cos\theta)\rangle$ is recognized directly as the order parameter introduced in Eq. (1.2) of Chap. 1.

(2) Uniaxial Phase

Due to the uniaxiality of the phase, $f(\Omega)$ is independent of ϕ:

$$f(\Omega) = f(\theta, \psi)/(2\pi) \quad . \tag{9.10}$$

The resulting orientational distribution function can be expanded in spherical harmonics:

$$f(\theta, \psi) = \sum_{L=0}^{\infty} \sum_{m=-L}^{L} \left(\frac{2L+1}{4\pi}\right)^{1/2} a_{Lm} Y_{Lm}(\theta, \psi) \quad . \tag{9.11}$$

Here the expansion coefficients a_{Lm} are given by

$$a_{Lm} = \left(\frac{2L+1}{4\pi}\right)^{-1/2} \langle Y^*_{Lm}(\theta,\psi)\rangle \quad , \tag{9.12}$$

i.e. for each L there are $2L+1$ order parameters. The distribution of the long molecular axis is described by the order parameters with $m=0$:

$$a_{L0} = \langle P_L(\cos\theta)\rangle \quad . \tag{9.13}$$

The first non-trivial order parameters appear for $L=2$ and $L=4$ and are $a_{20} = S_2$ and $a_{40} = S_4$ [see (9.9)].

The number of order parameters can be reduced considerably by assuming that the body representing the real molecule has an additional symmetry, namely two mirror planes that contain the ζ axis and are mutually perpendicular. A prolate ellipsoid with its shortest axis in the η direction, or a rectangular parallelepiped are examples of such a molecular model. This situation is shown in Fig. 9.1. Mathematically the additional symmetry is expressed by

$$f(\theta,\psi) = f(\theta,\pi-\psi) \quad , \quad f(\theta,\psi) = f(\theta,\pi+\psi) \quad . \tag{9.14}$$

Consequently for each L the number of order parameters is reduced from $2L+1$ to $1+L/2$. This means that only two order parameters are left for $L=2$, namely $a_{20} = \langle P_2(\cos\theta)\rangle$ and

$$a_{2-2} = a_{22} = (\tfrac{3}{8})^{1/2}\langle \sin^2\theta \cos 2\psi\rangle \quad . \tag{9.15}$$

For $L=4$ in addition to $a_{40} = \langle P_4(\cos\theta)\rangle$ only non-zero expressions result for $a_{4-2} = a_{42}$ and $a_{4-4} = a_{44}$ (Jen et al. 1977).

In the case of a uniaxial phase the orientational distribution function can also be expanded in terms of the projections of the director $\boldsymbol{n}$ on the axes of the molecule-fixed coordinate system. Such a procedure gives rise to a slightly alternative formulation of the order parameters. The components of $\boldsymbol{n}$ with respect to the molecule-fixed coordinate system are given by

$$\begin{aligned} n_\xi &= \xi_z = -\sin\theta\cos\psi \quad , \quad n_\eta = \eta_z = \sin\theta\sin\psi \quad , \\ n_\zeta &= \zeta_z = \cos\theta \quad , \end{aligned} \tag{9.16}$$

where $\boldsymbol{n}$ is along the z axis. The lowest order terms of such an expansion in the direction-cosines i_z with $i = \xi,\eta,\zeta$, are proportional to

$$\sum_{i,j} S_{ij} i_z j_z \quad , \quad i,j = \xi,\eta,\zeta \quad ,$$

where S_{ij} denotes the Saupe ordering matrix

$$S_{ij} = \tfrac{1}{2}\langle 3 i_z j_z - \delta_{ij} \rangle \quad . \tag{9.17}$$

The microscopic tensor order parameter S_{ij} is symmetric and traceless. Hence there are five independent elements, equivalent to the $2L+1$ order parameters for $L = 2$ as discussed above. After transformation to principal axes – which can be found immediately if the molecules contain two mirror-planes – only two elements are left. The first order parameter is given by

$$S_{\zeta\zeta} = \tfrac{1}{2}\langle 3\zeta_z^2 - 1 \rangle = \tfrac{1}{2}\langle 3 \cos^2\theta - 1 \rangle \quad . \tag{9.18}$$

As a second parameter either $S_{\xi\xi} = \frac{1}{2}\langle 3\xi_z^2 - 1\rangle$ or $S_{\eta\eta} = \frac{1}{2}\langle 3\eta_z^2 - 1\rangle$ can be used. It is convenient to take instead

$$D = S_{\xi\xi} - S_{\eta\eta} = \tfrac{3}{2}\langle \sin^2\theta \cos 2\psi \rangle \quad . \tag{9.19}$$

The order parameter D is related to the degree of "flatness" of the molecules. A finite D means that there is a difference in tendency of the two transverse molecular axes to project on the director. It does not mean that the phase is biaxial. There is no preference for either the ξ axis or the η axis of different molecules to be parallel. Comparison with (9.13) and (9.15) gives $S_{\zeta\zeta} = \langle P_2(\cos\theta)\rangle = a_{20}$ and $D = \sqrt{6}\, a_{22}$.

(3) General Expansion of $f(\Omega)$

The general expansion of $f(\phi, \theta, \psi)$ is not very useful for most mesophases of interest, because of their uniaxial symmetry. Nevertheless this case is mentioned here because of its relevance to the description of biaxial mesophases. The complete expansion of the orientational distribution function is given by

$$f(\Omega) = \sum_{L=0}^{\infty} \sum_{m,m'=-L}^{L} \frac{2L+1}{8\pi^2} a_{Lm'm} D^L_{m'm}(\Omega) \quad . \tag{9.20}$$

The functions $D^L_{m'm}(\Omega)$ are the so-called Wigner matrices or generalized spherical harmonics (Rose 1957; Zannoni 1979). Multiplying both sides by $D^{L*}_{m'm}(\Omega)$ and integrating over the angles it follows that

$$a_{Lm'm} = \langle D^{L*}_{m'm}(\Omega) \rangle \quad . \tag{9.21}$$

The averages, which completely define $f(\Omega)$, are just the orientational order parameters. In principle there are $(2L+1)^2$ order parameters for each L. This number is reduced if symmetry is present. The link with the previous expansions is made by noting that

$$D^L_{00}(\theta) = \left(\frac{2L+1}{4\pi}\right)^{-1/2} Y_{L0}(\theta) = P_L(\cos\theta) \quad . \tag{9.22}$$

Only a few of the microscopic order parameters discussed so far are accessible experimentally. In Sect. 9.2 it will be shown that second-rank tensor properties such as the magnetic susceptibility are related to $\langle P_2(\cos\theta)\rangle$ and D. In Sect. 9.4 a method to determine $\langle P_2(\cos\theta)\rangle$ as well as $\langle P_4(\cos\theta)\rangle$ will be discussed.

9.2 Magnetic Anisotropy

In the SI system of units, the magnetization $\boldsymbol{M}$ is defined in terms of the magnetic induction $\boldsymbol{B}$ and the magnetic field strength $\boldsymbol{H}$ by

$$\boldsymbol{M} = \mu_0^{-1}\boldsymbol{B} - \boldsymbol{H} \quad , \tag{9.23}$$

where μ_0 is the permeability of vacuum. The magnetic susceptibility of a system describes the response of the system to an external field of induction $\boldsymbol{B}$. For small values of the field the response is linear and can be written as

$$M_\alpha = \mu_0^{-1}\chi_{\alpha\beta}B_\beta \quad , \quad \alpha,\beta = x,y,z \quad , \tag{9.24}$$

where $\chi_{\alpha\beta}$ is an element of the magnetic susceptibility tensor $\tilde{\chi}$. For uniaxial phases like the nematic phase the tensor $\tilde{\chi}$ takes the form

$$\begin{pmatrix} \chi_\perp & 0 & 0 \\ 0 & \chi_\perp & 0 \\ 0 & 0 & \chi_\parallel \end{pmatrix} \quad ,$$

where the director is along the z axis. The average susceptibility is given by

$$\overline{\chi} = \tfrac{1}{3}(\chi_\parallel + 2\chi_\perp) \quad , \tag{9.25}$$

and the magnetic anisotropy is defined by

$$\Delta\chi = \chi_\parallel - \chi_\perp = \tfrac{3}{2}(\chi_\parallel - \overline{\chi}) \quad . \tag{9.26}$$

For an arbitrary angle between $\boldsymbol{B}$ and $\boldsymbol{n}$, the total magnetization can be written as

$$\boldsymbol{M} = \mu_0^{-1}\chi_\perp \boldsymbol{B} + \mu_0^{-1}\Delta\chi(\boldsymbol{B}\cdot\boldsymbol{n})\boldsymbol{n} \quad . \tag{9.27}$$

In addition to the volume susceptibility χ, one can introduce the mass susceptibility $\chi^{\mathrm{m}} = \chi/\varrho$, where ϱ is the density. The molar susceptibility $\chi^{\mathrm{M}} = \chi^{\mathrm{m}} M$ refers to a mole of substance, where M is the mass number. Like most organic substances liquid crystals are usually diamagnetic. Consequently $\chi_\parallel$ and $\chi_\perp$ are small and negative, of the order of 10^{-5} SI units. Although χ is dimensionless, it is a factor 4π larger in SI units than in CGS units, as the latter system is not rationalized. For χ^{m} and χ^{M} additional factors 10^{-3} occur due to the different units of density and mole of substance.

The anisotropy of a physical property such as the magnetic susceptibility has been used in Sect. 5.2 to define a macroscopic tensor order parameter $\tilde{\boldsymbol{Q}}$, by extracting the anisotropic part of $\tilde{\boldsymbol{\chi}}$. The elements $Q_{\alpha\beta}$ are given by

$$Q_{\alpha\beta} = \frac{\chi_{\alpha\beta}}{3\overline{\chi}} - \frac{1}{3}\delta_{\alpha\beta} \quad , \quad \alpha,\beta = x,y,z \quad . \tag{9.28}$$

$\tilde{\boldsymbol{Q}}$ is a second-rank tensor, that is diagonal if $\boldsymbol{n}$ is along the z axis, has zero trace, and vanishes in the isotropic phase. Therefore it suffices to consider only one element of $\tilde{\boldsymbol{Q}}$:

$$Q_{zz} = \frac{1}{3\overline{\chi}}(\chi_\parallel - \overline{\chi}) = \frac{2}{9\overline{\chi}}(\chi_\parallel - \chi_\perp) \quad . \tag{9.29}$$

All relevant information is contained in $\Delta\chi$ and $\overline{\chi}$.

In order to discuss the molecular interpretation of $\tilde{\boldsymbol{\chi}}$, and to relate $\tilde{\boldsymbol{Q}}$ to the microscopic order parameter tensor $\tilde{\boldsymbol{S}}$, the molecular magnetic polarizability $\tilde{\boldsymbol{\kappa}}$ is introduced, which is assumed to be diagonal in the molecule-fixed coordinate system ξ, η, ζ. When a magnetic field of induction $\boldsymbol{B}$ is applied along $\boldsymbol{n}$, a magnetic moment $\boldsymbol{m}$ is induced in a molecule with components:

$$\begin{aligned} m_\zeta &= \mu_0^{-1}\kappa_{\zeta\zeta} B \cos\theta = \mu_0^{-1}\kappa_{\zeta\zeta} B \zeta_z \\ m_\xi &= -\mu_0^{-1}\kappa_{\xi\xi} B \sin\theta\cos\psi = \mu_0^{-1}\kappa_{\xi\xi} B \xi_z \\ m_\eta &= \mu_0^{-1}\kappa_{\eta\eta} B \sin\theta\sin\psi = \mu_0^{-1}\kappa_{\eta\eta} B \eta_z \quad . \end{aligned} \tag{9.30}$$

The component of $\boldsymbol{m}$ along the z axis is

$$m_\parallel = \sum_i m_i i_z = \mu_0^{-1} B \sum_i \kappa_{ii} i_z^2 \quad , \quad i = \xi,\eta,\zeta \quad . \tag{9.31}$$

Taking the average over the orientation of N molecules leads to

$$\chi_{\|} = N \sum_i \kappa_{ii} \langle i_z^2 \rangle \quad . \tag{9.32}$$

Using (9.26) it follows that

$$\Delta\chi / N = [\kappa_{\zeta\zeta} - \tfrac{1}{2}(\kappa_{\xi\xi} + \kappa_{\eta\eta})]S + (\kappa_{\xi\xi} - \kappa_{\eta\eta})D \quad . \tag{9.33}$$

According to (9.29) the macroscopic order parameter Q_{zz} can be related to the microscopic order parameters S and D leading to

$$Q_{zz} = \frac{2N}{9\overline{\chi}} \left\{ \left[\kappa_{\zeta\zeta} - \frac{1}{2}(\kappa_{\xi\xi} + \kappa_{\eta\eta}) \right] S + \frac{1}{2}(\kappa_{\xi\xi} - \kappa_{\eta\eta}) D \right\} \quad . \tag{9.34}$$

Evidently a knowledge of $\Delta\chi$ is sufficient to determine the macroscopic order parameter Q_{zz} which distinguishes the nematic phase from the isotropic phase. However, (9.34) clearly shows that at least two parameters are needed to characterize the average degree of orientation of the molecules. These parameters cannot both be determined simultaneously from a single measurement of a second-rank tensor property.

In Fig. 9.2a measurements are shown of the susceptibilities of 7CB in the nematic and isotropic phase. Using the fact that $\overline{\chi}$ in the nematic phase

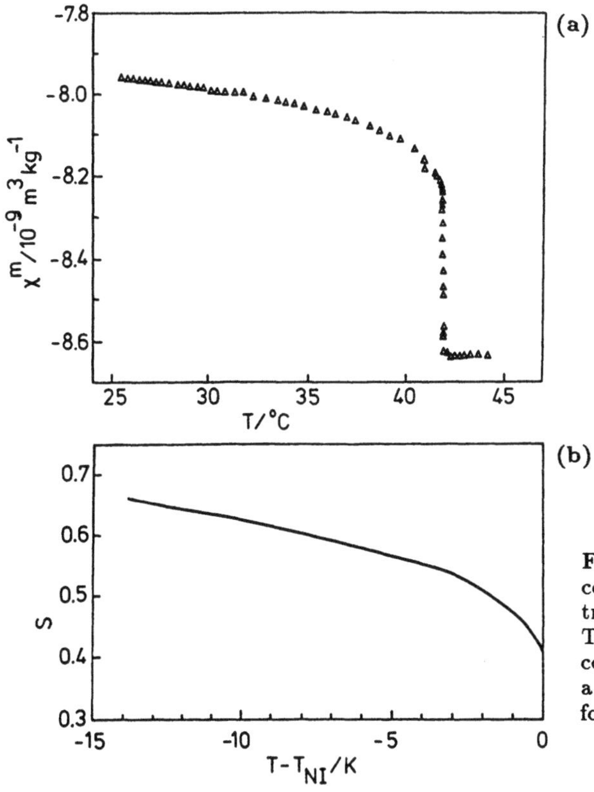

Fig. 9.2. **(a)** Diamagnetic susceptibility of 7CB. **(b)** Anisotropy $\Delta\chi$ as derived from **(a)**. The order parameter scale is constructed using (9.35) and a value of $1.92 \times 10^{-9}\,\mathrm{m^3kg^{-1}}$ for $\chi_l - \chi_t$

equals both $\overline{\chi}$ in a powdered solid and χ in the isotropic phase, $\Delta\chi$ can be calculated using Eq. (9.26) (Fig. 9.2b). In the interpretation of such experiments, the last term of (9.33) and (9.34) is often assumed to be small. This means that (9.33) can be written as

$$\chi_{\|} - \chi_{\perp} = (\chi_l - \chi_t)S \quad , \tag{9.35}$$

where $\chi_l = N\kappa_{\zeta\zeta}$ and $\chi_t = \frac{1}{2}N(\kappa_{\xi\xi} + \kappa_{\eta\eta})$. Equation (9.35) is known as Tsvetkov's expression for the order parameter S. To determine S the values of χ_l and χ_t are required. These can in principle be obtained from solid state measurements. For a discussion of the molecular factors that influence $\tilde{\chi}$, the reader is refered to De Jeu (1980). As far as information is available, it turns out that the factor $\kappa_{\xi\xi} - \kappa_{\eta\eta}$ that appears in (9.33) can be at least as large as the factor $\kappa_{\zeta\zeta} - \frac{1}{2}(\kappa_{\xi\xi} + \kappa_{\eta\eta})$. This means that the use of (9.35) can only be justified if $D \ll S$. As will be seen in the next section this often seems to be the case. Nevertheless neglecting the second term can introduce errors of the order of several percent. This is important if measurements are being discussed that are intrinsically more accurate than a few percent (Bunning et al. 1986).

The relations between macroscopic and microscopic order parameters derived in this section apply in principle to any second-rank tensor property. The magnetic susceptibility is a convenient choice because the diamagnetic moments of the molecules are very small, and consequently the interactions between these moments can be ignored. This can be concluded from the value of the relative diamagnetic permeability $\mu_r = 1 + \chi$ which differs only very little from unity. As a result, the field acting on a molecule can be taken to be equal to the externally applied field, as has been done.

9.3 Nuclear Magnetic Resonance

Nuclear magnetic resonance (NMR) gives information about the orientational order in various ways. The orientational order can be studied by considering (1) the coupling between nuclear dipole moments, (2) the coupling between a nuclear dipole and a nuclear quadrupole moment, and (3) the so-called chemical shift. Only the first method will be treated here in detail. A characteristic of the NMR spectrum of a liquid is that it consists of narrow lines: all direct magnetic interactions between the nuclear spins both of different molecules and within a molecule average to zero. This is due to the rapid translational diffusion as well as to the rotational tumbling of the molecules. In the nematic phase the first mechanism is still present: all *inter*molecular spin-spin interactions average to zero and the spectra still consist of narrow lines. However, compared with the isotropic phase, additional line splitting occurs since, due to the orientational order, the rotational symmetry is broken and *intra*molecular spin-spin coupling exists.

Consider two nuclear spins with $I = \frac{1}{2}$ (for example protons) within a molecule. The magnetic field and thus the director is taken along the z axis. The distance between the spins is denoted by $\boldsymbol{r} = |\boldsymbol{r}|\boldsymbol{u}$. The Hamiltonian of this system consists of the Zeeman term H_{Z} and the term H_{D} describing the interaction between the dipoles (Abragam 1961):

$$\begin{aligned} H &= H_{\mathrm{Z}} + H_{\mathrm{D}} \\ &= -\hbar\gamma B(I_{1z} + I_{2z}) \\ &\quad - (\hbar\gamma)^2 r^{-3}[3(\boldsymbol{I}_1 \cdot \boldsymbol{u})(\boldsymbol{I}_2 \cdot \boldsymbol{u}) - \boldsymbol{I}_1 \cdot \boldsymbol{I}_2] \quad , \end{aligned} \tag{9.36}$$

where $\hbar$ is Planck's constant divided by 2π and γ is the proton gyromagnetic ratio. The Larmor frequency, the frequency that is associated with the splitting by the magnetic field, is given by $\nu_0 = \gamma B/(2\pi)$ and is 42.5 MHz for a field of 1 T. Hence (9.36) can be averaged over the orientational fluctuations of the molecules, as these are much faster (see Chap. 11). Taking into account the uniaxial symmetry around the z axis i.e.

$$\langle u_x^2 \rangle = \langle u_y^2 \rangle = \tfrac{1}{2} - \tfrac{1}{2}\langle u_z^2 \rangle \quad ,$$

the averaging procedure results in

$$\langle H_{\mathrm{D}} \rangle = -(\hbar\gamma)^2 r^{-3} S_{uu}(3I_{1z}I_{2z} - \boldsymbol{I}_1 \cdot \boldsymbol{I}_2) \tag{9.37}$$

with $S_{uu} = \frac{1}{2}\langle 3u_z^2 - 1 \rangle$. The spectrum of the Hamiltonian (9.36) can be easily obtained by introducing the total spin $\boldsymbol{I} = \boldsymbol{I}_1 + \boldsymbol{I}_2$. Using the fact that both moments have spin $\frac{1}{2}$ (9.36) can be written as

$$\langle H \rangle = -\hbar\gamma B I_z - \tfrac{1}{2}(\hbar\gamma)^2 r^{-3} S_{uu}(3I_z^2 - \boldsymbol{I}^2) \quad , \tag{9.38}$$

where the constants are disregarded as they do not contribute to the energy intervals that determine the spectrum. For two protons I can only assume the values $I = 0$ and $I = 1$, corresponding to a singlet and a triplet state, respectively. In a resonance experiment only transitions between neighbouring triplet levels can be observed. These levels are characterized by $I_z = -1, 0, 1$. It follows directly from (9.38) that the relevant energy differences are given by

$$\Delta E = \hbar\gamma B \pm \tfrac{3}{2}(\hbar\gamma)^2 r^{-3} S_{uu} \quad . \tag{9.39}$$

Hence the NMR spectrum consists of a single line in the isotropic phase where $S_{uu} = 0$. This line splits into a doublet in the nematic phase, with a spacing (in frequency units)

$$\Delta\nu = \frac{3\hbar\gamma^2}{2\pi r^3} S_{uu} \quad . \tag{9.40}$$

If $\boldsymbol{u}$ is along the long molecular axis, i.e. $\boldsymbol{u} = \boldsymbol{\zeta}$, $\Delta\nu$ is a direct measure of the microscopic order parameter S. Using $u_z = u_\xi \xi_z + u_\eta \eta_z + u_\zeta \zeta_z$ it is easily verified that (9.40) can be rewritten as

$$\Delta\nu = \frac{3\hbar\gamma^2}{2\pi r^3} \left\{ \left[u_\zeta^2 - \frac{1}{2}(u_\xi^2 + u_\eta^2) \right] S + \frac{1}{2}(u_\xi^2 - u_\eta^2) D \right\} \quad . \tag{9.41}$$

The potential possibilities of this NMR method are clearly demonstrated by (9.41). S can be determined by choosing two particular protons such that $\boldsymbol{u}$ is approximately parallel to the long molecular axis. Next D can be obtained by taking another pair of protons such that $\boldsymbol{u}$ is at a certain angle with the ζ axis. In particular if this angle can be chosen to be close to 54.7° (the "magic angle") the first term in (9.41) will be small, and the influence of D will be large. In practice, however, the situation is not so simple, because the following experimental difficulties may interfere:

(i) The accuracy of the results for S and D depends strongly on the accuracy of the value and the direction of the relevant $\boldsymbol{r}$, due to the appearance of $\boldsymbol{r}$ to the third power. If the nuclei are separated by more than one bond, both the magnitude and the direction of $\boldsymbol{r}$ are very sensitive to variations in the bond angles. Usually, the accuracy of x-ray data of bond angles is insufficient.

(ii) Normal proton-NMR spectroscopy cannot be applied directly, as a great number of protons are usually present in a mesogenic molecule. They nearly all couple to each other due to their high gyromagnetic ratio, leading to very complicated spectra that cannot easily be analyzed.

The first problem can sometimes be overcome by making use of symmetry properties: for example, the *ortho* and *meta* protons attached to a benzene ring have internuclear vectors that are parallel or perpendicular to the axis through the *para* positions. In spite of the problems mentioned under (ii) ^{1}H-NMR has been useful, often in combination with analysis of the shape of the broad lines (see, for example, Limmer et al. 1981). The second problem can be dealt with by replacing all protons, except those in one specific part of the molecule, by deuterons. Due to their smaller magnetic moment the D-NMR signals are distinct from the ^{1}H signals, and any cou-

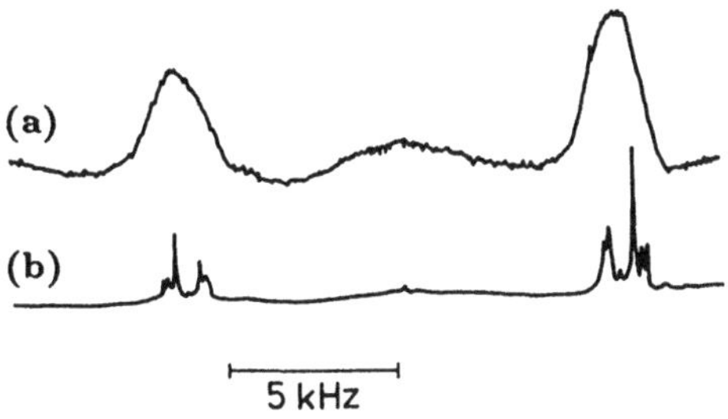

Fig. 9.3. ^{1}H-NMR spectrum (**a**) without and (**b**) with deuterium decoupling of partially deuterated 5CB in the nematic phase at 31°C (after Emsley et al. 1978)

pling between them is small. This results in a considerable simplification of the ^{1}H-spectrum. An example is shown in Fig. 9.3a where the ^{1}H-NMR spectrum is shown for

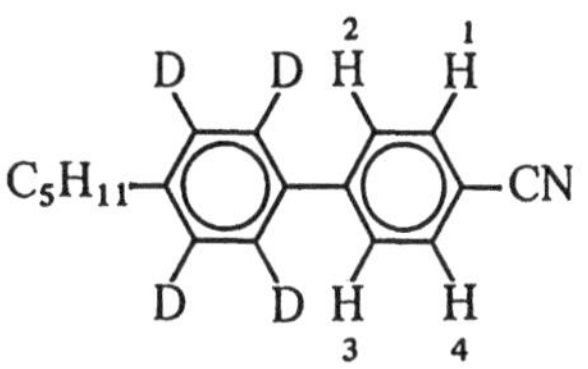

The spectrum closely approaches a doublet, as is expected for benzene. The width of the lines is due to residual coupling between the protons and the deuterons. Using proton-(deuterium) double resonance this coupling can be removed, producing the high-resolution spectrum of Fig. 9.3b. This spectrum can be fitted with specific values for the coupling between protons 1–2, 1–3, 1–4 and 2–3. This leads to $S = 0.56$ and $D = 0.05$ (at 31°C) *in a local coordinate system attached to the ring.* In order to transform this result to the principal axes of the molecule the latter axes have to be determined. Using information about the structure of closely related molecules Emsley et al. (1978) arrive at $S = 0.62$ and $D = -0.02$. If these results are typical for other mesogenic molecules the conclusion must be drawn that a description of the molecules as cylindrical objects is not unreasonable.

Isotopic substitution of one specific nucleus (D or ^{15}N) can be used in a similar way in combination with carbon-13 NMR. This isotope is present in the skeleton of a typical mesogenic molecule with a natural abundance of 1.1 %. Consequently the mutual dipolar coupling can be neglected. The coupling with neighbouring protons can be removed by irradiating the sample at the proton frequencies. Thus a spectrum is left dominated by the dipolar coupling between the various ^{13}C nuclei and the isotopically substituted nucleus at a well-chosen site. Though the experiment requires relatively elaborate equipment, it is one of the nicest ways to study dipolar coupling (Höhener et al. 1979).

Isotopic substitution with deuterium ($I = 1$) also offers the possibility to exploit the power of deuterium resonance. Nuclear spins with $I > \frac{1}{2}$ have a quadrupole moment, and the deuterium NMR spectrum is dominated by the electrostatic interaction of the quadrupole with the electric field gradients of the surrounding charge distribution of electrons. As already mentioned, the dipolar coupling is weak in this case and can be disregarded in a first approximation. The spin Hamiltonian can then be written as

$$H = H_Z + H_Q \tag{9.42}$$

where H_Z is the Zeeman term and H_Q the quadrupole Hamiltonian. H_Q depends on the electric quadrupole moment of the nucleus and on the electric field gradient. The latter is assumed to be axially symmetric around the chemical bond considered. The resulting splitting into a doublet is again of the form of (9.40) and (9.41), with $\hbar\gamma^2/(2\pi r^3)$ replaced by the static quadrupole coupling constant $\nu_Q = e^2qQ/h$, where eQ and eq denote the magnitude of the electric quadrupole moment and the electric field gradient, respectively. Of course, $\boldsymbol{u}$ is now along the direction of the electric field gradient, which in practice will be along a particular C-D bond.

In practice deuterium resonance is not very useful to obtain values for the order parameters, since the absolute accuracy is hampered by uncertainties in the value of ν_Q and in the exact direction of the electric field gradient. The method is very successful, however, in revealing relative variations in orientational order, for example within a molecule, because the spectra are well resolved. In particular, if a deuterated alkyl chain is investigated the spectrum shows doublets with different spacings for each CD_2 group. From the fact that the two deuterons of each CD_2 group are not distinct it must be concluded that the alkyl chain adds a symmetry plane to the system, while the long molecular axis must lie in that plane. In addition the results show that the chain cannot be rigid, as only one doublet would occur if all CD_2 groups were equivalent with respect to the ζ axis. Most probably this indicates isomeric rotations around the C-C bonds. Along the chain a decrease of the quadrupole splitting is always observed (Fig. 9.4). As ν_0 is constant along the chain this indicates a variation in S_{uu}, hence in the angle between the CD direction and the ζ axis. However, there has been considerable disagreement about the quantitative interpretation of the observed quadrupole splittings (Charvolin and Deloche 1979; Emsley and Luckhurst 1980). Detailed analysis of the results of specifically labelled mesogens has shown that a single set of order parameters that can be transformed to characterize the various segments, does not suffice to describe the orientational order of a non-rigid mesogenic molecule (Boden et al. 1981; Dong and Samulski 1982). In other words, the molecular orientation is not independent of the molecular conformation. Each conformation

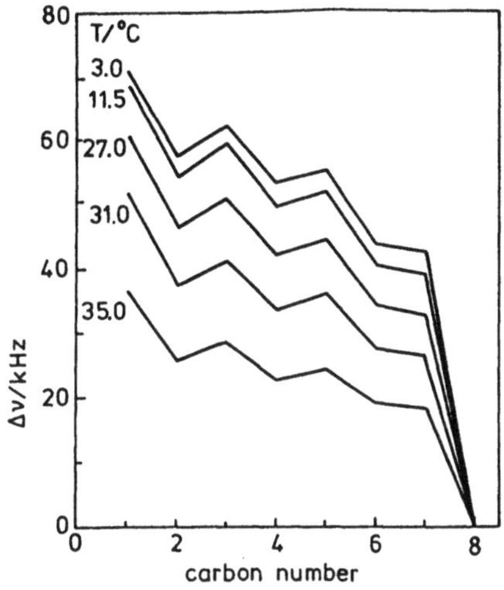

Fig. 9.4. Variation of deuterium quadrupole splittings $\Delta\nu$ with carbon number of the alkyl chain in 8CB (after Boden et al. 1981)

adopted by the molecules should in principle be described by its own set of order parameters. Schemes have been devised in which these sets are related by assuming that each tensor order parameter is diagonal in the frame that diagonalizes the moment of inertia tensor of the particular conformation (Samulski 1980; Samulski and Dong 1982).

Finally it should be mentioned that the orientational order is also related to the chemical shift. Due to the diamagnetism of the surrounding electrons a nucleus will be partly "shielded" from the external field. This means that the Zeeman part of the spin Hamiltonian can be written as

$$H_Z = -\gamma\hbar B(1 - \langle\sigma_{zz}\rangle)I_z \quad , \tag{9.43}$$

where $\tilde{\boldsymbol{\sigma}}$ is the shielding tensor, and the z axis is the direction of the magnetic field. The difference in chemical shift between the nematic and the isotropic phase can be evaluated analogously to (9.33) for the magnetic anisotropy, giving

$$\langle\sigma_{zz}\rangle - \overline{\sigma} = \tfrac{2}{3}\{[\sigma_{\zeta\zeta} - \tfrac{1}{2}(\sigma_{\xi\xi} + \sigma_{\eta\eta})]S + \tfrac{1}{2}(\sigma_{\xi\xi} - \sigma_{\eta\eta})D\} \quad . \tag{9.44}$$

For proton NMR the effect of the chemical shift is small, and in the case of liquid crystals not visible due to the dipolar coupling. Using carbon-13 NMR, however, the variation of the chemical shift with the order parameter can be observed easily. This method has been used to determine the temperature dependence of S under the assumption that D can be disregarded (Pines et al. 1974). In order to obtain the absolute scale for the order parameters, a knowledge of the diagonal elements of $\tilde{\boldsymbol{\sigma}}$ is required, which is, analogous to the case of the magnetic susceptibility, a complicated solid state problem (Höhener 1978).

9.4 Polarized Raman Scattering

So far, only the order parameters for $L = 2$ (see the expansion in Sect. 9.1) have come into play in the discussion of experiments. The reason is that most physical properties are described by a tensor of rank two and thus only give information on the second moment of the orientational distribution function. In this section Raman scattering will be considered, which is one of the few methods that provide information about $\langle P_4(\cos\theta)\rangle$.

To recall some basic aspects of Raman scattering in a simple way (Long 1978), consider scattering of light by particles small compared with the incident wavelength. The incident electric field $E = E_0 \cos(\omega_0 t)$ induces an electric moment

$$m = \alpha E_0 \cos(\omega_0 t) = m_0 \cos(\omega_0 t) \quad , \tag{9.45}$$

where α is the polarizability of the particle. According to classical electrodynamics, this time-dependent electric moment radiates light of frequency ω_0 (Rayleigh scattering) and of intensity $I \sim \omega_0^4 m_0^2$. Because of the molecular vibrations α itself depends on time. In the case of a single vibration with normal coordinate q_i and eigenfrequency ω_i, i.e.

$$q_i = Q_i \cos(\omega_i t) \quad , \tag{9.46}$$

α can be approximated for small amplitudes as

$$\alpha = \alpha_0 + \left(\frac{\partial\alpha}{\partial q_i}\right)_{q=0} q_i + \dots \quad . \tag{9.47}$$

Substitution in (9.45) gives:

$$\begin{aligned} m = {} & \alpha_0 E_0 \cos(\omega_0 t) + \frac{1}{2}\left(\frac{\partial\alpha}{\partial q_i}\right)_0 Q_i E_0 \cos\left[(\omega_0 + \omega_i)t\right] \\ & + \frac{1}{2}\left(\frac{\partial\alpha}{\partial q_i}\right)_0 Q_i E_0 \cos\left[(\omega_0 - \omega_i)t\right] \quad . \end{aligned} \tag{9.48}$$

Besides the first term, which leads to Rayleigh scattering, oscillations are found at frequencies $\omega_0 \pm \omega_i$ giving rise to Raman scattering. In a quantum-mechanical treatment the so-called Stokes Raman scattering at $\omega_0 - \omega_i$ originates from a lower energy level than the anti-Stokes Raman scattering at $\omega_0 + \omega_i$. Consequently the Stokes Raman lines are much more intense than the latter ones. The molecular factor that contributes to the inten-

sity of the Raman scattering is $(\partial\alpha/\partial q_i)_0^2 Q_i^2$. The generalization of (9.48) to anisotropic molecules is obtained by taking the tensorial nature of the Raman polarizability into account, i.e.

$$\tilde{\alpha}' = \left(\frac{\partial\tilde{\alpha}}{\partial q_i}\right)_0 Q_i \quad . \tag{9.49}$$

The resulting scattering intensities will be proportional to the squared elements of $\tilde{\alpha}'$.

Now the essentials of the application of Raman scattering to determine the orientational distribution in liquid crystals will be considered, without going into the details of the formulae for the scattering intensity. The starting point of the consideration is that it is always possible to choose a molecular coordinate system in which the Raman polarizability tensor associated with the selected vibration is diagonal:

$$\tilde{\alpha}' = \alpha_0' \begin{pmatrix} a & 0 & 0 \\ 0 & b & 0 \\ 0 & 0 & 1 \end{pmatrix} . \tag{9.50}$$

This particular frame is not necessarily the main molecular frame to which the order parameters refer. However, it will be assumed that the molecular symmetry is sufficiently known to guess the transformation which brings the two frames into coincidence. The experimental situation can be such that the z axis is chosen along the director of a thin uniform nematic layer, and the wave vector of the incident light is perpendicular to the layer. The incident light, which has been given a certain polarization, is scattered backwards and detected with a second polarizer either parallel or perpendicular to the first one. The various independent combinations that can be made for a uniaxial medium are illustrated in Fig. 9.5. In particular three independent depolarization ratios can be determined, defined by

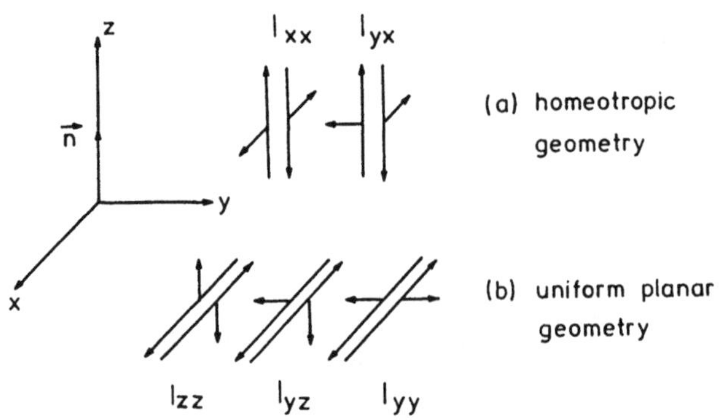

Fig. 9.5. The various configurations for depolarized Raman scattering

$$R_1 = I_{yz}/I_{zz} \quad , \quad R_2 = I_{zy}/I_{yy} \quad , \quad R_3 = I_{yx}/I_{xx} \quad ,$$

where $I_{\alpha\beta}$ denotes the scattered intensity when the incident polarization is along α and the outgoing polarization along β, where $\alpha, \beta = x, y, z$. The intensities are proportional to the squared elements of the Raman tensor, transformed to the laboratory coordinate system, and averaged over all molecular orientations. The results are

$$R_1 \sim \frac{\langle \alpha_{yz}'^2 \rangle}{\langle \alpha_{zz}'^2 \rangle} \quad , \quad R_2 \sim \frac{\langle \alpha_{zy}'^2 \rangle}{\langle \alpha_{yy}'^2 \rangle} \quad , \quad R_3 \sim \frac{\langle \alpha_{yx}'^2 \rangle}{\langle \alpha_{xx}'^2 \rangle} \quad . \tag{9.51}$$

These depolarization ratios can be directly related to the orientational statistics of the molecules. In this case both $\langle \cos^2 \theta \rangle$ and $\langle \cos^4 \theta \rangle$ will enter into the transformations, because squared elements of a second-rank tensor are involved. In addition to (9.51) the depolarization ratio R_{is} in the isotropic phase provides a relation between the elements a and b of the Raman tensor. Hence four measurements are available to determine four unknowns: thus a, b, $\langle \cos^2 \theta \rangle$ and $\langle \cos^4 \theta \rangle$ can be calculated, i.e. $\langle P_2(\cos \theta) \rangle$ as well as $\langle P_4(\cos \theta) \rangle$ can be obtained. Formulae relating the various quantities have been given by Jen et al. (1977). The sensitivity of the final results to changes in the parameters has been discussed in some detail by Dalmolen and De Jeu (1983).

Practical applications of the Raman method are hindered by several restrictions. Firstly an intense Raman band is required, well isolated from other bands in the spectrum. Secondly, the molecular vibration chosen should make a small angle (say $< 15°$) with the long molecular axis. Otherwise terms involving the Eulerian angle ψ come into play in addition to θ, resulting into an untractable situation. These considerations indicate that

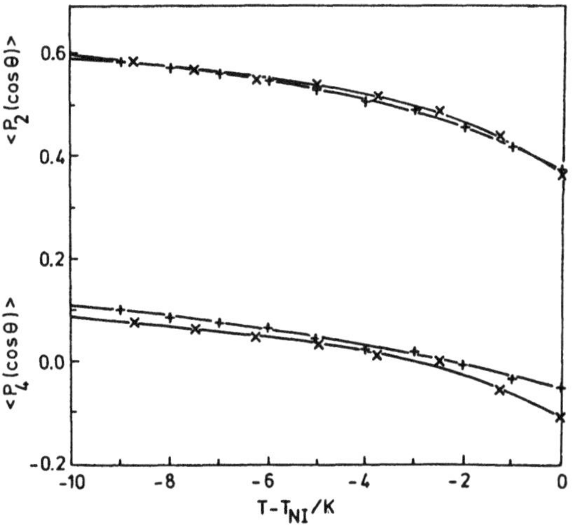

Fig. 9.6. Order parameters of 5CB (upper curves $\langle P_2(\cos \theta) \rangle$, lower curves $\langle P_4(\cos \theta) \rangle$) as determined from Raman scattering

reliable results can only be obtained if the Raman band is chosen in the aromatic core of the mesogenic molecule under investigation, and in addition the band must be symmetric with respect to the *para*-axis through one of the benzene rings. A Raman band that fulfils all these requirements, cannot be found in the majority of compounds.

In Fig. 9.6 $\langle P_2(\cos\theta)\rangle$ and $\langle P_4(\cos\theta)\rangle$ are shown for 5CB. Interestingly $\langle P_4(\cos\theta)\rangle$ is very small and even becomes negative, whereas $\langle P_2(\cos\theta)\rangle$ behaves very much as expected from other measurements. This suggests a strongly broadened distribution function, or even one with its maximum at some angle with $\boldsymbol{n}$. In order to understand the behaviour of $\langle P_4(\cos\theta)\rangle$ in the nCB-series a model has been proposed which attributes the low values to antiparallel dipole-dipole association (Dalmolen et al. 1985).

9.5 X-Ray Scattering

Consider a basic scattering experiment as shown in Fig. 9.7. The incident and scattered wave vectors have a magnitude $k_i = k_s = 2\pi/\lambda$, where λ is the wavelength. The scattering wave vector is given by

$$\boldsymbol{q} = \boldsymbol{k}_s - \boldsymbol{k}_i \quad , \quad q = |\boldsymbol{q}| = (4\pi \sin\alpha)/\lambda \quad , \tag{9.52}$$

where 2α is the angle between $\boldsymbol{k}_s$ and $\boldsymbol{k}_i$. The scattering of incident radiation by a centre at $\boldsymbol{r}$ is described (relative to the initial amplitude) by the scattering amplitude

$$f \exp(\mathrm{i}\boldsymbol{q}\cdot\boldsymbol{r}) \quad , \tag{9.53}$$

where f is the scattering power. Generalized to N centres the scattering amplitude is defined as

$$F(\boldsymbol{q}) = \sum_{j=1}^{N} f_j \exp(\mathrm{i}\boldsymbol{q}\cdot\boldsymbol{r}_j), \tag{9.54}$$

where $\boldsymbol{r}_j$ denotes the position of scattering centre j. A further generalization to a continuous distribution of scattering centres is the scattering amplitude

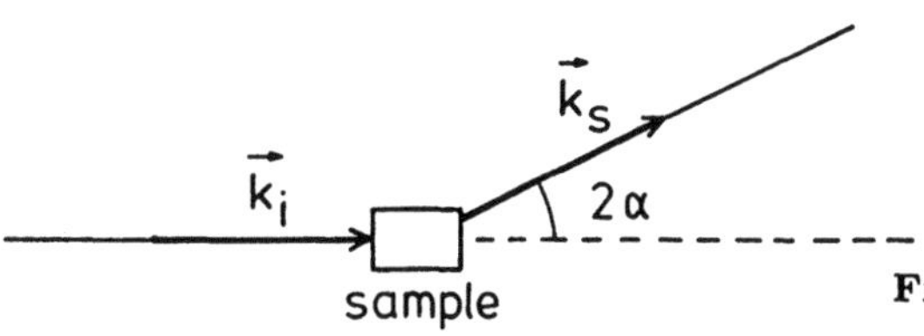

Fig. 9.7. Typical scattering geometry

$$F(\boldsymbol{q}) = \int \varrho(\boldsymbol{r}) \exp(\mathrm{i}\boldsymbol{q}\cdot\boldsymbol{r}) d\boldsymbol{r} \quad , \tag{9.55}$$

where $\varrho(\boldsymbol{r})$ is the time-averaged electron density. For the case of an isolated atom (9.55) is written as

$$f(\boldsymbol{q}) = \int \varrho_a(\boldsymbol{r}) \exp(\mathrm{i}\boldsymbol{q}\cdot\boldsymbol{r}) d\boldsymbol{r} \quad , \tag{9.56}$$

and is called the atomic scattering amplitude. It follows that $f(0) = Z$, the number of electrons. The function $f(\boldsymbol{q})$ falls off with increasing $\boldsymbol{q}$. Equation (9.56), in turn, can be generalized to a group of atoms, for which

$$\varrho(\boldsymbol{r}) = \sum_{j=1}^{N} \varrho_j(\boldsymbol{r} - \boldsymbol{r}_j) \quad . \tag{9.57}$$

This leads to

$$F(\boldsymbol{q}) = \sum_{j=1}^{N} f_j(\boldsymbol{q}) \exp(\mathrm{i}\boldsymbol{q}\cdot\boldsymbol{r}_j) \quad , \tag{9.58}$$

which is almost identical to (9.54). Therefore the diffraction by a set of atoms may be treated in terms of a diffraction by a set of points, provided the variation of the atomic scattering amplitude is accounted for.

So far only scattering amplitudes have been considered. In order to calculate the absolute value of the amplitude, and thus the experimentally relevant intensity, the amplitude of a wave scattered by a single point electron must be known. This amplitude can be calculated from classical electrodynamics. If all intensities are expressed in terms of the scattering intensity of such an electron, the following expression for the intensity results

$$I(\boldsymbol{q}) = |F(\boldsymbol{q})|^2 \quad , \tag{9.59}$$

where the factors that depend on the geometry of the experiment are not included.

The above formalism can be used to compute the scattering amplitude for any molecule, set of molecules, or a crystal. In the latter case the electron density is periodic in all three coordinates, and the Fourier integral of a general function, used above, must be replaced by the analogue of a periodic function. This leads to a $F(\boldsymbol{q})$ that has non-zero values only at certain points, the so-called reciprocal lattice. If disorder is present in a crystal, the scattering amplitude becomes a product of the perfect crystal structure amplitude and the transforms of the distributions for each atom. If the disorder is so large that the long-range order is destroyed, the structure can still be described by a periodic lattice, but convoluted with some damping function. Describing the damping in terms of a characteristic dimension ξ, it can also be said that there is positional order within a finite correlation length ξ. In this case the positions of the Bragg reflections remain unchanged, but the width of the reflections increases proportional to ξ^{-1}. This situation is clearly relevant for liquid crystals.

For molecular liquids it is convenient to separate the amplitude due to the molecular structure from the total scattering amplitude (Leadbetter 1979). Accordingly (9.58) is written as

$$F(\boldsymbol{q}) = \sum_{k,m} f_{km}(\boldsymbol{q}) \exp\left[\mathrm{i}\boldsymbol{q}\cdot(\boldsymbol{r}_k - \boldsymbol{R}_{km})\right] \tag{9.60}$$

where $\boldsymbol{r}_k$ gives the position of the centre of mass of molecule k and $\boldsymbol{R}_{km}$ the position of atom m within that molecule. Substitution of (9.60) into (9.59) leads to

$$I(\boldsymbol{q}) = \sum_{k,l,n,m} \langle f_{km}(\boldsymbol{q}) f^*_{ln}(\boldsymbol{q}) \times \exp\left[\mathrm{i}\boldsymbol{q}\cdot(\boldsymbol{r}_k - \boldsymbol{r}_l)\right] \exp\left[\mathrm{i}\boldsymbol{q}\cdot(\boldsymbol{R}_{ln} - \boldsymbol{R}_{km})\right]\rangle \quad , \tag{9.61}$$

where the brackets indicate that an average over the liquid is involved. Now the intensity can be written as

$$I(\boldsymbol{q}) = I_m(\boldsymbol{q}) + D(\boldsymbol{q}) \quad , \tag{9.62}$$

where the molecular structure factor $I_m(\boldsymbol{q})$ and the so-called interference function $D(\boldsymbol{q})$ are given by

$$\begin{aligned} I_m(\boldsymbol{q}) &= \sum_k \langle \sum_{m,n} f_{km}(\boldsymbol{q}) f^*_{kn}(\boldsymbol{q}) \exp\left[\mathrm{i}\boldsymbol{q}\cdot(\boldsymbol{R}_{kn} - \boldsymbol{R}_{km})\right]\rangle \\ &= N\langle |\sum_m f_{km} \exp(-\mathrm{i}\boldsymbol{q}\cdot\boldsymbol{R}_{km})|^2\rangle \quad , \end{aligned} \tag{9.63a}$$

$$D(\boldsymbol{q}) = \langle \sum_{k\neq l} \exp(\mathrm{i}\boldsymbol{q}\cdot\boldsymbol{r}_{kl}) \sum_{m,n} f_{km}(\boldsymbol{q}) f^*_{ln}(\boldsymbol{q}) \times \exp\left[\mathrm{i}\boldsymbol{q}\cdot(\boldsymbol{R}_{ln} - \boldsymbol{R}_{km})\right]\rangle \quad , \tag{9.63b}$$

where $\boldsymbol{r}_{kl} = \boldsymbol{r}_k - \boldsymbol{r}_l$. The term $I_m(\boldsymbol{q})$ gives the scattered intensity which would be observed from a very dilute gas of the same molecules. The function $D(\boldsymbol{q})$ is difficult to handle without making approximations. For dilute systems $D(\boldsymbol{q})$ only contains information on components larger than the molecular diameter. This means that $D(\boldsymbol{q})$ tends to zero at large values of q, where the small distance contributions are dominant. Assuming that $D(\boldsymbol{q})$ shows the same behaviour in the nematic region, $I_m(\boldsymbol{q})$ contains, in principle, the information concerning the distribution of molecular orientations if the scattering intensity is measured in this $\boldsymbol{q}$ range. For this reason PAA was investigated using neutron diffraction (Kohli et al. 1976). Although values for $\langle P_2(\cos\theta)\rangle$ and $\langle P_4(\cos\theta)\rangle$ were obtained, the results are not very accurate, mainly due to difficulties concerning the weakness of the scattering relative to the background.

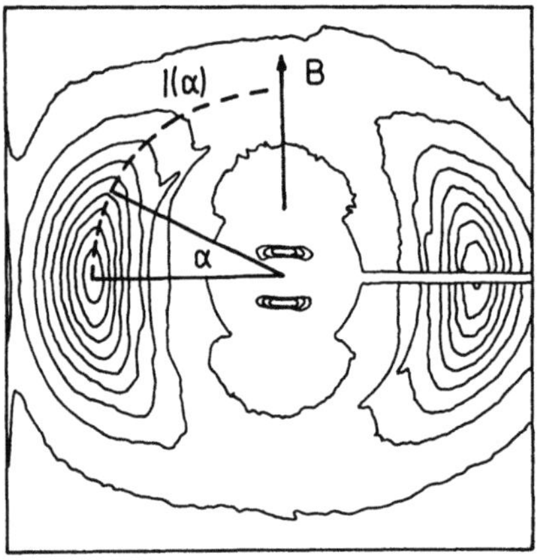

Fig. 9.8. Intensity contour map for x-ray scattering from the nematic phase of p,p'-dioctyloxyazoxybenzene at 120°C, aligned in a magnetic field (after Leadbetter and Norris 1979)

As a starting point for the discussion of the experimental results for x-ray scattering by nematic liquid crystals a typical x-ray diffraction pattern of an aligned nematic sample (uniform $\boldsymbol{n}$) is shown in Fig. 9.8. Although the diffraction pattern is liquid-like, two characteristic distances are evidently present, parallel and perpendicular to $\boldsymbol{n}$. For the situation $\boldsymbol{q} \parallel \boldsymbol{n}$ the characteristic distance d, which corresponds to the maximum of the scattering amplitude, is usually of the order of the molecular length. This diffraction peak must arise from correlations in the molecular arrangement along $\boldsymbol{n}$. In fact d is usually slightly smaller than l, typical values are $l - d \approx 2$ Å for $l \approx 17$ to 25 Å. This difference can be attributed to the orientational disorder of the molecules (Leadbetter 1979). For $\boldsymbol{q} \perp \boldsymbol{n}$ a strong peak in the diffraction pattern is found for a wide variety of compounds near $q \approx 1.4$ Å^{-1}, which corresponds to a distance of about 4.5 Å. This indicates an average molecular spacing, that lies in between the maximum (~6.5 Å) and the minimum (~3.5 Å) lateral dimension of a typical mesogenic molecule. Consequently strong local correlations in orientation about the long axis are present, i.e. the picture of molecules that are effectively uniaxial due to supposedly allowed averaging over all orientations about that axis, is certainly not correct on a local molecular level. This conclusion is supported further by x-ray scattering results from unaligned ("powder") samples (Leadbetter et al. 1975), that also show that the local molecular packing is quite similar in the nematic and isotropic phase. The latter is in agreement with the small density change at the NI phase transition.

More detailed information about $f(\theta)$, the orientational distribution function, can in principle be obtained from intensity profiles of an aligned sample as given in Fig. 9.8. To analyze these profiles it is noted that for pefect orientational order the scattering would be restricted to the equatorial plane $\boldsymbol{q} \perp \boldsymbol{n}$. The appearance of a pronounced arc around this plane [$I(\alpha)$ in Fig. 9.8] can be interpreted as arising from differently aligned clusters or coherence volumes of dimensions determined by the correlation length ξ. To calculate $D(\boldsymbol{q})$ in (9.63b) the orientational order within each cluster is taken to be perfect. In other words, locally perfect order is assumed, which decays with a correlation length much larger than ξ. In that situation the calculation of $I(\alpha)$ reduces to a geometrical problem (Leadbetter and Norris 1979). The width of the diffraction peaks indicates a perpendicular correlation length of the dimension of two or three molecules, so that the clusters involve at least ten molecules. The orientational distribution function $f'(\theta)$ for fluctuating clusters of this kind [as derived from $I(\alpha)$] is shown in Fig. 9.9. The true distribution function $f(\theta)$ can be expected to be the same or somewhat less ordered than $f'(\theta)$. The result of Fig. 9.9 agrees rather well with the predictions of the mean-field Maier-Saupe model. Unfortunately no systematic comparison is available between $\langle P_2(\cos\theta)\rangle$ and $\langle P_4(\cos\theta)\rangle$ derived from $f'(\theta)$ and values obtained from tensor properties as described in the previous sections of this chapter.

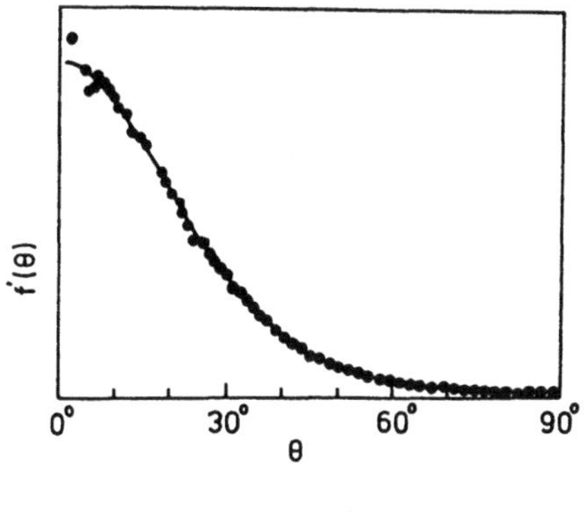

Fig. 9.9. Orientational distribution function $f'(\theta)$ derived from Fig. 9.8. The full line is the best fit to the data points according to Maier and Saupe's mean-field theory

10. Anisotropic Physical Properties

The occurrence of a fluid phase in combination with birefringence led to the discovery of liquid crystals. Nowadays the liquid crystalline nature of a given material is still most easily established by the anisotropy of a physical property such as the refractive index. A macroscopic anisotropy is found because the molecular anisotropy related to this property does not average out to zero as is the case in an isotropic phase. Consequently the orientational distribution function and the associated order parameters are an important part of the description. The present chapter can be considered to be an extension of the previous one, where the magnetic susceptibility was discussed as an example of an anisotropic physical property. In the following we shall extend our discussion to the refractive index and the dielectric permittivity, chosen because of their relevance to modern display applications of liquid crystals. This chapter ends with a section on anisotropic transport properties.

10.1 The Refractive Index

A uniaxial (liquid) crystal has two principal refractive indices, n_o and n_e. The first one, n_o, is observed for the ordinary ray, which is defined to be the light wave with the electric field vibrating perpendicular to the optical axis. The extraordinary index, n_e, is observed for a linearly polarized light wave where the electric field is parallel to the optical axis. In the case of a nematic or a uniaxial smectic liquid crystal the direction of the optical axis is given by the director. Using the subscripts $\|$ and $\perp$ for the directions parallel and perpendicular to the director, it follows that

$$n_o = n_\perp \quad ; \quad n_e = n_\| \quad . \tag{10.1}$$

The birefringence is defined by

$$\Delta n = n_e - n_o \quad , \tag{10.2}$$

which is thus equal to $n_\| - n_\perp$ for uniaxial nematic and smectic liquid crystals. For chiral nematics the situation is more complicated as the optical axis coincides with the helix axis. Thus the optical axis is perpendicular to the local director, i.e. $n_e = n_\perp$, and n_o is a function of both $n_\parallel$ and $n_\perp$ that depends on the relative magnitude of the wavelength with respect to the pitch. It appears that $n_\parallel > n_\perp$ for rod-like molecules. Thus a positive birefringence, $\Delta n > 0$, is found for conventional nematics with a value of Δn varying from close to zero to about 0.4 (see Fig. 10.1 for PAA), whereas chiral nematics show a negative birefringence. In the case of disc-like molecules the signs of Δn as mentioned are reversed.

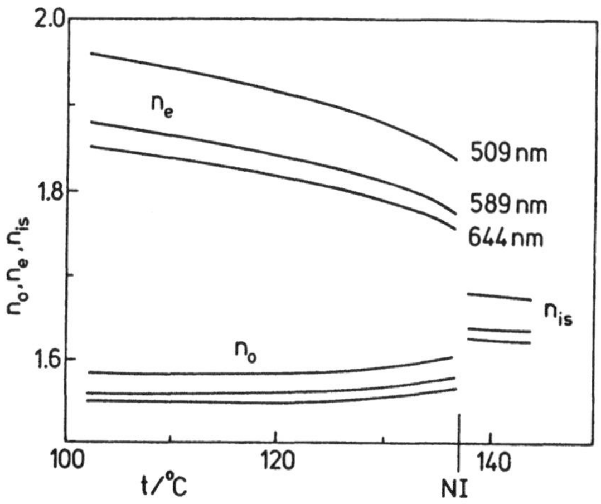

Fig. 10.1. Refractive indices of PAA in the nematic and isotropic phases as a function of the temperature (Chatelain and Germain 1964)

The refractive index is related to the response of matter to an electric field. On application of an electric field $\boldsymbol{E}$ an electric polarization $\boldsymbol{P}$ is induced that is related to $\boldsymbol{E}$ and the dielectric displacement $\boldsymbol{D}$ by

$$P_\alpha = D_\alpha - \varepsilon_0 E_\alpha \quad , \quad \alpha = x, y, z \quad , \tag{10.3}$$

where ε_0 is the permittivity of the vacuum. For small applied fields $\boldsymbol{P}$ is proportional to $\boldsymbol{E}$, i.e.

$$P_\alpha = \varepsilon_0 \chi^{\mathrm{e}}_{\alpha\beta} E_\beta \quad , \tag{10.4}$$

where $\chi^{\mathrm{e}}_{\alpha\beta}$ is an element of the electric susceptibility tensor $\tilde{\chi}^{\mathrm{e}}$. In the electric case the permittivity tensor $\tilde{\varepsilon}$ is mostly used instead of the susceptibility tensor. The permittivity tensor is defined by

$$\varepsilon_{\alpha\beta} = \delta_{\alpha\beta} + \chi^{\mathrm{e}}_{\alpha\beta} \quad . \tag{10.5}$$

In general $\varepsilon_{\alpha\beta}$ depends on the frequency and wavevector of the applied field. Choosing the z axis along the director, $\tilde{\varepsilon}$ is diagonal, and the following relations exist between the refractive indices and the elements of the permittivity tensor in the optical frequency range:

$$n_{\parallel} = \varepsilon_{zz}^{1/2} \quad ; \quad n_{\perp} = \varepsilon_{xx}^{1/2} = \varepsilon_{yy}^{1/2} \quad . \tag{10.6}$$

The connection with the microscopic theory is made by expressing the macroscopic permittivity tensor $\tilde{\varepsilon}$ in terms of the molecular polarizability $\tilde{\alpha}$. This polarizability consists of two contributions, namely, the electric field will induce (1) a displacement of the electrons relative to the nucleus in each atom (electronic polarization) and (2) a displacement of the nuclei relative to one another (ionic polarization). For organic liquids the second contribution is usually only 5–10 % of the first. Moreover, the ionic polarization plays a role up to frequencies in the infrared region, and can thus be neglected as far as the visible part of the spectrum is concerned. For this reason the refractive index, or rather the optical dielectric tensor $\tilde{\varepsilon}$, will be related to the molecular electronic polarizability $\tilde{\alpha}$. The polarization $\boldsymbol{P}$ is defined by

$$P_\beta = N\langle \alpha_{\beta\gamma} E^{\mathrm{i}}_\gamma \rangle \quad , \tag{10.7}$$

where N is the number of molecules per unit volume, and $\boldsymbol{E}^{\mathrm{i}}$ is the internal field, i.e. the average field that acts on a molecule. The internal field $\boldsymbol{E}^{\mathrm{i}}$ is equal to the sum of the external field and the field due to the induced moments of the surrounding particles. The brackets in (10.7) indicate an average over the orientations of the molecules. For an isotropic liquid (10.7) is usually written as

$$\boldsymbol{P} = N\overline{\alpha}\boldsymbol{E}^{\mathrm{i}} \quad , \tag{10.8}$$

where $\overline{\alpha} = \frac{1}{3}\alpha_{\beta\beta}$ and the average over the surrounding particles is thought to be included in $\boldsymbol{E}^{\mathrm{i}}$. The calculation of the internal field is the central problem of the theory of dielectrics, which is outside the scope of this section (see, for example, Böttcher and Bordewijk 1978).

When dealing with an anisotropic medium like a liquid crystal the internal field tensor $\tilde{\boldsymbol{K}}$ is often introduced for convenience. As the internal field is a linear function of the macroscopic field, its relation to the macroscopic field can be represented by

$$E^{\mathrm{i}}_\gamma = K_{\gamma\delta} E_\delta \quad , \tag{10.9}$$

where $\tilde{\boldsymbol{K}}$ is a second-rank tensor. Substitution of (10.9) in (10.7) gives

$$P_\beta = N\langle \alpha_{\beta\gamma} K_{\gamma\delta} \rangle E_\delta \quad . \tag{10.10}$$

Using (10.4) and (10.5) the macroscopic permittivity can be expressed in microscopic terms as

$$\varepsilon_{\beta\gamma} = \delta_{\beta\gamma} + (N/\varepsilon_0)\langle\alpha_{\beta\delta}K_{\delta\gamma}\rangle \quad . \tag{10.11}$$

This means that the dielectric anisotropy is given by

$$\Delta\varepsilon = n_{\parallel}^2 - n_{\perp}^2 = (N/\varepsilon_0)(\langle\alpha_{z\gamma}K_{\gamma z}\rangle - \langle\alpha_{x\gamma}K_{\gamma x}\rangle) \quad . \tag{10.12}$$

In order to proceed the internal field tensor $\tilde{\boldsymbol{K}}$ must be known. However, the required calculation is extremely difficult to perform starting from first principles. In the theory of dielectrics the simplest approximation in use is that of an electric field acting on a single molecule in a continuous environment with the macroscopic properties of the dielectric. Even with such a strong simplification, that ignores all short-range correlation, expression (10.12) is rather complicated to handle, and many inconsistent results can be found in the literature. In general $\tilde{\boldsymbol{K}}$ is influenced by two types of anisotropies.

i) The anisotropy of the molecular polarizability tensor $\tilde{\boldsymbol{\alpha}}$ must be taken into account.
ii) Any anisotropy of the continuum will influence $\tilde{\boldsymbol{K}}$.

Note that the first effect is relevant for any anisotropic molecule, whether in a nematic or in an isotropic phase. The second contribution is clearly absent in an isotropic liquid. Retaining all these anisotropies leads to untractable expressions. Results have only been obtained by making further approximations.

The most drastic approximation is to disregard all anisotropies as far as the internal field is concerned. Hence $\boldsymbol{E}^{\mathrm{i}}$ is the field in a spherical cavity in a homogeneous medium, the so-called Lorentz field:

$$\boldsymbol{E}^{\mathrm{i}} = \tfrac{1}{3}(\overline{\varepsilon} + 2)\boldsymbol{E} \quad , \tag{10.13}$$

where $\overline{\varepsilon} = \frac{1}{3}\varepsilon_{\alpha\alpha}$. Application of (10.13) to isotropic liquids, i.e. to (10.8), gives, in combination with (10.4) and (10.5), the well-known Lorenz-Lorentz equation

$$\overline{\alpha} = \frac{3\varepsilon_0(\overline{\varepsilon} - 1)}{N(\overline{\varepsilon} + 2)} \quad . \tag{10.14}$$

In case of an anisotropic medium (10.13) must be applied to (10.11) giving

$$\varepsilon_{\beta\gamma} = \delta_{\beta\gamma} + \frac{N}{3\varepsilon_0}(\overline{\varepsilon} + 2)\langle\alpha_{\beta\gamma}\rangle \quad . \tag{10.15}$$

The average $\langle \alpha_{\beta\gamma} \rangle$ can be evaluated by an analogous procedure to that leading to (9.32). Because of all the approximations already involved it is useless to go beyond a representation of the molecules in terms of cylindrically symmetric objects. Indicating the diagonal elements of $\tilde{\alpha}$ by $\alpha_l = \alpha_{\zeta\zeta}$ and $\alpha_t = \frac{1}{2}(\alpha_{\xi\xi} + \alpha_{\eta\eta})$ it follows that

$$\langle \alpha_{zz} \rangle = \tfrac{1}{3}[\alpha_l(1+2S) + \alpha_t(2-2S)] \tag{10.16a}$$

$$\langle \alpha_{xx} \rangle = \langle \alpha_{yy} \rangle = \tfrac{1}{3}[\alpha_l(1-S) + \alpha_t(2+S)] \quad . \tag{10.16b}$$

This means that (10.15) leads to

$$\frac{\varepsilon_{\parallel} - 1}{\overline{\varepsilon} + 2} = \frac{N}{9\varepsilon_0}[\alpha_l(1+2S) + \alpha_t(2-2S)] \quad , \tag{10.17a}$$

$$\frac{\varepsilon_{\perp} - 1}{\overline{\varepsilon} + 2} = \frac{N}{9\varepsilon_0}[\alpha_l(1-S) + \alpha_t(2+S)] \quad . \tag{10.17b}$$

If these equations are used to calculate $\overline{\alpha}$, (10.14) results. Hence (10.17) can be viewed as the generalization of the Lorenz-Lorentz equation to anisotropic liquids. The approximation of using the Lorentz-field also for anisotropic media is due to Vuks (1966).

Clearly (10.17) can be improved by taking the two types of anisotropies into account when calculating $\boldsymbol{E}^{\mathrm{i}}$. In the first instance the anisotropy of the continuum can be neglected as it is much smaller than the molecular anisotropy. Although the inclusion of the anisotropy of the molecules leads to further uncertainties in the theory, one point of interest should be mentioned (De Jeu and Bordewijk 1978). For cylindrically symmetric objects incorporation of the molecular anisotropy means that the tensor $\tilde{\boldsymbol{K}}$, which depends on molecular properties, has principal axes that coincide with those of $\tilde{\alpha}$. Therefore $\alpha_{\beta\delta}K_{\delta\gamma}$ can be considered to be a "dressed" or effective polarizability $\alpha^*_{\beta\gamma}$ with principal elements $\alpha_l K_l$ and $\alpha_t K_t$. Now (10.12) leads to

$$\Delta\varepsilon = n_{\parallel}^2 - n_{\perp}^2 = (N/\varepsilon_0)(\alpha_l K_l - \alpha_t K_t)S \quad . \tag{10.18}$$

As, according to (9.35) $S \sim \Delta\chi$, the assumptions above make sense if it is found that $\Delta\varepsilon \sim \Delta\chi$. This is indeed the case, as is shown in Fig. 10.2 for MBBA at various wavelengths.

In conclusion it seems reasonable to take the internal field into account in terms of an effective or "dressed" polarizability tensor. Such a procedure is useful for some other practical situations such as Raman scattering

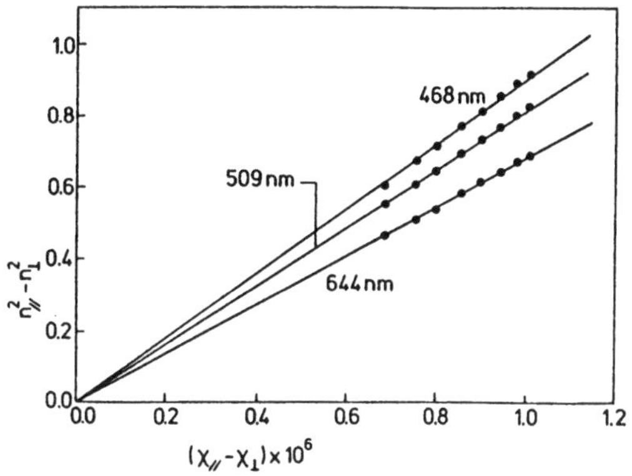

Fig. 10.2. The relation between $n_{\|}^2 - n_{\perp}^2$ and $\chi_{\|} - \chi_{\perp}$ as observed for MBBA (De Jeu 1980)

(Sect. 9.4). In order to calculate molecular polarizabilities from refractive index data further approximate models for the molecules must be introduced. For example, a molecule can be represented by an anisotropic homogeneously polarizable ellipsoid, filling up the cavity in an anisotropic polarized medium. The resulting formulas contain shape factors, that depend on the axial ratio of the ellipsoid. For a further discussion the reader is referred to De Jeu (1980).

Finally, it should be noted that the SI unit of α is Fm^2; the reduced polarizability α/ε_0 then has the dimensions of a volume (m^3) and equals $4\pi \times 10^{-6}$ times the polarizability in CGS units (cm^3). Replacing ε_0 by $1/4\pi$ in all equations the corresponding equations in CGS units are obtained.

10.2 The Dielectric Permittivity

Dielectric studies are concerned with the response of matter to the application of an electric field. For an anisotropic material the resulting polarization (per unit volume) is given by (10.4). Using (10.5) for the permittivity, this leads to

$$P_\alpha = \varepsilon_0(\varepsilon_{\alpha\beta} - \delta_{\alpha\beta})E_\beta \quad . \tag{10.19}$$

In a material consisting of non-polar molecules only an induced polarization occurs. For static fields this consists of two parts: the electronic polarization (which is also present at optical frequencies) and the ionic polarization. In materials with polar molecules, in addition to the induced polarization, an orientation polarization occurs, due to the tendency of the permanent dipole moments to orient themselves parallel to the field. In isotropic liquids and gases, the permittivity is isotropic, i.e. $\varepsilon_{\alpha\beta} = \varepsilon\delta_{\alpha\beta}$. In solids, on the other hand, the permittivity will usually be anisotropic. Then, however, the orientation polarization normally does not contribute significantly to the permittivity due to the rather fixed orientations of the molecules. In liquid crystals the complicated situation of an anisotropic permittivity in combination with a liquid-like behaviour occurs. This means that polar molecules can contribute significantly to the anisotropy of the permittivity due to the occurrence of an orientation polarization.

Consider a uniaxial liquid crystalline phase, i.e. a nematic or smectic A phase. As usual the macroscopic z axis is chosen along the director. The principal elements of $\tilde{\boldsymbol{\varepsilon}}$ are $\varepsilon_{zz} = \varepsilon_{\|}$ and $\varepsilon_{xx} = \varepsilon_{yy} = \varepsilon_{\perp}$. Figure 10.3 shows experimental data for $\varepsilon_{\|}$ and $\varepsilon_{\perp}$ of some nematics. The first case is a non-polar compound with $\Delta\varepsilon = \varepsilon_{\|} - \varepsilon_{\perp} > 0$. Once dipole moments are introduced, rather large values of $|\Delta\varepsilon|$ can be obtained (Fig. 10.3b,c) depending on the magnitude of the total permanent dipole moment $\boldsymbol{\mu}$ and the angle between $\boldsymbol{\mu}$ and the long molecular axis.

Theoretically the situation for non-polar molecules is relatively simple. Using the results from Fig. 10.3a it is found that for p, p'-diheptylazobenzene

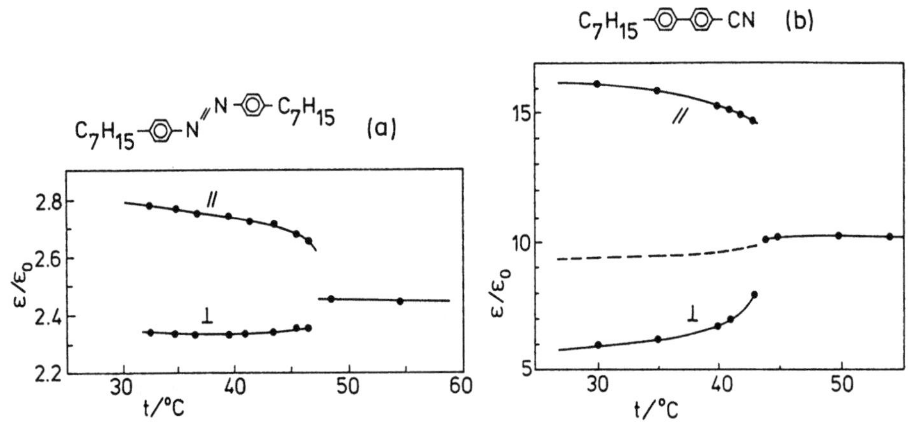

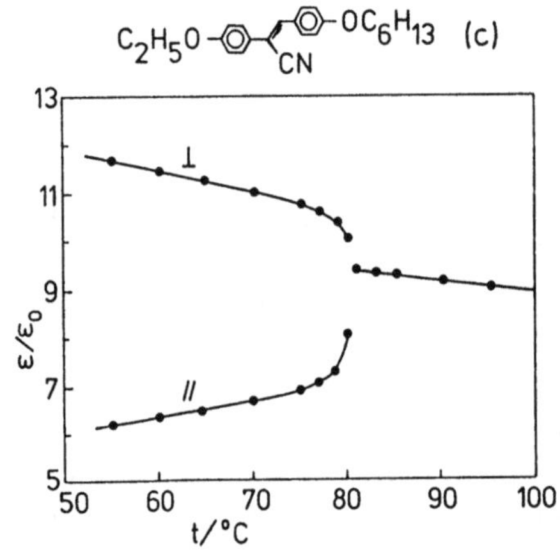

Fig. 10.3a–c. Static dielectric permittivity of some nematic liquid crystals. Examples are shown of **(a)** a non-polar molecule (De Jeu 1980), **(b)** a polar molecule giving rise to a positive anisotropy (Lippens et al. 1977) and **(c)** a polar molecule giving rise to a negative anisotropy (De Jeu 1980)

$\Delta\varepsilon \sim \Delta\chi$, analoguous to the situation for the optical frequency range where $n_{\|}^2 - n_{\perp}^2 \sim \Delta\chi$. This means that the whole reasoning of Sect. 10.1 can be applied. In the case of polar molecules the interactions between the dipole moments can be important, thus complicating the situation considerably.

In order to describe the dielectric permittivity of a medium with polar molecules, the orientation polarization must be included in (10.7). Now the polarization is defined as

$$P_\beta = N(\langle\alpha_{\beta\gamma}E_\gamma^{\mathrm{i}}\rangle + \langle\overline{\mu}_\beta\rangle) \quad , \tag{10.20}$$

where $\langle\overline{\boldsymbol{\mu}}\rangle$ indicates the average of the dipole moment $\boldsymbol{\mu}$ over the orientations of all molecules in the presence of a directing electric field $\boldsymbol{E}^{\mathrm{d}}$. The directing field is the effective field that tends to orient the permanent dipole moment. The fundamental equation for the permittivity of an anisotropic liquid follows directly from (10.19) and (10.20) and is given by

$$(\varepsilon_{\beta\gamma} - \delta_{\beta\gamma})E_\gamma = (N/\varepsilon_0)(\langle\alpha_{\beta\gamma}E_\gamma^{\mathrm{i}}\rangle + \langle\overline{\mu}_\beta\rangle) \quad . \tag{10.21}$$

The polarization $\langle\overline{\boldsymbol{\mu}}\rangle$ can be approximated as a linear function of the directing field because the field energy, $-\mu_\gamma E_\gamma^{\mathrm{d}}$, is small:

$$\exp\left(-\frac{\mu_\gamma E_\gamma^{\mathrm{d}}}{k_{\mathrm{B}}T}\right) \approx 1 - \frac{\mu_\gamma E_\gamma^{\mathrm{d}}}{k_{\mathrm{B}}T} \quad .$$

Now $\langle\overline{\mu}_\beta\rangle$ can be written as

$$\langle\overline{\mu}_\beta\rangle = \frac{1}{k_{\mathrm{B}}T}\langle\mu_\beta\mu_\gamma\rangle E_\gamma^{\mathrm{d}} \quad , \tag{10.22}$$

where $\langle\mu_\beta\mu_\gamma\rangle = 0$ if $\beta \neq \gamma$, as follows directly from the head-tail symmetry of the nematic phase.

Once (10.22) is substituted in the fundamental equation (10.21) the main problem that is left is (as in the theory of isotropic dielectrics) the relation between the externally applied field $\boldsymbol{E}$ and the internal and directing field, $\boldsymbol{E}^{\mathrm{i}}$ and $\boldsymbol{E}^{\mathrm{d}}$, respectively. According to Onsager the directing field differs from the internal field due to the presence of a reaction field. The dipole polarizes its surroundings, which in turn leads to a reaction field at the position of the dipole. As the reaction field is always parallel to the dipole it cannot direct the dipole, but it does contribute to $\boldsymbol{E}^{\mathrm{i}}$. The internal field is now given by the directing field $\boldsymbol{E}^{\mathrm{d}}$, plus the average reaction field. The various fields can only be calculated for certain models of the dielectric.

The simplest model is analogous to the one used in the previous section: a single molecule is considered in a spherical cavity, where its environment is taken as a continuum with the macroscopic properties of the dielectric (Böttcher 1973). This theory of Onsager has been extended to nematic liquid crystals by Maier and Meier (1961a, 1961b). As in the calculation in Sect. 10.1 they ignore all anisotropies as far as the calculation of $\boldsymbol{E}^{\mathrm{i}}$ and $\boldsymbol{E}^{\mathrm{d}}$ is concerned. According to their dielectric theory of nematics a molecule can be represented by a polarizability $\tilde{\boldsymbol{\alpha}}$ with principal elements α_l and α_t in a spherical cavity. Furthermore a permanent dipole moment $\boldsymbol{\mu}$ is present at an angle ω with the principal axis associated with α_l, i.e. $\mu_l = \mu \cos\omega$ and $\mu_t = \mu \sin\omega$. Just as in Onsager's theory, the relations between $\boldsymbol{E}$, $\boldsymbol{E}^{\mathrm{i}}$ and $\boldsymbol{E}^{\mathrm{d}}$ are expressed in terms of the cavity field factor h and the reaction field factor F. They are found to be

$$\boldsymbol{E}^{\mathrm{d}} = F\boldsymbol{E}^{\mathrm{i}} \quad , \quad \boldsymbol{E}^{\mathrm{i}} = hF\boldsymbol{E} \quad , \quad \text{where} \tag{10.23}$$

$$h = \frac{3\overline{\varepsilon}}{2\overline{\varepsilon}+1} \quad \text{and} \tag{10.24}$$

$$F = \left[1 - \frac{2N\overline{\alpha}(\overline{\varepsilon}-1)}{3\varepsilon_0(2\overline{\varepsilon}+1)}\right]^{-1} \tag{10.25}$$

with $\overline{\alpha} = \frac{1}{3}(\alpha_l + 2\alpha_t)$ and $\overline{\varepsilon} = \frac{1}{3}(\varepsilon_\| + 2\varepsilon_\perp)$. Choosing the z axis along the director the fundamental equation (10.21) for the underlying model is given by

$$\varepsilon_i - 1 = (NhF/\varepsilon_0)\left(\langle\alpha_i\rangle + F\frac{\langle\mu_i^2\rangle}{k_\mathrm{B}T}\right) \quad , \quad i = \|,\perp \quad . \tag{10.26}$$

The dependence of the quantities $\langle\alpha_i\rangle$ on the order parameter S is given by (10.16). Analogously $\langle\mu_i^2\rangle$ is found to be

$$\begin{aligned}\langle\mu_\|^2\rangle &= \tfrac{1}{3}[\mu_l^2(1+2S) + \mu_t^2(1-S)] \\ &= \tfrac{1}{3}\mu^2[1-(1-3\cos^2\omega)S] \quad ,\end{aligned} \tag{10.27a}$$

$$\begin{aligned}\langle\mu_\perp^2\rangle &= \tfrac{1}{3}[\mu_l^2(1-S) + \tfrac{1}{2}\mu_t^2(2+S)] \\ &= \tfrac{1}{3}\mu^2[1+\tfrac{1}{2}(1-3\cos^2\omega)S] \quad .\end{aligned} \tag{10.27b}$$

As expected, only the component μ_l contributes to $\langle\mu_\|^2\rangle$ and μ_t to $\langle\mu_\perp^2\rangle$ for the case $S = 1$.

The dielectric anisotropy follows directly from (10.26) and is given by

$$\Delta\varepsilon = (NhF/\varepsilon_0)\left[(\alpha_l - \alpha_t) - \frac{F\mu^2}{2k_\mathrm{B}T}(1-3\cos^2\omega)\right]S \quad . \tag{10.28}$$

Furthermore the following average dielectric constant is obtained

$$\overline{\varepsilon} = 1 + (NhF/\varepsilon_0)\left(\overline{\alpha} + \frac{F\mu^2}{3k_\mathrm{B}T}\right) \quad , \tag{10.29a}$$

which is equivalent to Onsager's equation for the permittivity of isotropic polar liquids. This can be seen by inserting the expression for h and F, and relating $\overline{\alpha}$ to the high-frequency permittivity $\overline{\varepsilon}_\infty$ using the Clausius-Mossotti equation. Equation (10.29a) can now be written as

$$\frac{(\overline{\varepsilon} - \overline{\varepsilon}_\infty)(2\overline{\varepsilon} + \overline{\varepsilon}_\infty)}{\overline{\varepsilon}(\overline{\varepsilon}_\infty + 2)^2} = \frac{N\mu^2}{9\varepsilon_0 k_\mathrm{B}T} \quad . \tag{10.29b}$$

This equation is often used to compute the value of the permanent dipole moment of a molecule once the low- and high-frequency permittivity and the density are known.

Clearly quantitatively correct results cannot be expected because of all the approximations involved. Nevertheless Maier and Meiers's equations satisfactorily account for many essential features of the permittivity of nematic liquid crystals consisting of polar molecules. This is best illustrated using (10.28) for $\Delta\varepsilon$. If $3\cos^2\omega = 1$ ($\omega \approx 55°$) the dipole moment contributes equally to $\varepsilon_\parallel$ and $\varepsilon_\perp$ and $\Delta\varepsilon$ is determined by the (positive) anisotropy of the polarizability. The dipole contribution to $\Delta\varepsilon$ is positive for $\omega<55°$ and negative for $\omega>55°$. In the latter case $\Delta\varepsilon$ itself may become negative depending on the magnitude of the dipole contribution. According to (10.28) the temperature dependence of $\Delta\varepsilon$ is governed by the induced polarization, which is proportional to S, and the orientation polarization, which varies as S/T. The remaining factors are only weakly dependent on temperature. In this way the following observations are nicely explained:

i) When $\Delta\varepsilon$ is negative or strongly positive, the S/T dependence of the dipole contribution to $\Delta\varepsilon$ is predominant over the whole temperature range. Consequently $|\Delta\varepsilon|$ increases with decreasing temperature (Fig. 10.3b,c).

ii) When $\Delta\varepsilon$ is positive and close to zero, the anisotropy of the induced polarization approximately equals that of the orientation polarization. Just below T_NI the temperature dependence of S is the decisive factor and $\Delta\varepsilon$ increases with decreasing temperature. At lower temperatures, however, the variation of S is small and the counteracting S/T dependence of the orientation polarization predominates. This means that $\Delta\varepsilon$ may be found first to increase and then after going through a maximum, to decrease with decreasing temperature.

It is difficult to test (10.26) more quantitatively. In principle, if information about $\overline{\alpha}$ and $\alpha_l - \alpha_t$ is available from optical measurements the quantities μ and ω can be obtained. The Onsager equation can be used to calculate μ from the value of the dielectric permittivity in the isotropic phase and the angle ω follows from $\Delta\varepsilon$ using (10.26). These results can then be compared with those of Kerr-effect measurements. Though reasonable agreement is often obtained, it should be realized that the interpretation of the Kerr constant suffers from the same internal field problem as the permittivity. Hence consistency does not necessarily mean that the theory can be trusted quantitatively. For further details, including more elaborate models, the reader is referred to De Jeu (1978). An extensive review of the relation between the permittivity and the molecular structure has been given by Kresse (1982).

Dielectric studies of liquid crystals have proven to be very useful as a source of information about specific intermolecular interactions. The latter may show up as deviations from the "normal" behaviour of the permittivity as discussed above. As an example the permittivity of 7CB can be considered (Fig. 10.3b). The relatively large dipole moment of the CN-group ($\sim$4.4 D) is approximately directed along the long molecular axis, i.e. $\omega \approx 0$. According to (10.28) this leads to a large positive dielectric anisotropy. In addition, the contribution of the induced polarization may be neglected in view of the magnitude of μ. Hence the behaviour of $\overline{\varepsilon}$ in the nematic phase and ε in the isotropic phase can be expected to be determined by the factor μ^2/T in (10.29a). A comparison with the experimental results shown in Fig. 10.3b leads to the following two observed deviations:

i) Instead of the expected trend ε decreases with decreasing temperature.
ii) The quantity μ^2 as calculated from (10.29b) is a factor of two smaller than expected for a CN-substituted compound.

These results can be interpreted as evidence for a strong antiparallel dipole-dipole correlation. This means that μ^2 must be replaced by $\mu^2_{\text{eff}} = g(T)\mu^2$ in the equations used so far, where $g(T)$ is the correlation factor which becomes smaller with decreasing temperature (Böttcher and Bordewijk 1978). Thus the factor $1/T$ in μ^2_{eff}/T can be more than compensated, leading to the observed behaviour. Further evidence for the correctness of this interpretation is obtained from the layer spacing d of some smectic compounds with a CN group. In many typical cases values $d \approx 1.4l$ are observed, where l is the molecular length. This is roughly compatible with a model where the antiparallel correlation is attributed to overlapping aromatic cores of the molecules (Brownsey and Leadbetter 1980). Sometimes this type of behaviour is accompanied by long-range antiferroelectricity at lower temperatures as is evident from a dramatic decrease of the permittivity. The corresponding S_{A2} phase then has a periodicity $d \approx 2l$ (compare Sect. 3.2).

A different type of dipole-dipole correlation is observed in smectic phases, where the interaction of a dipole moment with the dipoles of surrounding molecules turns out to be different from that in the nematic phase, due to the distribution of the centres of mass in layers. For dipoles situated in the central part of the molecules, the distance between the dipoles of molecules in different smectic layers is much greater than the distance between neighbouring dipoles in the same layer. This means that for the longitudinal component μ_l, antiparallel correlation with neighbours within the layers is dominant over parallel correlations with neighbours in adjacent layers. This leads, on average, to antiparallel correlation and thus to a decrease of $\varepsilon_{\parallel}$. Similarly the difference in distance between the two types of neighbours leads to a parallel correlation of μ_t and an increase of $\varepsilon_{\perp}$. This can even lead to a change in sign of $\Delta\varepsilon$ in some cases as is shown in Fig. 10.4.

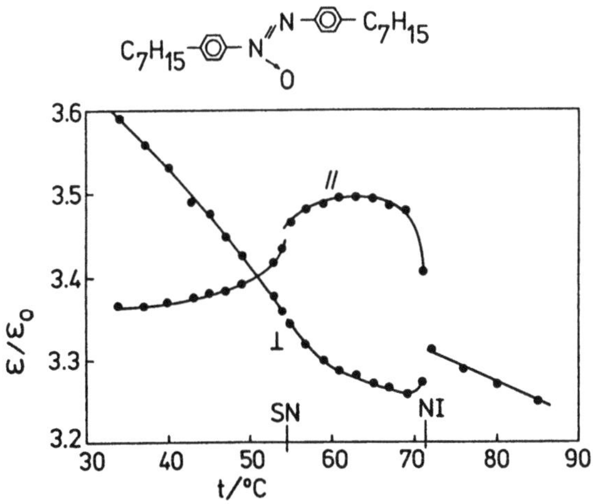

Fig. 10.4. Change of sign of $\Delta\varepsilon$ due to parallel correlation of μ_t and antiparallel correlation of μ_l in a layered structure (De Jeu et al. 1974)

10.3 Transport Properties

10.3.1 Mass Diffusion

As expected, the mass diffusion in liquid crystalline systems is anisotropic. Fick's law (Jost and Hauffe 1972) can be written as

$$J_\alpha = -D_{\alpha\beta}\partial_\beta c \quad , \tag{10.30}$$

where $\boldsymbol{J}$ is the diffusion flow, c the concentration, and $D_{\alpha\beta}$ an element of the diffusion tensor. For a uniaxial system with the z axis along the unique axis this tensor is given by

$$\tilde{D} = \begin{pmatrix} D_\perp & 0 & 0 \\ 0 & D_\perp & 0 \\ 0 & 0 & D_\parallel \end{pmatrix} \quad . \tag{10.31}$$

For the nematic phase it is found that $D_\parallel > D_\perp$, the actual values depending on both the mesomorphic medium and the type of diffusing molecules.

Some features of the diffusion constants, as for example the anisotropy, can be qualitatively understood by assuming that the short-range structure of the liquid can be described by a quasi-lattice. In such a model the diffusion process consists of a particle jumping into a neighbouring empty lattice site (hole). Disregarding the exact nature of this process D_i can be expected to have the following form

$$D_i \sim f_i \exp\left(-\frac{w}{k_\mathrm{B}T}\right) \quad , \tag{10.32}$$

where f_i is a structure factor related to the direct surroundings and w the activation energy to create a hole. According to (10.32) the temperature dependence of D_i is mainly due to this activation process. For ideal orientational order f_i will be inversely proportional to the area obtained by projecting a rod-like molecule onto a plane in the relevant direction. These areas are given by $A_\parallel = (\pi/4)W^2$ and $A_\perp = LW$, where L and W are the length and width of a molecule, respectively. Hence a maximum anisotropy can be expected of about

$$D_\parallel / D_\perp \simeq L/W \quad . \tag{10.33}$$

For imperfect orientational order ($S<1$), (10.33) can be expected to be modified analogously to the ratio $\langle\alpha_\parallel\rangle / \langle\alpha_\perp\rangle$ as discussed in the previous section [see e.g. (10.16)], i.e.

$$\frac{D_\parallel}{D_\perp} = \frac{(L/W)(1+2S)+(2-2S)}{(L/W)(1-S)+(2+S)} \quad . \tag{10.34}$$

For $L/W = 3$ and $S = 0.6$ (10.34) leads to $D_\parallel / D_\perp = 1.9$. Equation (10.34) has been derived more carefully by Chu and Moroi (1975).

The ratio $D_\parallel / D_\perp$ will be drastically changed in the smectic state because of the layered structure. An additional potential barrier ΔU will be present hampering the diffusion process perpendicular to the layer. Hence (10.33) must be modified to read

$$D_\parallel / D_\perp = (L/W) \exp\left(-\frac{\Delta U}{k_\mathrm{B}T}\right) \quad . \tag{10.35}$$

Thus in spite of the fact that $L/W>1$ a ratio $D_{\parallel}/D_{\perp}<1$ may be observed in smectics.

The diffusion constants can be determined experimentally in several ways. An elegant method is the injection of a dye at a given point in a uniform planar nematic layer. The observed coloured concentration profile takes the form of an ellipsoid with axes proportional to $D_{\parallel}^{1/2}$ and $D_{\perp}^{1/2}$ (Rondelez 1974; Schulz and Schumann 1978). Application of this method to MBBA with nitroso-dimethylaniline as dye results in

$$D_{\parallel} = 1.2 \times 10^{-6}\,\mathrm{cm^2 s^{-1}}, \quad D_{\perp} = 0.7 \times 10^{-6}\,\mathrm{cm^2 s^{-1}},$$

leading to an anisotropy ratio $D_{\parallel}/D_{\perp} = 1.7$. The injection of a chiral nematic liquid crystal changes the nematic uniform planar texture into a Grandjean texture. Now the concentration profile and thus $D_{\parallel}$ and $D_{\perp}$ can be deduced from the position of the disclination lines (Hakemi and Labes 1974). If the chiral nematic is injected into a racemic mixture of the same molecules, this method provides a measurement of the self-diffusion coefficients.

Pulsed nuclear magnetic resonance in the presence of a magnetic field gradient is a common technique to measure the diffusion constant in isotropic liquids. Due to the field gradient the precession frequency of the nuclear spins (Larmor frequency) changes with the position of the spins in the sample, and thus the translational diffusion is the principal mechanism that destroys the imposed transverse nuclear spin polarization. In liquid crystals, however, a competing mechanism exists: the nuclear magnetic dipole-dipole interactions which dominate over the effect of diffusion (see also Sect. 11.2). In order to avoid this problem, a spherically symmetric probe molecule (e.g. tetramethylsilane, TMS) can be used for which the dipole interactions approximately average out. In the nematic phases studied, only a small anisotropy $D_{\parallel}/D_{\perp}>1$ was found, and for the S_A phase $D_{\parallel}/D_{\perp}\ll 1$ (Krüger and Spiesecke 1973). Blinc et al. (1973) showed that the dipolar interactions can also be averaged out in a general way by using a special superposition of pulse sequences. In this way the self-diffusion coefficients can be measured which gives for nematic MBBA at room temperature (Zupančič et al. 1974):

$$D_{\parallel} = 0.7 \times 10^{-6}\,\mathrm{cm^2 s^{-1}}, \quad D_{\perp} = 0.5 \times 10^{-6}\,\mathrm{cm^2 s^{-1}}.$$

These values are somewhat smaller than those for the dye quoted above, and give $D_{\parallel}/D_{\perp} = 1.5$. This NMR method has been extensively used by Noack (1984). His data for MBBA are compared with results from other methods in Fig. 10.5. Apart from the differences in T_{NI} there are large dis-

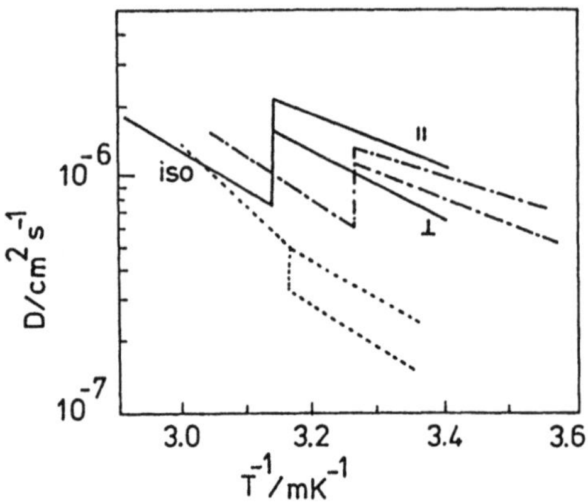

Fig. 10.5. Diffusion coefficients of MBBA from various sources (after Noack 1984). The dotted line is according to the dye-method, the other results are obtained with NMR. Note the differences in clearing temperature

crepancies. Surprisingly Noack finds at $T_{\rm NI}$ that $D_{\|}, D_{\perp} > D_{\rm iso}$. Furthermore a strict exponential temperature dependence is observed [compare (10.32)], without any indication of a dependence on S. These measurements have also been done for the PAA series, where for PAA itself discrepancies again exist with the results from other sources. The various results for diffusion coefficients have been reviewed by Krüger (1982), where references to more elaborate models can also be found. Experiments on the smectic A phase show $D_{\|}/D_{\perp} < 1$ as expected. Systematic measurements, however, are still scarce.

10.3.2 Electrical Conduction

The electrical conductivity of liquid crystals is mainly due to residual impurities, the nature of which is often unknown. Without taking special measures the specific electrical conductivity is usually in the order of 10^{-8} down to $10^{-12}\,\Omega^{-1}\,\rm cm^{-1}$. Precise measurements require ultra-pure liquid crystals that are subsequently doped with specific ionic impurities. The electrical conductivity can be measured using the standard methods employed for dielectric liquids in general. These measurements are not trivial, however, due to problems arising for example from electrode reactions, double layers leading to electrode polarization, separation of displacement and resistive current, etc. Early measurements of Svedberg (1914, 1915) in the nematic phase indicate an anisotropy $\sigma_{\|}/\sigma_{\perp}$ varying from 1.3 to 1.6. More recent systematic measurements still give similar values (Heppke et al. 1976).

The conductivity is directly related to the impurity diffusion via the Einstein relation. This relation can be derived as follows. In the presence of an electric field E along the x axis the density $n(x)$ of particles with charge q is given by Boltzman's distribution law

$$n(x) = n_0 \exp\left(\frac{qEx}{k_B T}\right) \quad . \tag{10.36}$$

The absence of a net current in the equilibrium state means that the electrical flow j_q and the diffusion flow j_D compensate each other, i.e.

$$j_q + j_D = 0 \quad , \quad \text{or}$$

$$\mu n E - D\frac{dn}{dx} = 0 \quad , \tag{10.37}$$

where μ is the mobility of the charge carriers. According to (10.37) the density $n(x)$ is given by

$$n(x) = n_0 \exp\left(\frac{\mu E x}{D}\right) \quad . \tag{10.38}$$

Equating the two exponents in (10.36) and (10.38) results in

$$\mu = \frac{qD}{k_B T} \quad \text{or} \tag{10.39}$$

$$\sigma = Nq\mu = \frac{Nq^2 D}{k_B T} \quad , \tag{10.40}$$

where N is the number of charge carriers. Consequently the ratio $\sigma_{\|}/\sigma_{\perp}$ can be expected to be approximately the same as $D_{\|}/D_{\perp}$. This is indeed observed experimentally.

Precise studies of the electrical conductivity of liquid crystals with well-controlled concentrations of impurity ions are relatively scarce. In general, however, no large variations of the ratio $\sigma_{\|}/\sigma_{\perp}$ are found for nematics. This ratio is of great importance because the anisotropy of the conductivity is the driving force for the hydrodynamic instabilities of a nematic layer in an electric field (see Sect. 8.4). The mobility of the charge carriers in nematics and thus the conductivity is influenced by frictional forces. The nature of these forces has been studied by Herino (1981).

Some typical results for $\sigma_{\|}/\sigma_{\perp}$ in a smectic A phase are shown in Fig. 10.6. In agreement with $D_{\|}/D_{\perp} < 1$ it is also found that $\sigma_{\|}/\sigma_{\perp} < 1$. In fact this result is already found in the nematic phase due to pretransitional smectic short-range order. In some strongly polar smectics with layer spacings $d \approx 1.4l$ a ratio $\sigma_{\|}/\sigma_{\perp} \approx 1$ is observed (Mircea-Roussel et al. 1975). According to (10.35) this behaviour can be explained by a rather weak potential barrier ΔU, only just balancing the anisotropy expected from the

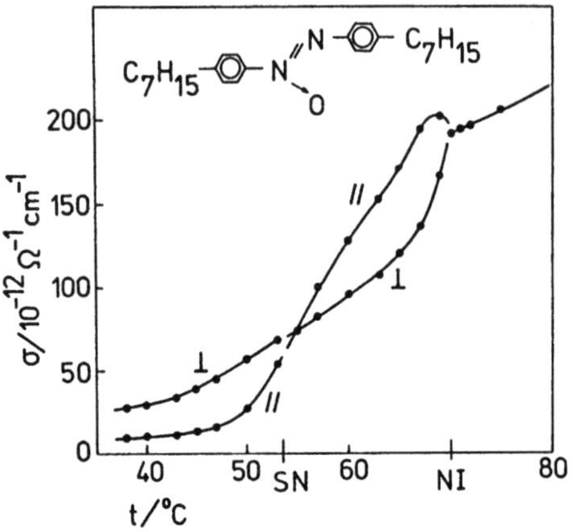

Fig. 10.6. Typical change of the sign of the anisotropy of conduction $\sigma_{\|} - \sigma_{\perp}$ around a nematic-smectic phase transition

L/W ratio. Finally it should be remarked that the biaxial symmetry of the smectic C phase introduces an additional anisotropy. A different value of $\sigma_{\perp}$ is obtained depending on whether it is measured parallel or perpendicular to the C-director. This difference has been measured and the relevant ratio amounts to about 1.08 (Heppke and Schneider 1974). Compared to the ratio $\sigma_{\perp}/\sigma_{\|} \approx 10$ this additional asymmetry is small.

11. Dynamic Behaviour of the Molecules

In this chapter we consider the dynamic behaviour of molecules in liquid crystals. As the experimental data are always time averages, relevant information on the dynamics of the molecules can only be obtained from experimental techniques involving short time scales. A knowledge of the molecular motion offers the important possibility of assessing the validity of the various approximate molecular models. First we discuss quasi-elastic incoherent neutron scattering. This tool is sensitive to the collective motions of groups of atoms in a molecule on a time scale of less than 10^{-8} s. Next we consider magnetic resonance of electrons, nuclear spins and nuclear quadrupoles – techniques that cover time scales from 10^{-8} s to 10^{-5} s. Finally we discuss the information obtained from dielectric relaxation, which ranges from time scales of 10^{-8} s to the order of 10^{-3} s.

11.1 Neutron Scattering

The analysis of quasi-elastic neutron scattering data is quite similar to the analysis of x-ray scattering data (see Fig. 9.7). The starting point is the assumption that the scattering amplitude $F(\boldsymbol{q})$ can be written as

$$F(\boldsymbol{q}) = \sum_{j=1}^{N} b_j \exp(\mathrm{i}\boldsymbol{q} \cdot \boldsymbol{r}_j) \quad , \tag{11.1}$$

where N denotes the number of nuclei, b_j the scattering length of the nucleus at site $\boldsymbol{r}_j$, and $\boldsymbol{q}$ the scattering wave vector. Comparison of (11.1) with (9.54) shows that both expressions are similar, with the scattering length b_j of a nucleus replacing the scattering power f_j for x-rays, which is due to the electron density. For the sake of simplicity the system is considered to consist of only one type of atom. Nevertheless different scattering lengths may appear due to different isotopes and different nuclear spin states. Clearly the calculation of the scattering intensity involves averaging over the possible scattering lengths and positions. Assuming that both averaging procedures are uncorrelated and that $\langle b_j b_k\rangle = \langle b_j\rangle\langle b_k\rangle$, $j{\neq}k$, the scattered intensity is given by

$$I(\boldsymbol{q}) = N\langle b^2\rangle + \sum_{j\neq k} \langle b\rangle^2 \langle \exp(\mathrm{i}\boldsymbol{q}\cdot\boldsymbol{r}_{jk})\rangle \quad , \tag{11.2}$$

where $\boldsymbol{r}_{jk} = \boldsymbol{r}_j - \boldsymbol{r}_k$. Usually the expression for the scattering amplitude is written as

$$I(\boldsymbol{q}) = I_{\mathrm{inc}}(\boldsymbol{q}) + I_{\mathrm{coh}}(\boldsymbol{q}) \quad \text{where} \tag{11.3}$$

$$I_{\mathrm{inc}}(\boldsymbol{q}) = N(\langle b^2\rangle - \langle b\rangle^2) \quad , \tag{11.3a}$$

$$I_{\mathrm{coh}}(\boldsymbol{q}) = \sum_{j,k}^{N} \langle b^2\rangle \langle \exp(\mathrm{i}\boldsymbol{q}\cdot\boldsymbol{r}_{jk})\rangle \quad . \tag{11.3b}$$

The coherent term provides structural information, and can be treated along similar lines as discussed in Sect. 9.5 for x-ray diffraction. The structure-independent incoherent term is, in practice, especially important for systems containing protons, as protons give rise to scattering cross-sections $4\pi\langle b^2\rangle = 10^{-26}\,\mathrm{m}^2$ and $4\pi\langle b\rangle^2 = 10^{-28}\,\mathrm{m}^2$, meaning that $I_{\mathrm{inc}}(\boldsymbol{q})$ is relatively large. As the coherent scattering intensity is concentrated in a few peaks, it is possible to interpret the incoherent scattering in terms of the motion of protons.

For a review of the theory of incoherent neutron scattering in liquid crystals the reader is referred to Leadbetter and Richardson (1979). Here only a few important features will be considered. In order to interpret the incoherent scattering spectrum the assumption is made that the motion of the protons can be decomposed into two independent motions, namely a diffusive or random motion and an oscillatory motion, giving rise to a quasi-elastic ($k_s \approx k_i$, see Fig. 9.7) and an inelastic ($k_s \neq k_i$) contribution to $I_{\mathrm{inc}}(\boldsymbol{q})$, respectively. Such a separation is empirically meaningful if the experimental spectra can be separated into a quasi-elastic and an inelastic part. The motion of various functional groups within the molecule can be studied by using partial deuteration. An example of the influence of partial deuteration on the spectrum of PAA is shown in Fig. 11.1. Both the quasi-elastic peak (underneath a well-resolved elastic peak) and the strong inelastic peak disappear completely if the methyl groups instead of the benzene rings are deuterated. This result indicates that the quasi-elastic scattering is due to a random rotational motion of the methyl groups which does not involve the aromatic core. It follows from additional model calculations that only the methyl group (and not the full methoxy group) is involved in this rotation, which can be described by a jump model with three-fold rotation symmetry. The rotational correlation time is about 5×10^{-12} s at 100°C.

Deuteration has been used to study molecular reorientation as well. Usually samples with deuterated end chains are taken, so that the motion

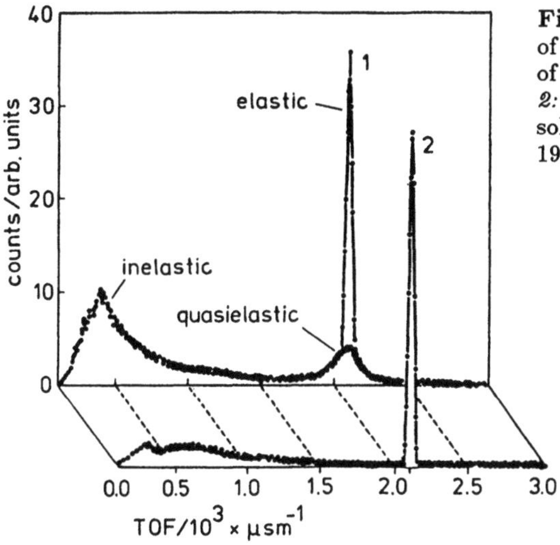

Fig. 11.1. Neutron scattering spectra of two partially deuterated derivatives of PAA (*1:* phenyl rings deuterated; *2:* methyl positions deuterated) in the solid state at 100°C (Hervet et al. 1976)

of the cores is seen. The results indicate in general that the molecules show a uniaxial rotation about their long axis, both in the nematic phase and in most smectic phases, the correlation time being of the order of 10^{-11} s to 10^{-10} s. The uniaxiality of this rotational motion only disappears in the most ordered smectic phases such as the smectic E phase, indicating an orientational order of the short axes of the molecules.

The separation of the elastic and the quasi-elastic peak is possible because the translational diffusion is slow compared to the rotations. Using the very highest resolution the broadening of the elastic peak at large $\boldsymbol{q}$, which is due to this translational diffusion, can also be measured. Diffusion constants result ranging from 10^{-11} to $10^{-10}\,\mathrm{m^2s^{-1}}$, in reasonable agreement with the data obtained by other means (Sect. 10.3.1). (See, for example, Richardson et al. 1980.)

11.2 Nuclear Magnetic Resonance and Relaxation

As discussed in Sect. 9.3 nuclear magnetic resonance (NMR) is used to determine the order parameter $\langle P_2(\cos\theta)\rangle$ [see (9.40)]. A prerequisite for such a calculation is the assumption that the line splitting $\Delta\nu$, which is due to the nuclear spin-spin interaction, can be considered to be an average value. This means that the correlation time τ associated with the orientational fluctuations must satisfy the relation

$$\Delta\nu\,\tau \ll 1 \quad . \tag{11.4}$$

Typically values of $\tau \lesssim 10^{-5}$ s are found, which is an upper limit for τ. The results that are obtained for $\langle P_2(\cos\theta)\rangle$ appear to be quite similar to those found by applying the method of electron paramagnetic resonance (EPR) of spin-probes dissolved in liquid crytals. Since the time scale of EPR is of the order of 10^{-8} s, the supplementary averaging involved in the NMR measurement does not appear to have any influence. This indicates a time scale of the orientational fluctuations $<10^{-8}$ s. Any effect of reorientation in the time window 10^{-8} to 10^{-5} s must have a small amplitude. The observed values of $\langle P_2(\cos\theta)\rangle$, however, require large amplitude reorientations so that the fluctuations of the long molecular axis must be quite rapid.

Besides orientational fluctuations, conformational changes within the molecules can be found, i.e. a second time scale is relevant. Deuteron magnetic resonance in particular, has been able to provide information about these conformational changes (Charvolin and Deloche 1979). For instance the DMR spectra of alkyl chains (see Sect. 9.3) indicate isomeric rotations about the C-C bonds. These rotations are fast on the time scale of DMR, which is 10^{-5} s. In a similar way, with appropriate deuteration, conformational changes within the aromatic core of TBBA have been established. It appears that the relevant time scales of these conformational changes are quite small. This means that for many purposes the concept of an average conformation of the chains makes sense. Such an average conformation is, however, not necessarily independent of temperature (see, for example, Jähnig 1979).

A dynamic picture of the molecules can be constructed from these results in combination with those of neutron scattering experments. The molecules show internal rotations and deformations on a time scale of 10^{-10}s, whereas the orientational fluctuation of the long molecular axis requires a time scale of about 10^{-9} s. This difference in time scales, though small, has often been used to support the convenient picture of an "averaged" molecule, subjected to orientational fluctuations. More recently, however, it has been realized that the difference in time scale is not a sufficient reason to consider the internal dynamics and the molecular reorientations to be independent. In addition the directions of the principal axes of the inertial tensor seem to be relevant. For a further discussion of the interpretation of the line splittings observed with magnetic resonance methods the reader is referred to Emsley and Luckhurst (1980), and Emsley (1985).

Information concerning the dynamics of the molecules can also be obtained from nuclear magnetic relaxation effects. The longitudinal relaxation rate $1/T_1$ is a particularly relevant quantity since the molecular motions cause the nuclear spin relaxation through the modulation of the interactions involving the nuclear spins (Abragam 1962). Here the discussion will be restricted to protons, i.e. to the magnetic dipole-dipole interaction. For isotropic liquids, the effect of the molecular motions on the relaxation process can often be described in terms of a single parameter, the correlation time τ_c. Consequently the dependence of $1/T_1$ on ν_0, the Larmor frequency (see Sect. 9.3), is also relatively simple. On the other hand several processes may contribute to the relaxation in liquid crystals. This leads, in general, to a complicated situation, which is particularly interesting if the frequency dependence of $1/T_1$ is considered over a wide range of Larmor frequencies.

For the sake of clarity the possible contributions to $1/T_1$ in a nematic will be discussed using MBBA as an example. The various possible relaxation mechanisms are shown in Fig. 11.2. Their relative importance depends on the temperature and on the difference between their inverse correlation time and the Larmor frequency at which the relaxation is measured. The various contributions to $1/T_1$ are caused by three classes of molecular motion:

i) internal rotations within the molecules;
ii) orientational fluctuations;
iii) self-diffusion.

In the following the effect of each of these mechanisms will be discussed briefly.

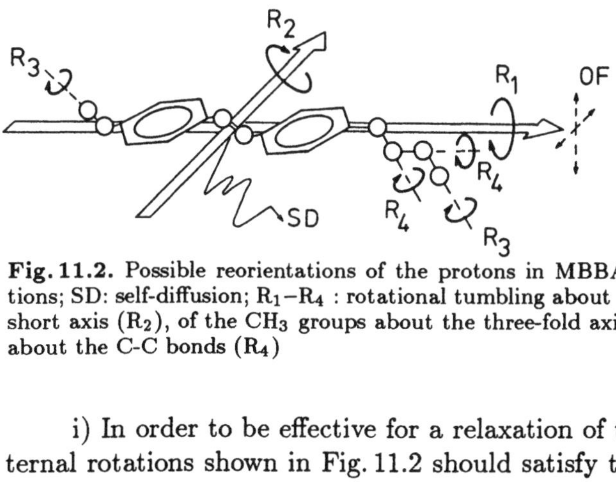

Fig. 11.2. Possible reorientations of the protons in MBBA; OF: orientational fluctuations; SD: self-diffusion; R_1–R_4 : rotational tumbling about the long axis (R_1), about the short axis (R_2), of the CH_3 groups about the three-fold axis (R_3), of the CH_2 segments about the C-C bonds (R_4)

i) In order to be effective for a relaxation of the nuclear spins the internal rotations shown in Fig. 11.2 should satisfy two requirements. In the first place the rotation must impose a modulation of the component along the molecular axis of the vector that connects the two protons considered. For example, rotation about the long molecular axis does not contribute to the relaxation of a dipole-dipole coupling parallel to that axis. Secondly, the inverse correlation time and the Larmor frequency should not differ too much. The latter requirement excludes any important influence of the rotations R_1, R_3 and R_4 in the example of MBBA (see Fig. 11.2). Their correlation times are too small (see Sect. 11.1) compared to the inverse Larmor frequency, which in practice is between 10^{-3} and 10^{-8} s. The rotation R_2 about a short axis, however, could be effective. This rotation has been studied with the method of dielectric relaxation (see the next section).

ii) Characteristic for liquid crystals is the appearance of an effective mechanism for relaxation via orientational fluctuations. Insofar as the long wavelength fluctuations are relevant these can be described in terms of the continuum theory (see Sect. 7.2). Applying Eq. (8.26), it follows that a fluctuation of wavevector $\boldsymbol{k}$ has a correlation time τ given by

$$\tau = \frac{\eta}{Kk^2} \quad , \tag{11.5}$$

where η is a typical viscosity and K a Frank constant [see also (8.58)]. In order to be effective the inverse correlation time and the Larmor frequency should be of comparable magnitude. This is indeed the case as can be verified by inserting typical values of $K \approx 10^{-11}$ N, $\eta \approx 10^{-2}$ Pa·s and a wavelength $\lambda = 2\pi/k \approx 60$ nm, which results in an inverse correlation time of the order of $10^7\ \mathrm{s}^{-1}$. Orientational fluctuations have been incorporated into the standard theory for relaxation by Pincus (1969) and the following functional behaviour of $1/T_1$ is found:

$$\frac{1}{T_1} \sim \frac{k_\mathrm{B}TS^2}{K}\left(\frac{K}{\eta}\right)^{-1/2} \nu_0^{-1/2} \quad . \tag{11.6}$$

Such a frequency dependence of $1/T_1$, which is relatively temperature independent, has been observed in nematics over certain temperature ranges. This behaviour is in strong contrast with the $1/T_1$ behaviour in isotropic liquids, and is the original cause of interest in NMR relaxation in liquid crystals (Wade 1977).

iii) The contribution of self-diffusion to nuclear spin relaxation originates from two effects. The first one is *intra*molecular and is due to the fact that a molecule, which diffuses through a nematic with "frozen" fluctuations, adjusts its long axis to the locally preferred direction. This means that there is still a contribution of the form (11.6) in the limit $K/\eta \to 0$. Boldly assuming that the orientational and diffusive motions of the molecules are uncoupled the *intra*molecular effect of diffusion modifies (11.6) into (Pincus 1969):

$$\frac{1}{T_1} \sim \frac{k_\mathrm{B}TS^2}{K}\left(\frac{K}{\eta} + D\right)^{-1/2} \nu_0^{-1/2} \quad , \tag{11.7}$$

where D is the diffusion constant. For a more thorough theoretical discussion the reader is referred to Ukleja et al. (1976). The second effect of self-diffusion concerns the modification of the dipolar interactions between nuclear spins belonging to different molecules (*inter*molecular effect). Such a process is well-established for isotropic liquids, where it leads, under certain conditions, to a frequency dependence of $1/T_1$ of the form

$$1/T_1 \sim \nu_0^{1/2} \quad . \tag{11.8}$$

The relevant theory does not change qualitatively when dealing with nematics instead of isotropic liquids (Žumer and Vilfan 1978). An extension of the theory to the S_A phase has also been given (Vilfan and Žumer 1980).

Figure 11.3 shows measurements of T_1 for MBBA over a wide frequency range (Graf et al. 1977). These data have been carefully analyzed in terms of the three contributions to $1/T_1$ discussed above. At temperatures close to the nematic-isotropic transition temperature all three effects are found to be important. At lower temperatures the contributions of the orienta-

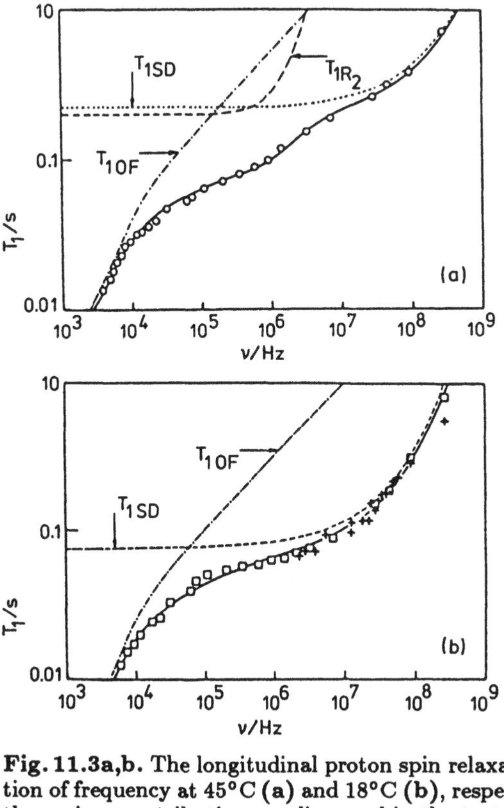

Fig. 11.3a,b. The longitudinal proton spin relaxation time in nematic MBBA as a function of frequency at 45°C (**a**) and 18°C (**b**), respectively. The data have been fitted using the various contributions as discussed in the text (Graf et al. 1977)

tional fluctuations and the self-diffusion suffice to explain the results. The correlation times and the activation energies of the three rotations involved appear to be in good agreement with those obtained from other spectroscopic methods.

11.3 Dielectric Relaxation

In Sect. 10.2 the properties of the dielectric permittivity of liquid crystals in static fields were discussed. In this section the dynamic properties are considered, i.e. the behaviour of the dielectric permittivity in the presence of alternating fields. The main difference between the static and dynamic permittivity is due to the process of (re)orientation of the permanent dipole moments caused by the changing field. Such a process involves a definite time interval as can be clearly seen after the removal of a static field, in which case the orientational polarization decays exponentially to its equilibrium value with characteristic time constant τ. In alternating fields this (re)orientation process leads to a time lag between the direction of the average orientation of the dipole moments and the field. This effect can be clearly observed at frequencies of about τ^{-1}. At higher frequencies the time lag is such that the orientation polarization hardly contributes to the permittivity, because the orientational polarization and the variations of the field are no longer correlated. The residual permittivity ε_∞ is due to the induced polarization only.

In the case of isotropic liquids the dielectric relaxation can be described in terms of a complex permittivity $\varepsilon(\omega) = \varepsilon'(\omega) - \mathrm{i}\varepsilon''(\omega)$, where ω denotes the angular frequency of the applied alternating field. According to Debye's classical theory (see, for example, Böttcher and Bordewijk 1978) $\varepsilon(\omega)$ is given by

$$\varepsilon(\omega) = \varepsilon_\infty + \frac{\varepsilon - \varepsilon_\infty}{1 + \mathrm{i}\omega\tau} \quad , \tag{11.9}$$

where $\varepsilon = \varepsilon(0)$ and $\varepsilon_\infty = \varepsilon(\infty)$ are the static and high-frequency permittivity, respectively. It follows directly from (11.9) that

$$\omega\tau = \frac{\varepsilon''(\omega)}{\varepsilon'(\omega) - \varepsilon_\infty} \quad , \tag{11.10a}$$

$$\frac{(\varepsilon - \varepsilon_\infty)[\varepsilon'(\omega) - \varepsilon_\infty]}{[\varepsilon'(\omega) - \varepsilon_\infty]^2 + [\varepsilon''(\omega)]^2} = 1 \quad . \tag{11.10b}$$

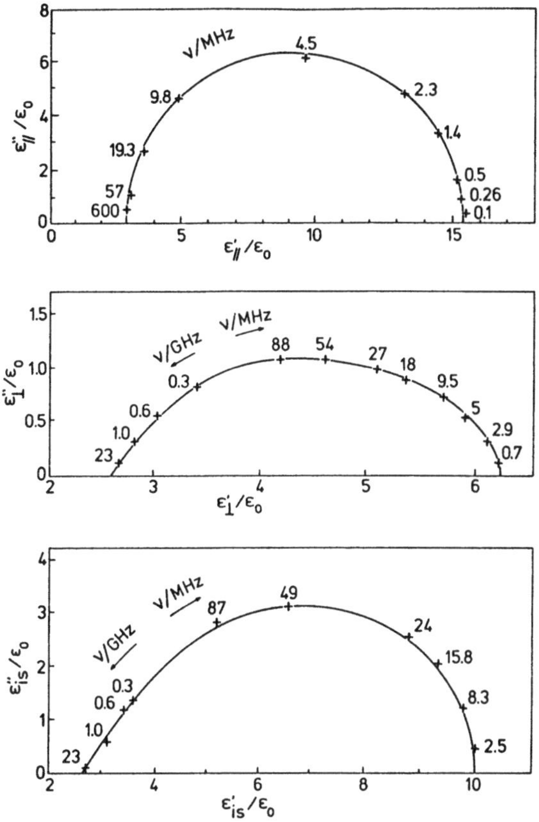

Fig. 11.4. Cole-Cole plots for the dielectric relaxation of 7CB (Lippens et al. 1977)

Consequently in Debye's theory according to (11.10b) a plot of $\varepsilon'(\omega)$ vs $\varepsilon''(\omega)$ must give a semicircle. In the case of liquid crystals (11.10b) can be checked for each of the components $\varepsilon_{\|}(\omega)$ and $\varepsilon_{\perp}(\omega)$. As an example the so-called Cole-Cole plots for 7CB are shown in Fig. 11.4. Only for $\varepsilon_{\|}(\omega)$ a true semicircle is found. The other cases show deviations indicating a more complicated relaxation to the equilibrium configuration, that cannot be described by a single relaxation time.

The influence of the permanent dipoles on the behaviour of $\varepsilon_{\|}(\omega)$ and $\varepsilon_{\perp}(\omega)$ can be understood by considering (10.27). It follows that the contributions of μ_l and μ_t are given by

$$\varepsilon_{\|}(\omega) - \varepsilon_{\|,\infty} \sim \langle\mu_{\|}^2\rangle = \tfrac{1}{3}[\mu_l^2(1+2S) + \mu_t^2(1-S)] \quad , \tag{11.11a}$$

$$\varepsilon_{\perp}(\omega) - \varepsilon_{\perp,\infty} \sim \langle\mu_{\perp}^2\rangle = \tfrac{1}{3}[\mu_l^2(1-S) + \tfrac{1}{2}\mu_t^2(2+S)] \quad . \tag{11.11b}$$

First consider the component μ_l. The behaviour of μ_l depends on the mutual direction of the director and the applied field. If they are parallel, i.e. in the case of $\varepsilon_{\parallel}(\omega)$, a reorientation must be accomplished by a rotation about the short axis (R_2 in Fig. 11.2). Consequently the reorientation process is considerably affected by the nematic order, leading to relatively large values of τ (low relaxation frequencies) that can be observed easily. If the field is perpendicular to the director, i.e. considering the case $\varepsilon_{\perp}(\omega)$, the reorientation of μ_l can be accomplished by a rotation over π, while keeping the angle θ between the long molecular axis and the director fixed. The relaxation time of such a process may be expected to decrease with increasing nematic order (smaller θ). However, this will be difficult to observe in general, as the contribution of μ_l to $\varepsilon_{\perp}(\omega)$ is small for typical values of S. The component μ_t can be reoriented by a rotation about the long molecular axis. In a first approximation such a rotation does not depend on the nematic order. As follows from (11.11) with increasing nematic order the importance of μ_t increases for $\varepsilon_{\perp}(\omega)$ but decreases for $\varepsilon_{\parallel}(\omega)$. In conclusion the dielectric relaxation of $\varepsilon_{\parallel}(\omega)$ and $\varepsilon_{\perp}(\omega)$ may be ascribed to the relaxation processes of μ_l and μ_t, respectively. Clearly the interesting differences between the dielectric relaxation in the isotropic and the nematic phase must be related to the reorientation of μ_l.

The effects discussed can be best observed in compounds with $\mu_t \ll \mu_l$. As an example for the behaviour of such a compound the relaxation frequencies of 7CB are shown in Fig. 11.5. The relaxation frequencies are found to shift in the expected directions by lowering the temperature below the clearing temperature. Clearly the presence of orientational order leads to a considerable increase, roughly a factor of 10, in the correlation time describing the reorientation about a short molecular axis, which by the way is

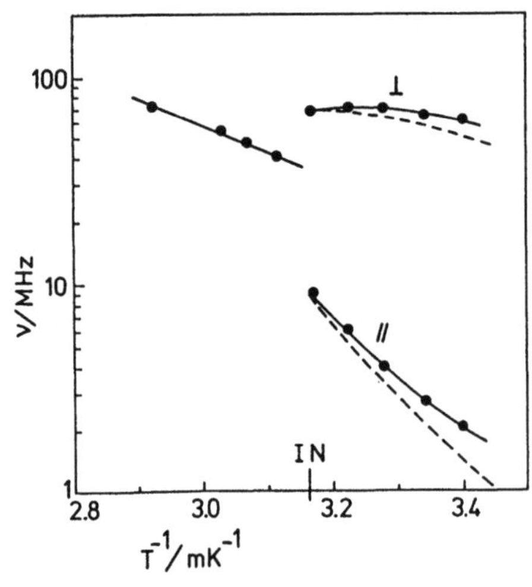

Fig. 11.5. Relaxation frequencies of 7CB (Davies et al. 1976); the broken line is a theoretical result (see text)

the slowest of the various molecular reorientations. Relaxation frequencies for $\varepsilon_{\parallel}(\omega)$ lower than 1 MHz ($\tau > 10^{-6}$ s) are quite possible.

The relaxation times introduced via (11.9) are macroscopic ones, and though the various contributions have been discussed in terms of particular molecular reorientations, the relation between each τ and the corresponding molecular correlation time τ_c depends on the choice of a particular model. Dielectric relaxation in isotropic liquids is usually described in terms of a rotational diffusion process being the result of many molecular collisions. This description can immediately be generalized to nematic liquids, as the short-range structure of the nematic and isotropic phase of a given compound are quite similar (see Sect. 9.5). The only modification concerns the introduction of an orienting potential due to the nematic phase (Martin et al. 1971). The broken line in Fig. 11.5 is obtained using this approach, and is in agreement with the qualitative discussion given above. Hard quantitative information, however, can hardly be expected. Another approach has been followed by Benguigui (1984) who showed that for a series of compounds the relaxation of $\varepsilon_{\parallel}(\omega)$ in both the nematic and smectic A phase can be described by

$$\nu_R = \nu_0 \exp\left(-\frac{\varepsilon S^2}{T - T_0}\right) , \qquad (11.12)$$

where T_0 represents a free volume factor (Zeller 1982a; Diogo and Martins 1981). This form has been verified by Zeller (1982b) for supercooled nematics over a very wide temperature range, down to the glass transition. Such a free volume effect is usually associated with a mechanism involving large jumps of the molecules, thus describing the relaxation of $\varepsilon_{\parallel}(\omega)$ as a single-particle process. However, the compatibility of such a model with the substantial information on strong local intermolecular correlations is not clear.

Concluding this chapter it should be remarked that the present analysis of spectroscopic data in terms of molecular motions is rather hybrid, because macroscopic concepts enter into the description. This means that the analysis may give rise to values for macroscopic quantities, for example the diffusion constants or viscosities, that may well differ from those obtained in other types of measurement.

Part IV

Liquid Crystalline Phases and Phase Transitions

12. Landau Theory of Liquid Crystalline Phases

In this chapter we deal with the phenomenological theory of phase transitions in liquid crystals, the so-called Landau-De Gennes theory. First of all the basic ideas of the Landau theory are discussed. The theory is then applied to the nematic-isotropic transition. Attention is paid to two pretransitional effects, namely induced birefringence or the Cotton-Mouton effect and light scattering. This latter phenomenon is calculated in terms of fluctuation theory. Next we discuss the smectic A-nematic phase transition. This transition is described first in terms of one order parameter: the amplitude of a density wave. The result is a second-order transition. We then show that the second-order character can be changed into a first-order one by incorporating either the influence of the amplitudes of other density waves or the effect of the orientational ordering.

12.1 The Basic Ideas of Landau Theory

The Landau theory (Landau and Lifshitz 1980) sets out to give a phenomenological description of phase transitions. These transitions involve a change of symmetry. Generally the more symmetrical (less ordered) phase corresponds to higher temperatures and the less symmetrical (more highly ordered) one to lower temperatures. In order to give a mathematical description of the transition, order parameters are introduced representing the difference in symmetry between the two phases. These order parameters, which are of a tensorial nature, are defined in such a way that they are zero in the more symmetric phase. The thermodynamic quantities of the less symmetric phase are obtained by expanding the thermodynamic potential in powers of the order parameters in the neighbourhood of the order-disorder transition point, assuming that such an expansion makes sense. The motivation for this procedure is derived from the continuity of the change of state at a phase transition of the second kind, i.e. the order parameters show arbitrarily small values near the transition point. Hence the procedure is in principle restricted to second-order phase transitions. The thermodynamic behaviour of the order parameters in the less symmetric phase is then determined from the condition that their values must minimize the postulated expansion of the thermodynamic potential. For the sake of clarity the idea of Landau will first be demonstrated for a simple example before applying the theory to liquid crystal phase transitions.

Consider a phase transition which is characterized by a scalar order parameter η. Such an order parameter may describe for instance the degree of ordering of a binary alloy AB. The completely ordered alloy (low temperature phase) has a simple cubic lattice with atoms of type A at the vertices and atoms of type B at the centres of the unit cell. The disordered alloy (high temperature phase) has a body-centred cubic lattice due to the completely random distribution of the atoms of type A and B over the lattice sites. The order parameter η is now simply the difference between the probabilities of finding an atom of type A or B at a given site of the lattice. Clearly $\eta = 0$ corresponds to the disordered alloy, i.e. the high temperature phase, whereas $\eta = \pm 1$ denotes the completely ordered alloy (low temperature phase). The general form of the thermodynamic potential $G(p, T, \eta)$ is postulated, in the neighbourhood of the transition point, to be given by

$$G(p,T,\eta) = G(p,T,0) + \alpha\eta + \tfrac{1}{2}A\eta^2 + \tfrac{1}{3}B\eta^3 + \tfrac{1}{4}C\eta^4 + \dots \quad , \qquad (12.1)$$

where the coefficients $\alpha, A, B, C, \dots$ are functions of the pressure p and the temperature T. Furthermore η reflects the difference in symmetry of the two phases, $\eta = 0$ in the high-temperature phase and $\eta \neq 0$ below the phase transition. In the latter situation the thermodynamic behaviour of the order parameter η follows from the stability conditions

$$\frac{dG}{d\eta} = 0 \quad , \quad \frac{d^2G}{d\eta^2} > 0 \quad . \qquad (12.2)$$

At the transition temperature T_c these stability conditions are

$$\frac{dG}{d\eta} = \alpha = 0 \quad , \quad \frac{d^2G}{d\eta^2} = A = 0 \quad , \qquad (12.3)$$

because of the coexistence of both phases. This means (1) $\alpha = 0$, (2) $A = a(T - T_c)$ with $a = (dA/dT)_{T_c} > 0$. The first conclusion follows from the requirement that the high temperature phase with $\eta = 0$ must give rise to an extreme value of $G(p,T,\eta)$. The second conclusion can be drawn from the different behaviour of the thermodynamic potential at $\eta = 0$ for T above and below the transition temperature T_c. The function $G(p,T,0)$ must have a minimum for $T > T_c$, i.e. $A > 0$, and a relative maximum for $T < T_c$, i.e. $A < 0$. Summarizing, the Landau theory postulates that the phase transition in the present case can be described by the following expression for the difference in thermodynamic potential of the two phases

$$\Delta G = G(p,T,\eta) - G(p,T,0)$$
$$= \tfrac{1}{2}a(T-T_c)\eta^2 - \tfrac{1}{3}B\eta^3 + \tfrac{1}{4}C\eta^4 + \ldots \quad (12.4)$$

where the negative sign in front of the coefficient B has been chosen for reasons of convenience.

The thermodynamic behaviour of the order parameter follows directly from the stability condition

$$\frac{dG}{d\eta} = 0 = a(T-T_c)\eta - B\eta^2 + C\eta^3 + \ldots \quad . \quad (12.5)$$

This equation has the following solutions near the transition point

$$(1) \quad \eta = 0 \quad , \quad \text{the high temperature phase} \quad , \quad (12.6a)$$

$$(2) \quad \eta = \frac{B \pm [B^2 - 4aC(T-T_c)]^{1/2}}{2C} \quad . \quad (12.6b)$$

Now the second-order nature of the phase transition, i.e. η is continuous at the transition, requires $B = 0$, i.e. the thermodynamic behaviour of η reads in the low temperature phase

$$\eta = \pm\left[\frac{a(T_c - T)}{C}\right]^{1/2} \quad . \quad (12.7)$$

This means that for reasons of stability the coefficient C must be positive.

Clearly the symmetry of the low temperature phase determines the form of the expansion (12.1). When dealing with a phase which is invariant for replacing η by $-\eta$, as for instance the mentioned example of the binary alloy AB, all the coefficients belonging to the uneven terms in η must be zero.

Finally it should be mentioned that the Landau theory can be extended to the case of phase transitions of the first kind. This extension will be discussed in the next sections. The main idea of this generalization can be expressed easily in terms of the present example of a scalar order parameter, where a first-order transition is obtained if the coefficient B is included in the description. If the symmetry of the low temperature phase does not allow for the appearance of odd powers in η, i.e. $B = 0$, a first-order transition can be obtained by taking C to be negative. In addition a positive sixth power term in η is then required in order to ensure stability of the ordered phase.

12.2 The Nematic-Isotropic Transition

The phase transition from the nematic to the isotropic phase is phenomenologically described in terms of an extension of the Landau theory of second-order phase transitions. According to Landau the free energy density function can be approximated by a low-order polynomial in the order parameter near the second-order transition as the order parameter will be small. De Gennes extended the Landau theory to first-order transitions in liquid crystals. However, the order parameter is not small at the transition. Consequently the De Gennes theory is only qualitatively correct from a molecular statistical point of view, as will also be shown later on. From a phenomenological point of view, however, the De Gennes theory provides a satisfactory description of the phase transition.

The nematic state is described by the symmetric tensor order parameter $\tilde{\boldsymbol{Q}}$ with zero trace, i.e. $Q_{\alpha\alpha} = 0$. In order to describe the first-order nematic-isotropic phase transition it is sufficient, according to the Landau-De Gennes theory (De Gennes 1971), to expand the thermodynamic potential up to the fourth order in the tensor order parameter $\tilde{\boldsymbol{Q}}$ near the transition. The experimentally observed discontinuity in the density at the transition is found to be small, about 0.3 % at atmospheric pressure. Consequently the related terms in the expansion of the thermodynamic potential can be neglected in a first approximation. Because of the scalar nature of the thermodynamic potential the expression is given by

$$\begin{aligned} g_{\mathrm{n}} = g_{\mathrm{i}} &+ \tfrac{1}{2} A_{\alpha\beta\gamma\delta} Q_{\alpha\beta} Q_{\gamma\delta} - \tfrac{1}{3} B_{\alpha\beta\gamma\delta\mu\nu} Q_{\alpha\beta} Q_{\gamma\delta} Q_{\mu\nu} \\ &+ \tfrac{1}{4} C_{\alpha\beta\gamma\delta\mu\nu\varrho\sigma} Q_{\alpha\beta} Q_{\gamma\delta} Q_{\mu\nu} Q_{\varrho\sigma} \quad , \end{aligned} \tag{12.8}$$

where g_{n} and g_{i} represent the Gibbs free energy density of the nematic and the isotropic phase, respectively. The negative sign in front of $\tilde{\boldsymbol{B}}$ is chosen for convenience. The tensors $\tilde{\boldsymbol{A}}$, $\tilde{\boldsymbol{B}}$ and $\tilde{\boldsymbol{C}}$ depend on the pressure and the temperature. The term linear in $Q_{\alpha\beta}$ does not appear due to the different symmetry of the two phases [analogous to Eq. (12.3)].

Usually uniaxial nematics are considered and expression (12.8) can be simplified considerably for the following reasons. First of all the tensor order parameter $\tilde{\boldsymbol{Q}}$ according to (5.16) becomes

$$Q_{\alpha\beta} = S(N_{\alpha\beta} - \tfrac{1}{3}\delta_{\alpha\beta}) \quad , \tag{12.9}$$

where the order parameter S with values between 0 and 1 is introduced for reasons of convention. Secondly, analogous to the theory of curvature elasticity (see Chap. 5), the tensors $\tilde{\boldsymbol{A}}$, $\tilde{\boldsymbol{B}}$ and $\tilde{\boldsymbol{C}}$ must be a combination of the tensors $\tilde{\boldsymbol{N}}$ and $\tilde{\boldsymbol{\delta}}$ with elements $N_{\alpha\beta}$ and $\delta_{\alpha\beta}$, respectively. In order to facilitate a comparison with molecular statistical calculations which are often performed at constant density, the Helmholtz free energy density f will be considered instead of g. For uniaxial nematics the expansion of f can be written as

$$f_{\mathrm{n}} = f_{\mathrm{i}} + \tfrac{1}{2}AQ_{\alpha\beta}Q_{\beta\alpha} - \tfrac{1}{3}BQ_{\alpha\beta}Q_{\beta\gamma}Q_{\gamma\alpha} + \tfrac{1}{4}C(Q_{\alpha\beta}Q_{\beta\alpha})^2 \quad , \qquad (12.10)$$

where f_{n} and f_{i} represent the free energy densities of the nematic and isotropic phase and the coefficients A, B and C depend on the temperature T. The change in pressure at the transition is small and is therefore neglected in the expansion. The Landau-De Gennes expansion (12.10) has been formulated in terms of the invariants $Q_{\alpha\beta}Q_{\beta\alpha}$ and $Q_{\alpha\beta}Q_{\beta\gamma}Q_{\gamma\alpha}$. All other invariants can always be reduced to forms consisting of these two invariants. For example the invariant $Q_{\alpha\beta}Q_{\beta\gamma}Q_{\gamma\delta}Q_{\delta\alpha}$ equals $\frac{1}{2}(Q_{\alpha\beta}Q_{\beta\alpha})^2$. This equality directly follows from the diagonal representation (5.10) of the tensor order parameter. It should be realized that the expansion (12.10) is not the most general one for a biaxial nematic.

The phase transition is described now by writing

$$A = a(T - T_{\mathrm{c}}^*) \quad , \qquad (12.11)$$

where T_{c}^* is a temperature slightly below the transition temperature T_{c} and a is a positive temperature-independent constant. Furthermore near the phase transition the temperature dependence of the coefficients B and C, where $C > 0$, is neglected. According to (12.9) it holds that

$$Q_{\alpha\beta}Q_{\beta\alpha} = \tfrac{2}{3}S^2 \quad ; \quad Q_{\alpha\beta}Q_{\beta\gamma}Q_{\gamma\alpha} = \tfrac{2}{9}S^3 \quad . \qquad (12.12)$$

Substitution of (12.11) and (12.12) into (12.10) gives the following free energy density for a nematic phase:

$$f_{\mathrm{n}} = f_{\mathrm{i}} + \tfrac{1}{3}a(T - T_{\mathrm{c}}^*)S^2 - \tfrac{2}{27}BS^3 + \tfrac{1}{9}CS^4 \quad . \qquad (12.13)$$

The equilibrium value of S is obtained by minimizing the free energy density (12.13) with respect to S. This means that S is determined by the equation

$$a(T - T_{\mathrm{c}}^*)S - \tfrac{1}{3}BS^2 + \tfrac{2}{3}CS^3 = 0 \quad . \qquad (12.14)$$

The solutions of (12.14) are

$$S = 0 \quad , \quad \text{the isotropic phase} \quad , \qquad (12.15a)$$

$$S_{\pm} = \frac{B}{4C}\left\{1 \pm \left[1 - \frac{24aC(T - T_c^*)}{B^2}\right]^{1/2}\right\} \quad . \tag{12.15b}$$

The correct solution describing the temperature dependence of the order parameter in the nematic phase is the S_+ solution. This follows directly from the calculation of the transition temperature T_c by using the condition $f_n = f_i$, i.e.

$$a(T_c - T_c^*)S_c^2 - \tfrac{2}{9}BS_c^3 + \tfrac{1}{3}CS_c^4 = 0 \quad , \tag{12.16}$$

and the second relation between S_c and T_c given by (12.14):

$$a(T_c - T_c^*)S_c - \tfrac{1}{3}BS_c^2 + \tfrac{2}{3}CS_c^3 = 0 \quad . \tag{12.17}$$

Combining (12.16) and (12.17) yields

$$\tfrac{1}{9}BS_c^3 = \tfrac{1}{3}CS_c^4 \quad . \tag{12.18}$$

This means that the following two solutions are possible:

$$(1) \quad S_c = 0 \quad , \quad T_c = T_c^* \quad , \tag{12.19a}$$

$$(2) \quad S_c = \frac{B}{3C} \quad , \quad T_c = T_c^* + \frac{B^2}{27aC} \quad . \tag{12.19b}$$

Clearly the S_- solution gives the result $S_c = 0$ at $T = T_c^*$, whereas the S_+ solution gives $S_c = B/(3C)$ at the higher transition temperature $T_c = T_c^* + B^2/(27aC)$. It follows that the S_+ solution is the thermodynamically stable solution. The solutions (12.15b) determine a third temperature $T_c^\dagger$ given by

$$T_c^\dagger = T_c^* + \frac{B^2}{24aC} \quad . \tag{12.20}$$

If $T > T_c^\dagger$ the solutions S_+ and S_- no longer apply because of their imaginary behaviour.

Concluding then, the Landau-De Gennes theory distinguishes four different temperature regions.

i) $T > T_c^\dagger$. Only the solution $S = 0$, i.e. the isotropic phase, exists.

ii) $T_c < T < T_c^\dagger$. The minimum of the free energy density function is still given by $S = 0$, i.e. the isotropic phase is the thermodynamically stable state. There is a relative minimum at $S = S_+$ and a relative maximum at $S = S_-$. Consequently an energy density barrier of height $f_n(S_-) - f_n(S_+)$ exists between the two minima $S = 0$ and $S = S_+$. This means that a metastable nematic phase can be obtained in this temperature region by overheating. At $T_c^\dagger$ the height of the barrier becomes zero, for $S_+(T_c^\dagger) = S_-(T_c^\dagger) = B/(4C)$, and the corresponding point in the free energy density curve is a point of inflection.

iii) $T_c^* < T < T_c$. Here $S = S_+$ gives rise to the lowest free energy density and $S = 0$ corresponds to a relative minimum. The difference in entropy density between the nematic phase and the isotropic phase at T_c is given by

$$\Delta\Sigma = -\left.\frac{\partial(f_n - f_i)}{\partial T}\right|_{T=T_c} = \frac{1}{3}aS_c^2 = \frac{aB^2}{27C^2} \quad . \tag{12.21}$$

This means that the latent heat per unit volume, ΔH, is given by

$$\Delta H = \frac{aB^2 T_c}{27C^2} \quad . \tag{12.22}$$

Table 12.1. Values of the parameters in the Landau expansion for MBBA

Parameter	Value
a	$42 \times 10^3\ \mathrm{Jm^{-3}\,K^{-1}}$
B	$64 \times 10^4\ \mathrm{Jm^{-3}}$
C	$35 \times 10^4\ \mathrm{Jm^{-3}}$

Consequently the parameters a, B and C can be determined by measuring S_c, T_c and ΔH. As an example their values for MBBA are given in Table 12.1. The value of T_c^* follows from the behaviour of light scattering in the isotropic phase. The nematic phase is the thermodynamically stable state in this region. The $S = S_-$ solution still corresponds to a relative maximum of the free energy density. The height of the barrier is given by $f_n(S_-) - f_i$. At T_c^* the height becomes zero, because $S_- = 0$. This means that the isotropic phase, which can be obtained by undercooling, is metastable in the temperature region between T_c^* and the clearing point T_c.

iv) $T < T_c^*$. The S_+ solution has the lowest free energy density, i.e. the nematic phase is the thermodynamically stable one. Now the S_- solution corresponds to a relative minimum, whereas $S = 0$ gives rise to a relative maximum. The S_- solution describes a state, where the molecules tend to

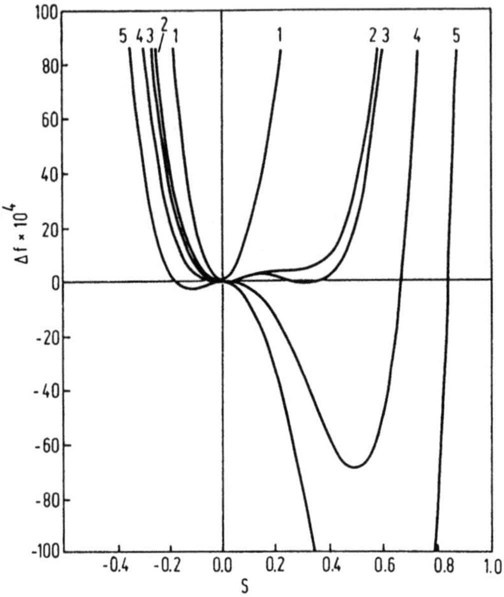

Fig. 12.1. A plot of the difference in free energy density [see (12.13)]

$$\Delta f = \frac{f_n - f_i}{\frac{1}{3}aT_c^*} = \left(\frac{T}{T_c^*} - 1\right)S^2 - \frac{2}{9}S^3 + \frac{1}{3}S^4 \quad ,$$

with $B = C = aT_c^*$, versus S for five different values of the temperature. The curves *1* to *5* refer respectively to the temperatures, $T = 1.2T_c^* > T_c^\dagger$, $T = T_c^\dagger$, $T = T_c$, $T = T_c^*$ and $T = 0.95T_c^*$. Similar plots in the isotropic phase (1), the metastable nematic phase (2), the metastable isotropic phase (3) and the nematic phase (4) lie in regions bounded respectively by curve *2*, curves *2* and *3*, curves *3* and *4*, and curve *4*

orient perpendicular to the uniaxial axis. There is, however, no way of obtaining this metastable state. The height of the energy barrier between the S_- and S_+ solution is given by $f_i - f_n(S_-)$.

In Fig. 12.1 the functional dependence of $\Delta f = f_n - f_i$ on the parameter S is given for five different temperatures, namely $T_1 > T_c^\dagger$, $T_c^\dagger$, T_c, T_c^* and $T_2 < T_c^*$. The height h of the energy barrier at $T = T_c$ between the isotropic state, $S = 0$, and the nematic state, $S = S_c$, is given by

$$h = \frac{B^4}{11664C^3} \quad . \tag{12.23}$$

The four different temperature regions can be clearly distinguished.

The influence of external fields on the physical properties of a given system can be easily taken into account within the framework of the Landau theory. Consider the effect of a magnetic field $\boldsymbol{B}$ on a nematic compound. According to (6.5) the presence of such a field leads to the following extra orientation dependent term in the free energy density

$$f_{\mathrm{M}} = -\tfrac{1}{2}\mu_0^{-1}\Delta\chi_{\alpha\beta}B_\alpha B_\beta \quad . \tag{12.24}$$

The dependence of this free energy density on the tensor order parameter $\tilde{\boldsymbol{Q}}$ is obtained from (5.9) with $T_{\alpha\beta} = \chi_{\alpha\beta}$, i.e.

$$\chi_{\alpha\beta} = \overline{\chi}\delta_{\alpha\beta} + 3\overline{\chi}Q_{\alpha\beta} \quad , \tag{12.25}$$

where $\overline{\chi} = \frac{1}{3}\chi_{\alpha\alpha}$. Assuming that the coupling with the magnetic field does not destroy the uniaxial symmetry of the nematic phase, expression (12.9) for $Q_{\alpha\beta}$ can be used giving

$$\chi_{\alpha\beta} = \overline{\chi}\delta_{\alpha\beta} + \Delta\chi_{\max}S(N_{\alpha\beta} - \tfrac{1}{3}\delta_{\alpha\beta}) \quad , \tag{12.26}$$

where $\Delta\chi_{\max}$ is the anisotropy in the perfectly aligned phase. The sign of $\Delta\chi_{\max}$ is extremely important; it must be positive to stabilize the uniaxial symmetry. A negative sign means that the ordering of the nematic is such that the director is perpendicular to the field. Consequently the field direction introduces a second axis and the phase is biaxial, meaning that the complete Landau theory of biaxial nematics must be used. Therefore, only the case of a positive anisotropy will be considered here. When $\boldsymbol{n}$ and $\boldsymbol{B}$ are parallel, the field gives rise to the following additional free energy density

$$f_{\mathrm{M}} = -\tfrac{1}{3}\mu_0^{-1}\Delta\chi_{\max}SB^2 \quad . \tag{12.27}$$

Clearly the field induces order in the originally disordered high-temperature phase. Mathematically speaking the order in this phase is the result of the term proportional to S. Note that the symmetries of both phases do not differ any more due to the presence of the field. This means that the term proportional to S cannot be excluded on symmetry grounds.

The order induced by a magnetic (electric) field in the originally isotropic liquid, leads to a weak birefringence called the Cotton-Mouton (Kerr) effect. Above T_{c} the order parameter S is small and the free energy density f is given by

$$f = f_{\mathrm{i}} - \tfrac{1}{3}\mu_0^{-1}\Delta\chi_{\max}SB^2 + \tfrac{1}{3}a(T - T_{\mathrm{c}}^*)S^2, \tag{12.28}$$

where the cubic and quartic terms in Eq. (12.13) are neglected because of their small values. Minimization of f with respect to S gives the value of the order parameter

$$S = \frac{\mu_0^{-1} \Delta\chi_{\max} B^2}{2a(T - T_c^*)} \quad . \tag{12.29}$$

According to electrodynamics the birefringence Δn can be expressed by

$$\Delta n = n_{\|} - n_{\perp} = \sqrt{\varepsilon_{\|}} - \sqrt{\varepsilon_{\perp}} \quad , \tag{12.30}$$

where $\varepsilon_{\|}$ and $\varepsilon_{\perp}$ are the elements of the dielectric tensor at optical frequencies parallel and perpendicular to the director. This tensor has the same form as the magnetic susceptibility tensor and is given by

$$\varepsilon_{\alpha\beta} = \overline{\varepsilon}\delta_{\alpha\beta} + \Delta\varepsilon_{\max} S(N_{\alpha\beta} - \tfrac{1}{3}\delta_{\alpha\beta}) \quad , \tag{12.31}$$

where $\overline{\varepsilon} = \frac{1}{3}\varepsilon_{\alpha\alpha}$ and $\Delta\varepsilon_{\max}$ is the anisotropy in the perfectly aligned phase. It directly follows from the parallel alignment of $\boldsymbol{n}$ and $\boldsymbol{B}$ and the small value of the order parameter S that

$$n_{\|} = \sqrt{\varepsilon_{\|}} = \left(\overline{\varepsilon} + \frac{2}{3}\Delta\varepsilon_{\max} S\right)^{1/2} \simeq \sqrt{\overline{\varepsilon}} + \frac{\Delta\varepsilon_{\max} S}{3\,\overline{\varepsilon}^{1/2}} \quad ,$$

$$n_{\perp} = \sqrt{\varepsilon_{\perp}} = \left(\overline{\varepsilon} - \frac{1}{3}\Delta\varepsilon_{\max} S\right)^{1/2} \simeq \sqrt{\overline{\varepsilon}} - \frac{\Delta\varepsilon_{\max} S}{6\,\overline{\varepsilon}^{1/2}} \quad ,$$

i.e. the induced birefringence is

$$\Delta n = \frac{\Delta\varepsilon_{\max} S}{2\,\overline{\varepsilon}^{1/2}} \quad . \tag{12.32}$$

The ratio $\Delta n/(\mu_0^{-1}B^2)$ is called the Cotton-Mouton constant. Substitution of (12.29) in (12.32) expresses this constant in terms of the material parameters and the temperature resulting in

$$\frac{\Delta n}{\mu_0^{-1}B^2} = \frac{\Delta\varepsilon_{\max}\Delta\chi_{\max}}{4a\,\overline{\varepsilon}^{1/2}(T - T_c^*)} \quad . \tag{12.33}$$

Clearly this "constant" diverges at $T = T_c^*$ and decreases according to the power law $(T - T_c^*)^{-1}$ for $T > T_c^*$. This behaviour is shown in Fig. 12.2. The observed temperature dependence near the nematic-isotropic transition deviates slightly from (12.33). This is hardly surprising in view of the molecular field character of the Landau-De Gennes theory. Possible causes for these discrepancies have been discussed in the review of Gramsbergen et al. (1986).

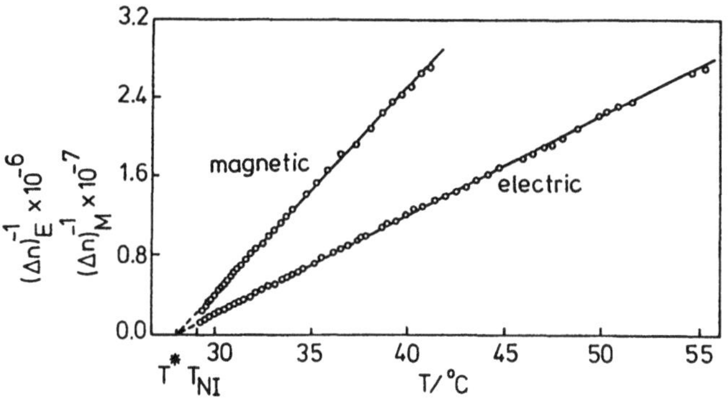

Fig. 12.2. Reciprocal magnetic and electric field induced birefringence as a function of temperature in the isotropic phase of 6CB (Ratna et al. 1973)

12.3 Fluctuations in the Isotropic Phase

A second pretransitional effect in nematics concerns the light scattering in the isotropic phase. Figure 12.3 schematically shows the observed behaviour of the intensity of the scattered light as a function of temperature. A divergence is found at the temperature T_c^*. Because of the first-order nematic-isotropic transition at the temperature $T_c > T_c^*$ this divergence is not fully realized. According to the theory of light scattering (see Sect. 7.3) a proper description of the underlying phenomenon is obtained by calculating the relevant thermal fluctuations in the isotropic phase. These fluctuations are related to the thermal behaviour of the elements of the local order pa-

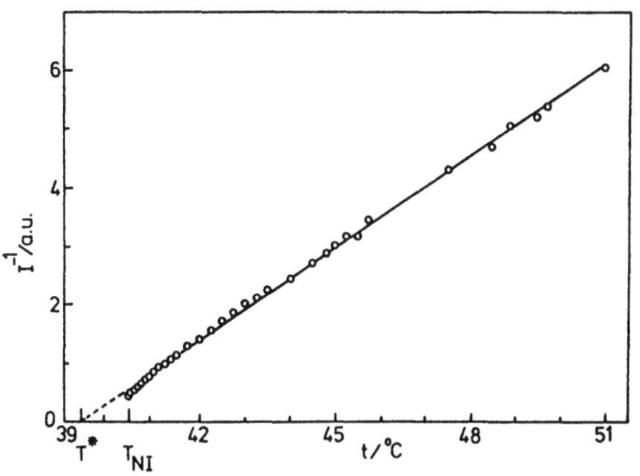

Fig. 12.3. The observed behaviour of the inverse intensity of the scattered light in the isotropic phase as a function of temperature for 8CB (Zink and De Jeu 1985)

rameter and are nothing more than the deviations of the mean value of each of these elements. Clearly the mean values of all elements of the local order parameter tensor $\tilde{\boldsymbol{Q}}(\boldsymbol{r})$ are zero in the isotropic phase. Consequently the elements $Q_{\alpha\beta}(\boldsymbol{r})$ of the local order parameter tensor describe the orientational fluctuations in the isotropic phase. They satisfy the constraint $Q_{\alpha\alpha}(\boldsymbol{r}) = 0$.

The temperature dependent behaviour of the fluctuations can be calculated in an approximate way by once more resorting to the Landau-De Gennes theory. According to this theory the fluctuations can be taken into account by means of an appropriate generalization of the original expression (12.10) for the free energy density of the system. For this purpose the position-dependent order parameter tensor $\tilde{\boldsymbol{Q}}(\boldsymbol{r})$ is used instead of $\tilde{\boldsymbol{Q}}$, and terms are incorporated that describe the variation of $\tilde{\boldsymbol{Q}}(\boldsymbol{r})$ with position. The variation with position is assumed to be small, meaning that it suffices to consider only spatial derivative terms of the lowest order, i.e. $\partial_\alpha Q_{\alpha\gamma}$ and $\partial_\alpha Q_{\beta\gamma}$. The introduction of gradient terms is essential for the calculation of the relevant correlations. In addition only the simplest gradient terms are taken into account. Terms like $Q_{\alpha\beta}(\boldsymbol{r})[\partial_\alpha Q_{\gamma\delta}(\boldsymbol{r})][\partial_\beta Q_{\gamma\delta}(\boldsymbol{r})]$ for example, are neglected for the simple reason that they are hard to deal with. As a result an expression is obtained that can describe the effect of the local fluctuations in the isotropic phase:

$$\begin{aligned} f(\boldsymbol{r}) = f_{\mathrm{i}} &+ \tfrac{1}{2}a(T - T_{\mathrm{c}}^{*})Q_{\alpha\beta}(\boldsymbol{r})Q_{\beta\alpha}(\boldsymbol{r}) - \tfrac{1}{3}BQ_{\alpha\beta}(\boldsymbol{r})Q_{\beta\gamma}(\boldsymbol{r})Q_{\gamma\alpha}(\boldsymbol{r}) \\ &+ \tfrac{1}{4}C[Q_{\alpha\beta}(\boldsymbol{r})Q_{\beta\alpha}(\boldsymbol{r})]^2 + \tfrac{1}{2}L_1[\partial_\alpha Q_{\beta\gamma}(\boldsymbol{r})][\partial_\alpha Q_{\beta\gamma}(\boldsymbol{r})] \\ &+ \tfrac{1}{2}L_2[\partial_\alpha Q_{\alpha\gamma}(\boldsymbol{r})][\partial_\beta Q_{\beta\gamma}(\boldsymbol{r})] \quad , \end{aligned} \tag{12.34}$$

where f_{i} represents the free energy density of the isotropic phase without local fluctuations and L_1 and L_2 are constants that are closely related to the elastic constants. This directly follows from the fact that the order parameter S does not change with an elastic deformation of the phase. Consequently the distortion free energy density up to order S^2 is given by

$$\begin{aligned} f_{\mathrm{d}} &= L_1S^2[\partial_\alpha n_\beta(\boldsymbol{r})][\partial_\alpha n_\beta(\boldsymbol{r})] \\ &\quad + \tfrac{1}{2}L_2S^2\{n_\alpha(\boldsymbol{r})n_\beta(\boldsymbol{r})[\partial_\alpha n_\gamma(\boldsymbol{r})][\partial_\beta n_\gamma(\boldsymbol{r})] + [\partial_\alpha n_\alpha(\boldsymbol{r})]^2\} \\ &= (L_1 + \tfrac{1}{2}L_2)S^2\{[\operatorname{div}\boldsymbol{n}(\boldsymbol{r})]^2 + [\boldsymbol{n}(\boldsymbol{r}) \times \operatorname{curl}\boldsymbol{n}(\boldsymbol{r})]^2\} \\ &\quad + L_1S^2[\boldsymbol{n}(\boldsymbol{r}) \cdot \operatorname{curl}\boldsymbol{n}(\boldsymbol{r})]^2 \quad . \end{aligned} \tag{12.35}$$

Thus, up to order S^2 the following relations hold

$$K_1 = K_3 = 2(L_1 + \tfrac{1}{2}L_2)S^2 \quad , \tag{12.36a}$$

$$K_2 = 2L_1S^2 \quad . \tag{12.36b}$$

As the elastic constants are positive this means that

$$L_1 > 0 \quad , \quad L_1 + \tfrac{1}{2}L_2 > 0 \quad . \tag{12.37}$$

Note that K_1 and K_3 are not equal anymore if a term of the form $Q_{\alpha\beta}(\boldsymbol{r})[\partial_\alpha Q_{\gamma\delta}(\boldsymbol{r})]\,[\partial_\beta Q_{\gamma\delta}(\boldsymbol{r})]$ is introduced because this term gives rise to $2S^3\{[\operatorname{div}\boldsymbol{n}(\boldsymbol{r})]^2 + [\boldsymbol{n}(\boldsymbol{r})\cdot\operatorname{curl}\boldsymbol{n}(\boldsymbol{r})]^2 + 2[\boldsymbol{n}(\boldsymbol{r})\times\operatorname{curl}\boldsymbol{n}(\boldsymbol{r})]^2\}$.

Neglecting the cubic and quartic terms in the elements of the fluctuation tensor it directly follows from (12.34) that local fluctuations in the isotropic phase give rise to a local free energy density of the form

$$\begin{aligned} f(\boldsymbol{r}) = f_{\mathrm{i}} &+ \tfrac{1}{2}a(T-T_{\mathrm{c}}^*)Q_{\alpha\beta}(\boldsymbol{r})Q_{\beta\alpha}(\boldsymbol{r}) + \tfrac{1}{2}L_1[\partial_\alpha Q_{\beta\gamma}(\boldsymbol{r})][\partial_\alpha Q_{\beta\gamma}(\boldsymbol{r})] \\ &+ \tfrac{1}{2}L_2[\partial_\alpha Q_{\alpha\gamma}(\boldsymbol{r})][\partial_\beta Q_{\beta\gamma}(\boldsymbol{r})] \quad . \end{aligned} \tag{12.38}$$

Next the local fluctuations $Q_{\alpha\beta}(\boldsymbol{r})$ are expanded in a Fourier series

$$Q_{\alpha\beta}(\boldsymbol{r}) = \sum_{\boldsymbol{k}} Q_{\alpha\beta}(\boldsymbol{k})\exp(\mathrm{i}\boldsymbol{k}\cdot\boldsymbol{r}) \quad \text{with} \tag{12.39}$$

$$Q_{\alpha\beta}(\boldsymbol{k}) = \frac{1}{\Omega}\int d\boldsymbol{r}\, Q_{\alpha\beta}(\boldsymbol{r})\exp(-\mathrm{i}\boldsymbol{k}\cdot\boldsymbol{r}), \tag{12.40}$$

where Ω represents the volume of the system. As $Q_{\alpha\beta}(\boldsymbol{r})$ is real it follows that

$$Q_{\alpha\beta}^*(\boldsymbol{k}) = Q_{\alpha\beta}(-\boldsymbol{k}) \quad . \tag{12.41}$$

Substitution of (12.39) into (12.38) and integration of the resulting expression over the volume of the system gives the following contribution of the fluctuations to the total free energy

$$\begin{aligned} F_{\mathrm{f}} = \tfrac{1}{2}\Omega a(T-T_{\mathrm{c}}^*)\sum_{\boldsymbol{k}} Q_{\alpha\beta}^*(\boldsymbol{k})Q_{\alpha\beta}(\boldsymbol{k}) &+ \tfrac{1}{2}\Omega L_1 \sum_{\boldsymbol{k}} k^2 Q_{\alpha\beta}^*(\boldsymbol{k})Q_{\alpha\beta}(\boldsymbol{k}) \\ &+ \tfrac{1}{2}\Omega L_2 \sum_{\boldsymbol{k}} k_\alpha k_\beta Q_{\alpha\gamma}^*(\boldsymbol{k})Q_{\beta\gamma}(\boldsymbol{k}) \end{aligned} \tag{12.42}$$

with $k^2 = k_\alpha k_\alpha$. Clearly the value of F_{f} strongly depends on the values of the amplitudes $Q_{\alpha\beta}(\boldsymbol{k})$. Basic to the Landau-De Gennes theory is the postulate that the appearance of a given set of amplitudes $\{Q_{\alpha\beta}(\boldsymbol{k})\}$ is described by a probability distribution of the form

$$P[\{Q_{\alpha\beta}(\boldsymbol{k})\}] = \frac{1}{Z}\exp(-\beta F_{\mathrm{f}}) \quad , \tag{12.43}$$

where $\beta = 1/k_B T$ with k_B Boltzmann's constant and T the temperature. Z is a normalization constant such that the probability distribution function is normalized to one. For $\boldsymbol{k} \neq 0$ the amplitudes $Q_{\alpha\beta}(\boldsymbol{k})$ are complex quantities, i.e.

$$Q_{\alpha\beta}(\boldsymbol{k}) = r_{\alpha\beta}(\boldsymbol{k}) + \mathrm{i} s_{\alpha\beta}(\boldsymbol{k}) \tag{12.44}$$

where $r_{\alpha\beta}(\boldsymbol{k})$ and $s_{\alpha\beta}(\boldsymbol{k})$ are real variables. These variables satisfy the relations

$$r_{\alpha\beta}(\boldsymbol{k}) = r_{\alpha\beta}(-\boldsymbol{k}) \quad , \quad s_{\alpha\beta}(\boldsymbol{k}) = -s_{\alpha\beta}(-\boldsymbol{k}) \tag{12.45}$$

because of condition (12.41) and

$$r_{\alpha\beta}(\boldsymbol{k}) = r_{\beta\alpha}(\boldsymbol{k}) \quad , \quad s_{\alpha\beta}(\boldsymbol{k}) = s_{\beta\alpha}(\boldsymbol{k}) \tag{12.46}$$

because of the symmetry of the fluctuation tensor. So far the discussion of the fluctuations has been completely parallel to the treatment in Sect. 7.2. At this stage a difference arises due to the fact that the trace of the fluctuation tensor $\tilde{\boldsymbol{Q}}(\boldsymbol{k})$ is zero. This leads to

$$r_{\alpha\alpha}(\boldsymbol{k}) = s_{\alpha\alpha}(\boldsymbol{k}) = 0 \quad . \tag{12.47}$$

The analogy with Sect. 7.2 can be maintained by incorporating the constraint (12.47) into the probability distribution by means of the Dirac delta function, represented here as

$$\delta(x) = \lim_{\varepsilon \to 0} (2\pi\varepsilon)^{-1/2} \exp(-x^2/2\varepsilon) \tag{12.48}$$

with $x = r_{\alpha\alpha}(\boldsymbol{k})$ or $x = s_{\alpha\alpha}(\boldsymbol{k})$. In this way the variables $r_{xx}(\boldsymbol{k})$, $r_{yy}(\boldsymbol{k})$ and $r_{zz}(\boldsymbol{k})$ or $s_{xx}(\boldsymbol{k})$, $s_{yy}(\boldsymbol{k})$ and $s_{zz}(\boldsymbol{k})$ can be treated as independent variables and the full symmetry between these variables can be satisfied in an elegant way. For simplicity the elastic constant L_2 is assumed to be zero. The influence of the corresponding term will be discussed briefly at the end of this section. The probability distribution of the really independent variables is now given by

$$\begin{aligned} P = {} & \frac{1}{Z} \exp[-\beta F_{\mathrm{f}}(0)]\, \delta[Q_{\alpha\alpha}(0)] \\ & \times \prod_{\boldsymbol{k}}{}' \exp[-\beta F_{\mathrm{f}}(\boldsymbol{k})]\, \delta[r_{\gamma\gamma}(\boldsymbol{k})]\, \delta[s_{\mu\mu}(\boldsymbol{k})] \end{aligned} \tag{12.49}$$

where $\prod_{\boldsymbol{k}}'$ is a product over all independent variables except the variables labelled by $\boldsymbol{k} = 0$, whereas

$$\begin{aligned} F_{\mathrm{f}}(0) = {} & \tfrac{1}{2}\Omega a(T - T_{\mathrm{c}}^*)[Q_{xx}^2(0) + Q_{yy}^2(0) + Q_{zz}^2(0) \\ & + 2Q_{xy}^2(0) + 2Q_{xz}^2(0) + 2Q_{yz}^2(0)] \end{aligned} \tag{12.50}$$

and

$$F_{\mathrm{f}}(\boldsymbol{k}) = \Omega[a(T-T_{\mathrm{c}}^*) + L_1k^2][r_{xx}^2(\boldsymbol{k}) + r_{yy}^2(\boldsymbol{k}) + r_{zz}^2(\boldsymbol{k}) + 2r_{xy}^2(\boldsymbol{k}) + 2r_{xz}^2(\boldsymbol{k}) + 2r_{yz}^2(\boldsymbol{k}) + s_{xx}^2(\boldsymbol{k}) + s_{yy}^2(\boldsymbol{k}) + s_{zz}^2(\boldsymbol{k}) + 2s_{xy}^2(\boldsymbol{k}) + 2s_{xz}^2(\boldsymbol{k}) + 2s_{yz}^2(\boldsymbol{k})] \quad . \tag{12.51}$$

For reasons of clarity the relevant thermal averages based upon the tensor elements $Q_{\alpha\beta}(0)$ will be calculated first. The probability distribution for these variables is

$$P_0 = \frac{1}{Z_0\sqrt{2\pi\varepsilon}} \exp\{-\beta F_{\mathrm{f}}(0) - \frac{1}{2\varepsilon}[Q_{xx}(0) + Q_{yy}(0) + Q_{zz}(0)]^2\} \quad , \tag{12.52}$$

where Z_0 is such that the probability distribution P_0 is normalized to one. According to a theorem of Landau and Lifshitz [see also (7.31) and (7.32)] the thermal averages $\langle Q_{xx}^2(0)\rangle$, $\langle Q_{yy}^2(0)\rangle$ and $\langle Q_{zz}^2(0)\rangle$ follow directly from the inverse of the matrix

$$A = \begin{pmatrix} \alpha + 1/\varepsilon & 1/\varepsilon & 1/\varepsilon \\ 1/\varepsilon & \alpha + 1/\varepsilon & 1/\varepsilon \\ 1/\varepsilon & 1/\varepsilon & \alpha + 1/\varepsilon \end{pmatrix} \quad \text{with} \tag{12.53}$$

$\alpha = \beta\Omega a(T - T_{\mathrm{c}}^*)$, whereas

$$\langle Q_{xy}^2(0)\rangle = \langle Q_{xz}^2(0)\rangle = \langle Q_{yz}^2(0)\rangle = \frac{1}{2\alpha} \quad . \tag{12.54}$$

It is easily verified that the inverse matrix A^{-1} is given by

$$A^{-1} = \begin{pmatrix} \frac{\alpha\varepsilon+2}{\alpha^2\varepsilon+3\alpha} & \frac{-1}{\alpha^2\varepsilon+3\alpha} & \frac{-1}{\alpha^2\varepsilon+3\alpha} \\ \frac{-1}{\alpha^2\varepsilon+3\alpha} & \frac{\alpha\varepsilon+2}{\alpha^2\varepsilon+3\alpha} & \frac{-1}{\alpha^2\varepsilon+3\alpha} \\ \frac{-1}{\alpha^2\varepsilon+3\alpha} & \frac{-1}{\alpha^2\varepsilon+3\alpha} & \frac{\alpha\varepsilon+2}{\alpha^2\varepsilon+3\alpha} \end{pmatrix} \quad . \tag{12.55}$$

Consequently the relevant thermal averages are

$$\langle Q_{xx}^2(0)\rangle = \langle Q_{yy}^2(0)\rangle = \langle Q_{zz}^2(0)\rangle = \frac{\alpha\varepsilon + 2}{\alpha^2\varepsilon + 3\alpha} \quad , \tag{12.56}$$

or in the limit $\varepsilon \to 0$

$$\langle Q_{xx}^2(0)\rangle = \langle Q_{yy}^2(0)\rangle = \langle Q_{zz}^2(0)\rangle = \frac{4}{3}\langle Q_{xy}^2(0)\rangle = \frac{4}{3}\langle Q_{xz}^2(0)\rangle$$
$$= \frac{4}{3}\langle Q_{yz}^2(0)\rangle = \frac{2k_\mathrm{B}T}{3\Omega a(T-T_\mathrm{c}^*)} \quad . \qquad (12.57)$$

The relevant thermal averages $\langle r_{\alpha\beta}^2(\boldsymbol{k})\rangle$ and $\langle s_{\alpha\beta}^2(\boldsymbol{k})\rangle$ are obtained in a similar way. It follows directly that $\langle r_{\alpha\beta}^2(\boldsymbol{k})\rangle = \langle s_{\alpha\beta}^2(\boldsymbol{k})\rangle$. The thermal averages $\langle r_{xx}^2(\boldsymbol{k})\rangle$, $\langle r_{yy}^2(\boldsymbol{k})\rangle$ and $\langle r_{zz}^2(\boldsymbol{k})\rangle$ follow from the inverse of the matrix

$$\begin{pmatrix} \alpha(k)+\gamma k_x^2+1/\varepsilon & 1/\varepsilon & 1/\varepsilon \\ 1/\varepsilon & \alpha(k)+\gamma k_y^2+1/\varepsilon & 1/\varepsilon \\ 1/\varepsilon & 1/\varepsilon & \alpha(k)+\gamma k_z^2+1/\varepsilon \end{pmatrix} \quad , \qquad (12.58)$$

where

$$\alpha(k) = 2\beta\Omega[a(T-T_\mathrm{c}^*)+L_1k^2] \quad . \qquad (12.59)$$

In the limit $\varepsilon \to 0$ these thermal averages become

$$\langle r_{xx}^2(\boldsymbol{k})\rangle = \langle r_{yy}^2(\boldsymbol{k})\rangle = \langle r_{zz}^2(\boldsymbol{k})\rangle = \frac{2}{3\alpha(k)} \quad . \qquad (12.60)$$

The thermal averages $\langle r_{xy}^2(\boldsymbol{k})\rangle$, $\langle r_{xz}^2(\boldsymbol{k})\rangle$ and $\langle r_{yz}^2(\boldsymbol{k})\rangle$ are obtained in a trivial way, because they follow from a matrix, which is inverse to a diagonal matrix with all diagonal elements equal to $2\alpha(k)$. Consequently it holds that

$$\langle r_{xy}^2(\boldsymbol{k})\rangle = \langle r_{xz}^2(\boldsymbol{k})\rangle = \langle r_{yz}^2(\boldsymbol{k})\rangle = \frac{1}{2\alpha(k)} \quad . \qquad (12.61)$$

Correlation functions of the type $\langle Q_{xx}(0)Q_{xx}(\boldsymbol{R})\rangle$ and $\langle Q_{xy}(0)Q_{xy}(\boldsymbol{R})\rangle$ can now be calculated in a straightforward way. The distance-dependence of these correlation functions is described by the concept of a correlation length. This length can be found without resorting to an explicit calculation of the correlation functions, because this length appears as a pole of the functions $\alpha^{-1}(k)$ in the complex k-plane. This pole is given by

$$\xi = \left[\frac{L_1}{a(T-T_\mathrm{c}^*)}\right]^{1/2} \quad . \qquad (12.62)$$

As a simple example the correlation function $\langle Q_{xx}(0)Q_{xx}(\boldsymbol{R})\rangle$ is calculated here. Using (12.39) and (12.41) it follows that

$$\langle Q_{xx}(0)Q_{xx}(\boldsymbol{R})\rangle = \sum_{\boldsymbol{k}_1\boldsymbol{k}_2} \langle Q^*_{xx}(\boldsymbol{k}_1)Q_{xx}(\boldsymbol{k}_2)\rangle \exp(\mathrm{i}\boldsymbol{k}_2\cdot\boldsymbol{R})$$
$$= \sum_{\boldsymbol{k}} \langle Q^*_{xx}(\boldsymbol{k})Q_{xx}(\boldsymbol{k})\rangle \exp(\mathrm{i}\boldsymbol{k}\cdot\boldsymbol{R}) \quad , \tag{12.63}$$

because $\langle Q^*_{xx}(\boldsymbol{k}_1)Q_{xx}(\boldsymbol{k}_2)\rangle = \delta_{\boldsymbol{k}_1\boldsymbol{k}_2}$ due to the independence of the fluctuations referring to different $\boldsymbol{k}$-modes. Using (12.44), i.e.

$$\langle Q^*_{xx}(\boldsymbol{k})Q_{xx}(\boldsymbol{k})\rangle = \langle r^2_{xx}(\boldsymbol{k})\rangle + \langle s^2_{xx}(\boldsymbol{k})\rangle$$

and (12.61), the following expression is found

$$\langle Q_{xx}(0)Q_{xx}(\boldsymbol{R})\rangle = \frac{2k_\mathrm{B}T}{3\Omega}\sum_{\boldsymbol{k}} \frac{\exp(\mathrm{i}\boldsymbol{k}\cdot\boldsymbol{R})}{a(T-T^*_\mathrm{c})+L_1k^2} \quad . \tag{12.64}$$

Replacing $\sum_{\boldsymbol{k}}$ by $\Omega/(2\pi)^3 \int d\boldsymbol{k}$ and changing over to polar coordinates (12.64) becomes

$$\langle Q_{xx}(0)Q_{xx}(\boldsymbol{R})\rangle = \frac{k_\mathrm{B}T}{6\pi^2}\int_0^\pi \sin\theta\, d\theta \int_0^{k_\mathrm{D}} dk \frac{k^2\exp(\mathrm{i}kR\cos\theta)}{a(T-T^*_\mathrm{c})+L_1k^2} \quad , \tag{12.65}$$

where the cut-off k_D is due to the fact that not all Fourier components are independent [see also (7.41)]. If k_D is replaced by ∞ the following approximate expression results

$$\langle Q_{xx}(0)Q_{xx}(\boldsymbol{R})\rangle = \frac{k_\mathrm{B}T}{6\pi L_1 R}\exp(-R/\xi) \quad , \tag{12.66}$$

where the correlation length ξ is defined in (12.62). The correlation length is a measure for the distance over which the local fluctuations are correlated. This distance is zero at infinitely high temperatures and diverges as $T \to T^*_\mathrm{c}$. Near $T = T^*_\mathrm{c}$ this divergent behaviour of ξ gives rise to the so-called pretransitional phenomena, e.g. the strong increase in the light scattering cross-section. The expression for this cross-section is given in (7.64). It directly follows from this expression that the depolarization ratio, i.e. the ratio between the intensities of the scattered light with polarizations parallel and perpendicular to the incident light at a scattering wave vector $\boldsymbol{q} = 0$, must be

$$\frac{I_{\parallel}(0)}{I_{\perp}(0)} = \frac{\langle Q^2_{xx}(0)\rangle}{\langle Q^2_{xy}(0)\rangle} = \frac{4}{3} \quad . \tag{12.67}$$

The introduction of the L_2 term complicates the calculation considerably; the inverse of a 6×6 matrix must be calculated in order to obtain the thermal averages $\langle r_{\alpha\beta}(\boldsymbol{k}) r_{\gamma\delta}(\boldsymbol{k})\rangle$. The calculation of the correlation functions requires some care (Govers and Vertogen 1984). Three poles are now present in the thermal averages. This means that three correlation lengths exist; they are found to be given by

$$\xi_1 = \left[\frac{L_1}{a(T - T_c^*)}\right]^{1/2} \quad , \quad \xi_2 = \left[\frac{L_1 + \frac{1}{2}L_2}{a(T - T_c^*)}\right]^{1/2} \quad ,$$
$$\xi_3 = \left[\frac{L_1 + \frac{2}{3}L_2}{a(T - T_c^*)}\right]^{1/2} \quad . \tag{12.68}$$

This immediately implies that, due to the positive sign of the correlation length, L_1 and L_2 must satisfy the relation

$$L_1 + \tfrac{2}{3}L_2 > 0 \quad . \tag{12.69}$$

It is obvious that only the largest distance makes sense. This means that ξ_1 is the correlation length, if $L_2 < 0$, whereas ξ_3 is the correlation length if $L_2 > 0$. It should be stressed here that the appearance of only one correlation length does not necessarily mean that the correlation functions $\langle Q_{\alpha\beta}(0) Q_{\gamma\delta}(\boldsymbol{R})\rangle$ are isotropic. In general they depend on the direction of $\boldsymbol{R}$ as well, but this dependence cannot be described in terms of different correlation lengths.

Finally it should be emphasized that the fluctuation theory presented here can hardly be expected to give a quantitatively correct description of the pretransitional phenomena. Because of the rather rough approximations involved, only a qualitative or a semi-quantitative understanding can be hoped for. The molecular-statistical analogue of the fluctuation theory is called the random phase approximation (Vertogen and Van der Meer 1979a).

12.4 The Smectic-Nematic Transition

The smectic phase, already extensively discussed in Chap. 3, distinguishes itself from the nematic phase by its layered structure. This means that additional order parameters must be introduced in order to describe the thermodynamic behaviour of this state. Here only the "simplest" smectic-nematic transition is considered, namely the transition from the smectic A ($\mathrm{S_A}$) to the nematic (N) phase. Naturally the Landau-De Gennes formalism is perfectly able to describe more complicated transitions like the

S_C (smectic C)-N transition. Such a description involves, compared to the S_AN transition, an additional order parameter that takes the biaxiality of the S_C phase into account. Also transitions from a given smectic phase to another smectic phase, e.g. the S_CS_A transition, can be dealt with without conceptual difficulties. A full treatment of all these transitions, however, lies outside the scope of this book. Moreover, the descriptions of the nematic-isotropic and the S_AN transition are more than sufficient to demonstrate this phenomenological approach.

The S_A phase can be thought of as consisting of so-called smectic layers or planes. The molecules are free to move within the layers, i.e. a smectic layer can roughly be considered to be a two-dimensional liquid. The molecules are on average aligned with their axis normal to the layers. Consequently the additional order parameter which distinguishes the smectic state from the nematic or isotropic state, is the deviation of the centre of mass density from uniform behaviour. Defining $\varrho(\boldsymbol{r})$ and ϱ_0, the centre of mass densities of the smectic and uniform phase respectively, and assuming that the average mass densities of both phases are equal, the relevant order parameter is found to be

$$\delta\varrho(\boldsymbol{r}) = \varrho(\boldsymbol{r}) - \varrho_0 \quad . \tag{12.70}$$

In the S_A phase this density is periodic perpendicular to the smectic planes and uniform parallel to the planes. If the direction perpendicular to the smectic planes is called the z direction the position-dependent order parameter $\delta\varrho(\boldsymbol{r})$ can be expanded in the following Fourier series

$$\delta\varrho(\boldsymbol{r}) = \sum_{n>0} \varrho_n \cos(2\pi nz/d + \phi) \quad , \tag{12.71}$$

where d is the interplanar spacing and ϕ an arbitrary phase angle. In the following this angle is chosen to be zero, i.e. the origin of the coordinate system is situated in a smectic plane. The Fourier expansion boils down to replacing a position-dependent order parameter $\delta\varrho(\boldsymbol{r})$ by an infinite set of order parameters $\{\varrho_n\}$. These are the amplitudes of the components $\cos(2\pi nz/d)$. According to the Landau-De Gennes theory, the average free energy density of the system may be postulated to be given up to fourth order in $\delta\varrho(\boldsymbol{r})$, by

$$\begin{aligned} f = f_0 &+ \tfrac{1}{2}\int A(z-z_1)\,\delta\varrho(z)\,\delta\varrho(z_1)\,dz\,dz_1 \\ &- \tfrac{1}{3}\int B(z-z_1,z-z_2)\,\delta\varrho(z)\,\delta\varrho(z_1)\,\delta\varrho(z_2)\,dz\,dz_1dz_2 \\ &+ \tfrac{1}{4}\int C(z-z_1,z-z_2,z-z_3)\,\delta\varrho(z)\,\delta\varrho(z_1)\,\delta\varrho(z_2)\,\delta\varrho(z_3) \\ &\times dz\,dz_1dz_2dz_3 \quad , \end{aligned} \tag{12.72}$$

where the functions $A(z-z_1)$, $B(z-z_1, z-z_2)$ and $C(z-z_1, z-z_2, z-z_3)$ describe the smectic phase, and f_0 represents the free energy density of the uniform phase.

Substitution of the Fourier series (12.71) into (12.72) yields the expression of the free energy density in terms of the infinite set of order parameters $\{\varrho_n\}$. Clearly there is no need to write down the resulting expression of the free energy density, because this expression is extremely difficult to handle. Further approximations must be introduced in order to continue the discussion.

The simplest approach to this transition is to consider only one Fourier component, i.e. the position-dependent order parameter is approximated by

$$\delta\varrho(z) = \varrho_1 \cos(2\pi z/d) \tag{12.73}$$

and to assume that the orientational order is perfect in the smectic phase and does not change at the transition. Substitution of (12.73) into (12.72) gives an average free energy density of the form

$$f = f_0 + \tfrac{1}{2}A_1\varrho_1^2 + \tfrac{1}{4}C_1\varrho_1^4 \quad , \quad \text{where} \tag{12.74}$$

$$A_1 = \int A(z-z_1) \cos\left(\frac{2\pi}{d}z\right) \cos\left(\frac{2\pi}{d}z_1\right) dz\, dz_1 \quad ,$$

$$B_1 = \int B(z-z_1, z-z_2) \cos\left(\frac{2\pi}{d}z\right) \cos\left(\frac{2\pi}{d}z_1\right) \cos\left(\frac{2\pi}{d}z_2\right) \times dz\, dz_1 dz_2 = 0 \quad ,$$

$$C_1 = \int C(z-z_1, z-z_2, z-z_3) \times \cos\left(\frac{2\pi}{d}z\right) \cos\left(\frac{2\pi}{d}z_1\right) \cos\left(\frac{2\pi}{d}z_2\right) \cos\left(\frac{2\pi}{d}z_3\right) \times dz\, dz_1 dz_2 dz_3 \quad .$$

The coefficient B_1 is zero, because the function $B(z-z_1, z-z_2)$, as well as $A(z-z_1)$ and $C(z-z_1, z-z_2, z-z_3)$, is symmetric in its variables. The phase transition is described by assuming that A_1 depends on the temperature T in a linear way

$$A_1 = a_1(T - T_1) \quad , \quad (a_1 > 0) \tag{12.75}$$

whereas the temperature dependence of C_1 is neglected. It is assumed here that $C_1 > 0$, i.e. a second-order phase transition occurs. If $C_1 < 0$ a term of the sixth power in ϱ_1 must be added in order to ensure stability and the phase transition becomes a first-order transition. The temperature dependence of ϱ_1 is obtained by minimizing expression (12.74) with respect to ϱ_1. It follows directly from $\partial f/\partial \varrho_1 = 0$ that

$$\varrho_1[a_1(T-T_1)+C_1\varrho_1^2]=0 \quad . \tag{12.76}$$

The solutions of this equation are

$$\varrho_1 = 0 \quad , \quad \text{the uniform phase} \quad , \tag{12.77a}$$

$$\varrho_1 = \pm\left[\frac{a_1(T_1-T)}{C_1}\right]^{1/2} \quad . \tag{12.77b}$$

Clearly the phase transition is a second-order one at a transition temperature T_1.

In order to obtain a first-order transition two possibilities exist in addition to the one already mentioned. The first possibility is to incorporate higher order Fourier components in the description while still assuming perfect orientational order both in the smectic phase and in the uniform state just above the transition (Meyer and Lubensky 1976). The second possibility consists of taking the effect of the orientational order into account, while still approximating the position-dependent order parameter by its first Fourier component (De Gennes 1972, 1973; McMillan 1972). Of course both possibilities can also be considered simultaneously giving rise to a very flexible description.

The effect of the incorporation of higher order Fourier components into the description is most easily seen by considering the influence of the second Fourier component, i.e. the position-dependent order parameter is approximated by

$$\delta\varrho(z) = \varrho_1 \cos(2\pi z/d) + \varrho_2 \cos(4\pi z/d) \quad . \tag{12.78}$$

Substitution of (12.78) into (12.72) gives the following average free energy density

$$\begin{aligned} f = f_0 &+ \tfrac{1}{2}A_1\varrho_1^2 + \tfrac{1}{2}A_2\varrho_2^2 - B\varrho_1^2\varrho_2 \\ &+ \tfrac{1}{4}C_1\varrho_1^4 + \tfrac{1}{4}C_2\varrho_2^4 + C_{12}\varrho_1^2\varrho_2^2 \quad , \end{aligned} \tag{12.79}$$

where the coefficients can be expressed in terms of the functions $A(z-z_1)$, $B(z-z_1, z-z_2)$ and $C(z-z_1, z-z_2, z-z_3)$. It is not necessary, however, to make these results explicit. In order to understand the change from a second-order to a first-order transition it suffices to observe that the introduction of the second Fourier component gives rise to a free energy density as given in (12.79). The phase transition is described by assuming that A_2 also depends on the temperature T in a linear way, i.e.

$$A_2 = a_2(T - T_2) \quad , \quad (a_2 > 0) \tag{12.80}$$

whereas the remaining coefficients hardly depend on the temperature. The order parameter ϱ_1 is still thought to be the dominant one, i.e. $T_1 > T_2$. The order parameters ϱ_1 and ϱ_2 are obtained by minimizing f with respect to them. In order to understand the influence of ϱ_2 on the behaviour of the transition, the order parameter ϱ_2 is expressed in terms of ϱ_1 and the resulting expression for the free energy density is compared with the original one (12.74) without the ϱ_2 term. It follows directly from $\partial f/\partial \varrho_2 = 0$ that

$$A_2\varrho_2 + C_2\varrho_2^3 - B\varrho_1^2 + 2C_{12}\varrho_1^2\varrho_2 = 0 \quad . \tag{12.81}$$

Using (12.81) ϱ_2 can be expressed in terms of ϱ_1. Assuming that ϱ_1 remains small near the phase transition, the influence of the second and fourth term can be neglected and ϱ_2 is approximated by

$$\varrho_2 = \frac{B}{A_2}\varrho_1^2 \quad , \tag{12.82}$$

i.e. the sign of ϱ_2 near the transition is determined by the sign of B. Substitution of (12.82) into (12.79) yields

$$f = f_0 + \frac{1}{2}A_1\varrho_1^2 + \frac{1}{4}\left(C_1 - \frac{2B^2}{A_2}\right)\varrho_1^4 + \frac{1}{6}D_1\varrho_1^6 \quad , \tag{12.83}$$

where the term $\frac{1}{6}D\varrho_1^6$ is added for reasons of stability. Now a situation arises, where a first-order transition is found if

$$C_1 - \frac{2B^2}{A_2} < 0 \quad , \tag{12.84}$$

whereas the transition is second-order otherwise. The transition is called tricritical on the border of a first- and second-order phase transition, i.e. for $C_1 = 2B^2/A_2$. Clearly the transition temperature of a first-order phase transition is higher than T_1.

The incorporation of the effect of the orientational order parameter proceeds in the following way. The free energy density of the uniform nematic phase, f_0, determines the value of the orientational order parameter $S_0(T)$ by the requirements

$$\left.\frac{\partial f}{\partial S}\right|_{S_0} = 0 \quad ; \quad \left.\frac{\partial^2 f}{\partial S^2}\right|_{S_0} = A_2 > 0 \quad . \tag{12.85}$$

In the smectic phase the tendency of the molecules to order in the smectic planes has an additional effect on the orientational order parameter. Calling the orientational order parameter in the smectic phase $S(T)$, the influence of the smectic layering on the orientational order parameter $S(T)$ can be described by an additional term in the free energy density, namely

$$-B\varrho_1^2(S - S_0) \quad , \quad (B > 0) \quad , \tag{12.86}$$

the implicit assumption being that ϱ_1 is small near the phase transition. Note the analogy between $(S - S_0)$ and ϱ_2. Substitution of S in the expression of the free energy density of a nematic phase gives

$$\begin{aligned} f(S) &= f_0 + \left.\frac{\partial f_0}{\partial S}\right|_{S_0}(S - S_0) + \frac{1}{2}\left.\frac{\partial^2 f_0}{\partial S^2}\right|_{S_0}(S - S_0)^2 \\ &= f_0(S_0) + \tfrac{1}{2}A_2(S - S_0)^2 \quad . \end{aligned} \tag{12.87}$$

Consequently the average free energy density is now given by

$$f = f_0(S_0) + \tfrac{1}{2}A_1\varrho_1^2 + \tfrac{1}{2}A_2(S - S_0)^2 - B\varrho_1^2(S - S_0) + \tfrac{1}{4}C_1\varrho_1^4 \quad . \tag{12.88}$$

In order to study the orienting effect of the smectic state, i.e. $S - S_0$, on the transition, the difference $S - S_0$ is expressed in term of ϱ_1 and the resulting expression is compared with the original one (12.74). It follows from $\partial f/\partial S = 0$ that

$$S - S_0 = \frac{B}{A_2}\varrho_1^2 \quad . \tag{12.89}$$

Substitution of (12.89) into (12.88) gives

$$f = f_0(S_0) + \frac{1}{2}A_1\varrho_1^2 + \frac{1}{4}\left(C_1 - \frac{2B^2}{A_2}\right)\varrho_1^4 + \frac{1}{6}D_1\varrho_1^6 \quad , \tag{12.90}$$

where the sixth-order term is added in order to ensure stability. Clearly the order of the transition depends critically on the sign of $C_1 - 2B^2/A_2$.

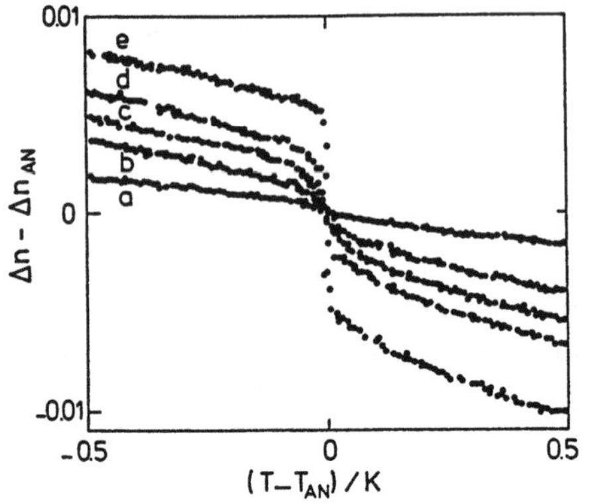

Fig. 12.4. Birefringence of mixtures of two members of the series of p, p'-dialkylazoxybenzenes (nAB, see also Fig. 6.4) around the S_AN phase transition. Different curves correspond to different mole fractions x of 8AB in 7AB: (a) $x = 0$, (b) $x = 0.345$, (c) $x = 0.464$, (d) $x = 0.573$, (e) $x = 0.689$. The tricritical point is approximately reached for case (c) with $T_{AN}/T_{NI} = 0.976$

To conclude, the effect of a coupling between ϱ_1 and other order parameters like ϱ_2 and $S - S_0$ increases the tendency for the S_AN transition to become first order. In fact this feature is quite general, as follows from the present analysis which stresses the analogy between the effects of ϱ_2 and $S - S_0$. Experimentally the situation is such that most S_AN transitions are second-order. For a few cases the existence of a tricritical point has been demonstrated (Ocko et al. 1984, Thoen et al. 1984). For the nCB series this point is found around $T_{AN}/T_{NI} \approx 0.99$, above which value the transition is first order. The occurrence of a tricritical point so close to the NI phase transition (the region where S varies strongly) indicates that from the two mechanisms mentioned the coupling between the smectic order parameter and S is probably the most relevant one. This is illustrated for another series in Fig. 12.4, where the change from a second-order to a first-order S_AN phase transition is reflected in the variation in the birefringence at T_{AN}. As discussed in Sect. 10.1 the latter quantity is proportional to S, so that these results give a direct verification of the coupling between ϱ_1 and S.

13. Molecular Statistical Theory of the Nematic Phase

The counterpart of the phenomenological theory of liquid crystals is the molecular-statistical description. In this chapter we discuss the molecular-statistical theory of nematic and chiral nematic liquid crystals. For reasons of clarity we reformulate the existing theories in such a way that they become accessible to simple analytical treatment. Thus we hope to ensure that the reader may gain clear insight into the status of the present theories with a minimum of effort. Because of the qualitative character of all these theories we do not feel that we jeopardize the real meaning of them by this simplified treatment. First of all we deal briefly with the theory of simple liquids. Then we generalize this theory to the case of anisotropic fluids and we derive an equation of state for nematics. The assumptions involved in this derivation are emphasized in order to avoid the danger that they are obscured by the mathematics. Starting from this equation of state for nematics two well-known approaches are discussed. The first approach is due to Onsager and ascribes the origin of nematic ordering to the anisotropic shape of the molecules, i.e. to repulsive interactions. The second approach, which was formulated by Maier and Saupe, states that the nematic ordering essentially originates from the anisotropic attractive interactions. Next a theory of the chiral nematic phase is presented starting from a model consisting of a Maier-Saupe type interaction and a twist interaction. Finally we give a brief discussion of the relevance of molecular models.

13.1 Introduction to the Theory of Simple Liquids

The molecular-statistical theory of liquids and liquid crystals sets out to understand the physical behaviour of these materials in terms of their components, the molecules. A prerequisite for a satisfactory molecular theory is accordingly a knowledge of the intermolecular interactions. Such a knowledge, however, is almost entirely lacking. Fortunately this does not necessarily mean that a molecular approach is out of the question. Model potentials may be introduced in order to compensate for this ignorance. This means that the most relevant characteristics of the molecules and their mutual interactions are represented in terms of simple models. In this way a qualitative understanding of the physical behaviour of liquids and possibly liquid crystals can be hoped for from a molecular point of view.

The molecular-statistical calculations themselves are prohibitively difficult to perform analytically, even for models based upon intermolecular interactions of a simple form. Consequently approximation methods have to be used. In order to obtain analytical expressions for the relevant physical quantities as far as is possible, the so-called spherical version of the models will be considered and most of the calculations will be performed in the mean field approximation. The spherical version is, to put it briefly, the replacement of a unit vector by a vector of variable length, but such that its thermally averaged length is always equal to one. The mean field approximation replaces the system of interacting molecules by a system of free molecules interacting with a field that is generated by the other molecules. This so-called mean or molecular field must then be determined self-consistently. Any unnecessary mathematical complications will be avoided in the following. As the theory itself is such that only qualitative results may be expected, mathematical effort spent in obtaining better approximations of simple models usually does not contribute to a better quantitative understanding of the behaviour of liquid crystals.

A molecular-statistical theory of the nematic phase and the nematic-isotropic transition must allow for the fact that nematics are fluids. For this reason the molecular-statistical theory of simple fluids is considered here first of all (Barker and Henderson 1976). Simple fluids are described satisfactorily by an equation of state of the following form

$$p = p_{\mathrm{HC}} - \tfrac{1}{2} J_0 \varrho^2 \quad , \tag{13.1}$$

where p represents the pressure, p_{HC} the so-called hard-core pressure, and $\varrho = N/\Omega$ the density of the system with N and Ω being the total number of molecules and the volume of the system, respectively. The constant J_0 is related to the attractive part of the inter-particle potential. The pressure consists of two terms, a positive and a negative contribution. Such an equation of state is the direct result of separating the interaction between the particles into a repulsive and an attractive part. In order to calculate the terms in (13.1) the free energy must be known. The free energy of the system, in turn, can be derived from the partition function, which is based on the kinetic and potential energy of the molecules. In the following the repulsive part of the potential between the particles i and j will be represented by $u(r_{ij})$, and the attractive part by $-v(r_{ij})$, where r_{ij} is the distance between the centres of mass of the particles i and j. The partition function of a system of N spherically symmetric particles of mass m is (Landau and Lifshitz 1980)

$$Z = \frac{\lambda^{-3N}}{N!} \int d\boldsymbol{r}_1 \ldots d\boldsymbol{r}_N \prod_{i<j} e_{ij} \exp\left[\frac{1}{2}\beta \sum_{k \neq l} v(r_{kl})\right] \quad , \tag{13.2}$$

with $\lambda = h/(2\pi m k_B T)^{1/2}$, h being Planck's constant, $\beta = 1/k_B T$ and

$$e_{ij} = \exp[-\beta u(r_{ij})] \quad .$$

The integral in (13.2) is called the configurational integral. Any rotation of the particles is neglected. The separation into a repulsive and an attractive part is brought about by rewriting (13.2) in the form

$$Z = Z_0 \left\langle \exp\left[\frac{1}{2}\beta \sum_{k \neq l} v(r_{kl})\right] \right\rangle_0 \quad , \tag{13.3}$$

where Z_0 is the partition function of the reference system solely consisting of particles interacting through the repulsive potential $u(r_{ij})$, i.e.

$$Z_0 = \frac{\lambda^{-3N}}{N!} \int d\boldsymbol{r}_1 \ldots d\boldsymbol{r}_N \prod_{i<j} e_{ij} \quad , \quad \text{and} \tag{13.4}$$

$$\left\langle \exp\left[\frac{1}{2}\beta \sum_{k \neq l} v(r_{kl})\right] \right\rangle_0 = \frac{\int d\boldsymbol{r}_1 \ldots d\boldsymbol{r}_N \prod_{i<j} e_{ij} \exp[\frac{1}{2}\beta \sum_{k \neq l} v(r_{kl})]}{\int d\boldsymbol{r}_1 \ldots d\boldsymbol{r}_N \prod_{i<j} e_{ij}} \quad , \tag{13.5}$$

i.e. the thermal average of

$$\exp\left[\frac{1}{2}\beta \sum_{k \neq l} v(r_{kl})\right]$$

is taken with respect to the repulsive reference system. Obviously the calculation breaks up into two parts, namely, (1) the partition function (13.4) and (2) the thermal average (13.5).

(1) *The Partition Function.* In general the reference system is chosen to be a collection of hard spheres, i.e. the repulsive part $u(r_{ij})$ of the potential is approximated by

$$\begin{aligned} u(r_{ij}) &= \infty \,, \quad \text{i.e.} \quad e_{ij} = 0 \quad , \quad r_{ij} \leq 2r_0 \\ u(r_{ij}) &= 0 \quad , \quad \text{i.e.} \quad e_{ij} = 1 \quad , \quad r_{ij} > 2r_0 \quad , \end{aligned} \tag{13.6}$$

where r_0 is the radius of the hard sphere. The calculation of the partition function Z_0 is a formidable problem and one must resort to approximation techniques.

The simplest approximation, which gives a qualitatively correct description of the properties of a hard-sphere system, is the so-called Van der Waals approximation. The basic idea of this approximation is the following. The calculation of the configuration integral (13.4) is trivial in the case of an ideal gas, i.e. $r_0 = 0$ (point particles). The partition function is then given by

$$Z_0 = \frac{\lambda^{-3N}}{N!} \int d\boldsymbol{r}_1 \dots d\boldsymbol{r}_N = \frac{\lambda^{-3N} \Omega^N}{N!} \quad , \tag{13.7}$$

i.e. each particle separately gives rise to a factor Ω. Next the finite volume $v_0 = \frac{4}{3}\pi r_0^3$ of each of the spheres is taken into account. This means that the accessible volume for the centre of mass of a sphere becomes less than the total volume Ω. Consider the two hard spheres as represented in Fig. 13.1.

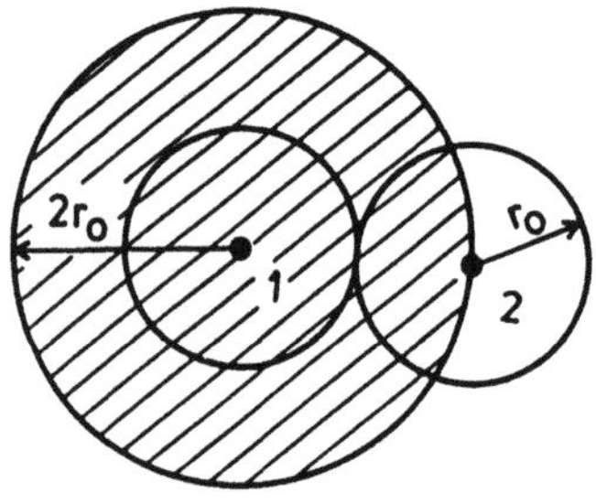

Fig. 13.1. Touching hard spheres. The shaded area is the excluded volume v_{excl} for the second sphere

The mere presence of the first sphere diminishes the available volume for the centre of mass of the second sphere with an amount $v_{\text{excl}} = 8v_0$, where v_{excl} is called the excluded volume. This means that, on average, the centre of mass of each of both spheres has an available volume $\Omega - 4v_0$ instead of Ω (being the ideal gas situation). According to the Van der Waals approximation N spheres give rise to an accessible volume $\Omega - 4Nv_0$ for the centre of mass of each sphere separately, i.e. N spheres give rise to an excluded volume $4Nv_0$. Consequently the partition function is approximated by

$$Z_0 = \frac{\lambda^{-3N} \Omega^N (1 - 4\eta)^N}{N!} \quad , \tag{13.8}$$

where $\eta = \varrho v_0$ is the packing fraction. The partition function (13.8) gives rise to the Helmholtz free energy

$$F = -k_{\mathrm{B}}T \ln Z_0 = Nk_{\mathrm{B}}T[3\ln\lambda + \ln\varrho - 1 - \ln(1-4\eta)] \quad , \tag{13.9}$$

and a hard-core pressure

$$p_{\mathrm{HC}} = -\frac{\partial F}{\partial \Omega} = \frac{\varrho k_{\mathrm{B}}T}{1-4\eta} \quad . \tag{13.10}$$

Clearly this expression for the hard-core pressure can be only qualitatively correct, because the term already diverges for $\eta = 1/4$, i.e. for a packing fraction, that is much lower than is found experimentally. Further insight into the Van der Waals approximation can be obtained by comparing the low density expansion of (13.10) with the exact low density or virial expansion of a system of hard spheres, which is well-known. The expansion of (13.10) gives

$$p_{\mathrm{HC}} = \varrho k_{\mathrm{B}}T(1 + 4\eta + 16\eta^2 + 64\eta^3 + 256\eta^4 + 1024\eta^5 + \dots) \quad , \tag{13.11}$$

whereas the exact virial expansion of a system of hard spheres is

$$p_{\mathrm{HC}} = \varrho k_{\mathrm{B}}T(1 + 4\eta + 10\eta^2 + 18.36\eta^3 + 28.2\eta^4 + 40\eta^5 + \dots) \quad . \tag{13.12}$$

A comparison of these two expansions clearly shows that the Van der Waals approximation overestimates the pressure considerably except for very low packing fractions, where both expansions yield the same result.

The entropy Σ of the system, in the Van der Waals approximation, is given by

$$\Sigma = -\frac{\partial F}{\partial T} = Nk_{\mathrm{B}} \ln(1-4\eta) + \Sigma_0 \quad , \tag{13.13}$$

where Σ_0 represents the entropy of the ideal gas. It immediately follows that the effect of the finite volume of the spheres, manifesting itself in the excluded volume per sphere 4η, is to reduce the entropy. This means that, as far as entropy is concerned, the system tries to reduce the excluded volume per sphere. It is obvious that such a reduction is impossible in the present case of spheres. However, the situation is different for non-spherical objects, e.g. rods, where this tendency plays a prominent role because of the orientational dependence of the excluded volume.

The system of hard spheres has been extensively studied using computer methods. The numerical results are described quite well by the following equation of state (Carnahan and Starling, 1969)

$$p_{\mathrm{HC}} = \varrho k_{\mathrm{B}}T \frac{1 + \eta + \eta^2 + \eta^3}{(1-\eta)^3} \quad . \tag{13.14}$$

This follows by comparing the exact virial expansion (13.12) with the low density expansion of (13.14) which is given by

$$p_{\mathrm{HC}} = \varrho k_{\mathrm{B}} T (1 + 4\eta + 10\eta^2 + 18\eta^3 + 28\eta^4 + 40\eta^5 \ldots) \quad . \tag{13.15}$$

On the other hand it is obvious that this phenomenological equation of state has its limitations as well. The hard-core pressure diverges for $\eta \to 1$, which is in contradiction with the fact that hard spheres cannot fill space completely ($\eta < 3/4$ for hard spheres). The Helmholtz free energy is now given by

$$F = N k_{\mathrm{B}} T \left[3 \ln \lambda + \ln \varrho + \frac{3 - 2\eta}{(1-\eta)^2} - 4 \right] \quad , \tag{13.16}$$

and the entropy is, compared to the ideal gas situation, diminished by an amount

$$\Sigma - \Sigma_0 = -N k_{\mathrm{B}} \frac{\eta(4 - 3\eta)}{(1-\eta)^2} \quad . \tag{13.17}$$

It directly follows from (13.17) that the Van der Waals approximation overestimates the entropy reduction.

(2) *The Thermal Average.* The quantity (13.5) is also extremely hard to calculate, and thus some kind of an approximation must be made. A procedure used quite often is the following decoupling:

$$\left\langle \exp\left[\frac{1}{2} \beta \sum_{k \neq l} v(r_{kl}) \right] \right\rangle_0 \simeq \exp\left[\frac{1}{2} \beta \left\langle \sum_{k \neq l} v(r_{kl}) \right\rangle_0 \right] \quad . \tag{13.18}$$

It further holds that

$$\begin{aligned} \left\langle \sum_{k \neq l} v(r_{kl}) \right\rangle_0 &= \sum_{k \neq l} \frac{\int d\boldsymbol{r}_1 \ldots d\boldsymbol{r}_N v(r_{kl}) \prod_{i>j} e_{ij}}{\int d\boldsymbol{r}_1 \ldots d\boldsymbol{r}_N \prod_{i>j} e_{ij}} \\ &= N \varrho \int d\boldsymbol{r}\, v(r) g_{\mathrm{HC}}(r) \quad . \end{aligned} \tag{13.19}$$

Here $g_{\mathrm{HC}}(r)$ is the hard-sphere radial distribution function, which is defined by

$$g_{\mathrm{HC}}(\boldsymbol{r}_1 - \boldsymbol{r}_2) = \frac{\Omega^2 \int d\boldsymbol{r}_3 \ldots d\boldsymbol{r}_N \prod_{i>j} e_{ij}}{\int d\boldsymbol{r}_1 \ldots d\boldsymbol{r}_N \prod_{i>j} e_{ij}} \quad . \tag{13.20}$$

Thus the relevant thermal average is approximated by

$$\left\langle \exp\left[\frac{1}{2}\beta\sum_{k\neq l} v(r_{kl})\right]\right\rangle_0 = \exp\left[\frac{1}{2}\beta N\varrho J_0\right] \quad , \tag{13.21}$$

where the constant J_0 is directly related to the attractive part of the interparticle potential, namely

$$J_0 = \int d\boldsymbol{r}\, v(r) g_{\mathrm{HC}}(r) \quad . \tag{13.22}$$

(3) *The Equation of State.* The calculation of the equation of state (13.1) starting from the partition function (13.3) now proceeds in the following way. The free energy of the system is

$$\begin{aligned} F &= -k_{\mathrm{B}}T\ln Z \\ &= -k_{\mathrm{B}}T\ln Z_0 - k_{\mathrm{B}}T\ln\left\langle \exp\left[\frac{1}{2}\beta\sum_{k\neq l} v(r_{kl})\right]\right\rangle_0 \quad , \end{aligned} \tag{13.23}$$

or, using the necessary approximations, i.e. (13.9) and (13.21),

$$F = Nk_{\mathrm{B}}T[3\ln\lambda + \ln\varrho - 1 - \ln(1-4\eta)] - \tfrac{1}{2}N\varrho J_0 \quad . \tag{13.24}$$

Consequently the equation of state now becomes

$$p = \frac{\varrho k_{\mathrm{B}}T}{1-4\eta} - \frac{1}{2}J_0\varrho^2 , \tag{13.25}$$

which is the well-known equation of state of Van der Waals. As remarked previously, this equation of state only gives a qualitative description of simple fluids. A satisfactory description is obtained if the hard-core pressure of Van der Waals (13.10) is replaced by the phenomenological hard-core pressure of Carnahan and Starling (13.14).

13.2 The Equation of State of Nematics

In order to generalize the description of simple liquids to nematics and to deal with the nematic-isotropic transition which occurs in the fluid phase, the non-spherical character of the molecules must be taken into account. In view of the mathematical difficulties already outlined, it is very hard to arrive at a reliable equation of state for a simple model of nematics starting from first principles. Therefore only a qualitatively correct equation of state is discussed here.

Three lines of approach exist in order to incorporate the effect of the non-spherical shape of the molecules and the consequent orientation-dependent intermolecular interaction into the equation of state. The first approach is due to Onsager (1949). It incorporates the relevant effect only into the repulsive part of the interaction, i.e. the shape manifests itself in the hard-core pressure. According to the second approach, which was suggested by Maier and Saupe (1959, 1960), the shape manifests itself in the attractive part of the interaction and thus contributes to the negative part of the pressure. Finally the third approach consists of taking the effect of the shape into account in both the repulsive and attractive part of the orientation-dependent intermolecular interaction, meaning that both pressure terms are affected. Here the third approach will be followed and the Onsager and Maier-Saupe approach will be dealt with as limiting cases of the general approach (Cotter 1977; Ypma and Vertogen 1978).

The desired equation of state for nematics must be based on a model system, because the intermolecular interactions are largely unknown. For this reason the intermolecular interaction again consists of an anisotropic repulsive interaction and an anisotropic attractive interaction. Usually the anisotropic repulsive intermolecular interaction is approximated by the steric hindering between hard rods. In analogy with the successful description of simple liquids the hard-rod model is now chosen as the reference system for the calculation of the relevant thermal averages. Generally the attractive part of the interaction between the molecules i and j depends both on the mutual distance vector $\boldsymbol{r}_{ij}$ between their centres of mass and on their orientation. For the sake of simplicity the molecules are assumed to be cylindrically symmetric with respect to their long molecular axis, with an orientation determined by the unit vector $\boldsymbol{a}$. Consequently the attractive part of the intermolecular interaction between the molecules i and j can be written as $-v(\boldsymbol{r}_{ij}, \boldsymbol{a}_i, \boldsymbol{a}_j)$. In complete analogy to the theory of simple liquids, in particular the expressions (13.3) and (13.5), the partition function of the present model system is given by

$$Z = Z_{\mathrm{HR}} \left\langle \exp\left[\frac{1}{2}\beta \sum_{k \neq l} v(\boldsymbol{r}_{kl}, \boldsymbol{a}_k, \boldsymbol{a}_l)\right]\right\rangle_0 \quad , \tag{13.26}$$

where Z_{HR} denotes the partition function of the hard-rod model and the thermal average is taken with respect to the hard-rod system. Thus the calculation of the equation of state for the present model separates into (1) the calculation of the partition function of the hard-rod model and (2) the calculation of the thermal average for a given attractive intermolecular interaction.

(1) *The Partition Function of the Hard-Rod Model.* The hard rods commonly used have the shape of a spherocylinder, a cylinder or an ellipsoid. The original approach of Onsager (1949) starts from hard rods that have the shape of a spherocylinder, being a cylinder of length L and diameter D capped at both ends with hemispheres of the same diameter. Clearly the system of hard spherocylinders becomes identical to a system of hard spheres in the limit $L \to 0$. In order to calculate the partition function of the system the Hamiltonian of the system must be formulated. The Hamiltonian of a system consisting of N cylindrical hard rods is given by

$$H = \sum_{k=1}^{N} T_k + \frac{1}{2} \sum_{k,l=1}^{N} V_{kl} \quad , \tag{13.27}$$

where T_k denotes the kinetic energy of rod k and V_{kl} the steric interaction between the rods k and l. This interaction is of the form

$$\begin{aligned} V_{kl} &= \infty \; , \quad \text{if } k \text{ and } l \text{ overlap,} \\ V_{kl} &= 0 \quad , \quad \text{if } k \text{ and } l \text{ do not overlap,} \\ V_{kk} &= 0 \; . \end{aligned} \tag{13.28}$$

The kinetic energy T of a cylinder reads

$$T = \frac{p_\theta^2}{2I_1} + \frac{(p_\phi - p_\psi \cos\theta)^2}{2I_1 \sin^2\theta} + \frac{p_\psi^2}{2I_3} + \frac{\boldsymbol{p}^2}{2m} \quad , \tag{13.29}$$

where θ, ϕ and ψ are the Eulerian angles; p_θ, p_ϕ and p_ψ are their canonical conjugate momenta (see any textbook on analytical mechanics); I_1 and I_3 are the principal moments of inertia of the cylinder, where I_3 is the moment of inertia about the cylinder axis; m and $\boldsymbol{p}$ are respectively the mass and the linear momentum of the cylinder. Consequently the hard-rod problem consists of the calculation of the following partition function

$$Z_{\mathrm{HR}} = \frac{1}{N! h^{6N}} \int \prod_{i=1}^{N} d\boldsymbol{r}_i d\boldsymbol{p}_i dp_{\theta_i} dp_{\phi_i} dp_{\psi_i} d\theta_i d\phi_i d\psi_i \exp(-\beta H) \quad . \tag{13.30}$$

It follows easily that

$$\begin{aligned} &\int \prod_{i=1}^{N} d\boldsymbol{p}_i dp_{\theta_i} dp_{\phi_i} dp_{\psi_i} \exp\left(-\beta \sum_{k=1}^{N} T_k\right) \\ &= (64\pi^6 k_{\mathrm{B}}^6 T^6 I_1^2 I_3 m^3)^{N/2} \prod_{k=1}^{N} \sin\theta_k \quad . \end{aligned} \tag{13.31}$$

Because of symmetry the integrand of (13.30) does not depend on the Eulerian angle ψ, which means that the integration over this angle can be carried out yielding a factor 2π. The problem thus boils down to the calculation of

$$Z_{\rm HR} = \frac{(\lambda\tau)^{-3N}}{N!} \int \prod_{i=1}^{N} d\boldsymbol{a}_i d\boldsymbol{r}_i \exp\left(-\frac{1}{2}\beta \sum_{k,l=1}^{N} V_{kl}\right) \quad , \tag{13.32}$$

where $d\boldsymbol{a}_i = \sin\theta_i d\theta_i d\phi_i$ and the constant τ is given by

$$\tau = \frac{h}{[(2\pi)^{5/3} k_{\rm B} T I_1^{2/3} I_3^{1/3}]^{1/2}} \quad . \tag{13.33}$$

In view of the preceding discussion of the hard-sphere model it probably does not come as a surprise that the calculation of the partition function (13.32) is still an unsolved problem. Neither the integration over the position coordinates $\boldsymbol{r}_i$ nor the integration over the Eulerian angles θ_i and ϕ_i can be carried out analytically. The usual way is to first deal with the integration over the position coordinates. According to the Van der Waals approximation the integral over the position coordinates

$$\frac{1}{N!} \int \prod_{i=1}^{N} d\boldsymbol{r}_i \exp\left(-\frac{1}{2}\beta \sum_{k,l=1}^{N} V_{kl}\right)$$

can be approximated in a qualitative sense by

$$Z(\boldsymbol{a}_1, \boldsymbol{a}_2, \ldots, \boldsymbol{a}_N) = \frac{1}{N!} \prod_{i=1}^{N} \left[\Omega - \frac{1}{2} \sum_{j=1;j\neq i}^{N} v_{\rm excl}(\boldsymbol{a}_i, \boldsymbol{a}_j)\right] \quad , \tag{13.34}$$

where $v_{\rm excl}(\boldsymbol{a}_i, \boldsymbol{a}_j)$ is the volume from which the centre of mass of a cylinder with an orientation parallel to the unit vector $\boldsymbol{a}_i$ is excluded, when it collides with a second cylinder with orientation $\boldsymbol{a}_j$. The excluded volume of two spherocylinders as well as two cylinders has been calculated by Onsager (1949). Although his calculation is an interesting mathematical exercise, it will not be repeated here because of all the approximations that are already involved in the calculation of the partition function. An approximation of the excluded volumes suffices.

The simplest approximation of the excluded volume of two spherocylinders or two cylinders is obtained by postulating

$$v_{\rm excl}(\boldsymbol{a}_1, \boldsymbol{a}_2) = A_0 - A_1 P_2(\boldsymbol{a}_1 \cdot \boldsymbol{a}_2) \quad . \tag{13.35}$$

The coefficients A_0 and A_1 are then determined by calculating the two simplest excluded volumes, namely the excluded volume belonging to two parallel rods and the one belonging to two mutually perpendicular rods. Both spherocylinders and cylinders will be considered here for the following reason. Spherocylinders have a sphere as limit, i.e. in this limit all quantities must reduce to their hard-sphere values. Cylinders are of interest because they describe both oblong and disc-like molecules simply by varying their length-to-breadth ratio. First of all the approximate expression for the excluded volume of two spherocylinders is calculated. It is immediately found that the parallel situation gives rise to

$$A_0 - A_1 = \tfrac{4}{3}\pi D^3 + 2\pi L D^2 \quad ,$$

whereas the mutually perpendicular situation gives

$$A_0 + \tfrac{1}{2}A_1 = \tfrac{4}{3}\pi D^3 + 2\pi L D^2 + 2L^2 D \quad .$$

Both situations and corresponding excluded volumes are shown in Fig. 13.2.

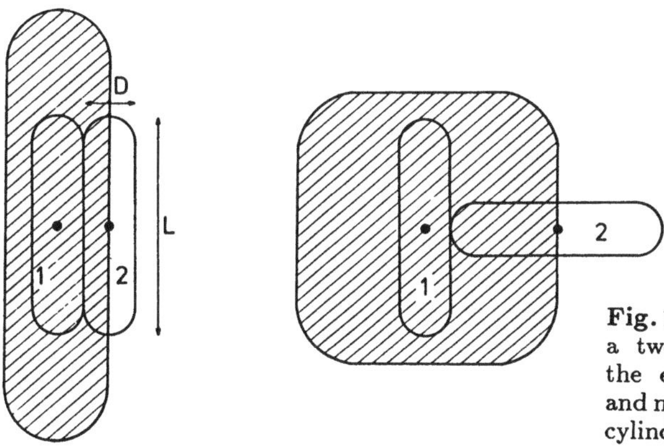

Fig. 13.2. The shaded area shows a two-dimensional intersection of the excluded volume for parallel and mutually perpendicular spherocylinders

As the volume of the spherocylinder is given by

$$v_0 = \tfrac{1}{6}\pi D^3 + \tfrac{1}{4}\pi L D^2 \quad ,$$

it follows that

$$v_{\mathrm{excl}}(\boldsymbol{a}_1, \boldsymbol{a}_2) = 8v_0\{1 + \gamma[1 - P_2(\boldsymbol{a}_1 \cdot \boldsymbol{a}_2)]\} \quad , \tag{13.36}$$

where

$$\gamma = \frac{2(x-1)^2}{\pi(3x-1)}$$

with $x = (L+D)/D$, i.e. x is the length-to-diameter or axial ratio. The parameter γ is proportional to this ratio provided that this ratio is large. Furthermore the Van der Waals theory for hard spheres is obtained in the limit $\gamma = 0$. The exact Onsager result is

$$v_{\mathrm{excl}}(\boldsymbol{a}_1, \boldsymbol{a}_2) = 8v_0[1 + \tfrac{3}{2}\gamma\,|\sin(\boldsymbol{a}_1, \boldsymbol{a}_2)|\,] \quad , \tag{13.37}$$

where $\sin(\boldsymbol{a}_1, \boldsymbol{a}_2)$ is the sine of the angle between the unit vectors $\boldsymbol{a}_1$ and $\boldsymbol{a}_2$. Consequently the present approximation of the excluded volume boils down to replacing $|\sin(\boldsymbol{a}_1, \boldsymbol{a}_2)|$ in the exact expression by $\sin^2(\boldsymbol{a}_1, \boldsymbol{a}_2)$.

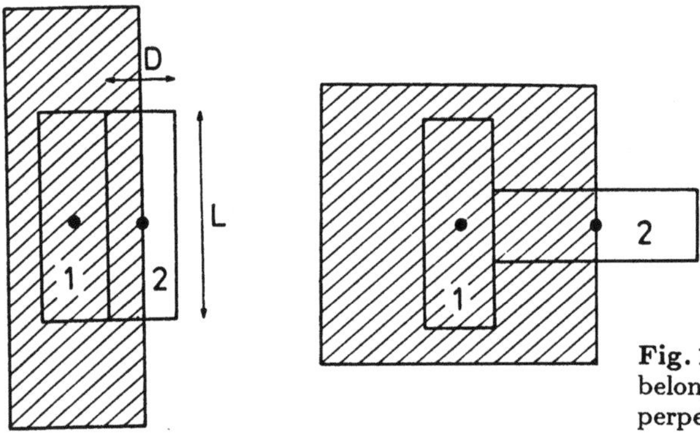

Fig. 13.3. The excluded volume belonging to parallel and mutually perpendicular cylinders

Next the excluded volume belonging to two cylinders will be calculated approximately. The excluded volumes belonging to the two extreme situations being considered are shown in Fig. 13.3. It is easily found that the excluded volume belonging to two parallel cylinders is given by

$$A_0 - A_1 = 2\pi L D^2 \quad ,$$

whereas the one belonging to two mutually perpendicular cylinders is

$$A_0 + \tfrac{1}{2}A_1 = 2DL(L+D) + \tfrac{1}{2}\pi D^2(L+D) \quad .$$

Now the following approximate expression for the excluded volume of two cylinders results

$$v_{\mathrm{excl}}(\boldsymbol{a}_1, \boldsymbol{a}_2) = 8v_0\left[\frac{2x}{3\pi} + \frac{1}{6x} + \frac{2}{3\pi} + \frac{1}{2} - \left(\frac{2x}{3\pi} + \frac{1}{6x} + \frac{2}{3\pi} - \frac{1}{2}\right)P_2(\boldsymbol{a}_1 \cdot \boldsymbol{a}_2)\right] \quad , \tag{13.38}$$

where $x = L/D$, the axial ratio of the cylinder, and $v_0 = \frac{1}{4}\pi D^2 L$, i.e. v_0 is the volume of a cylinder. The exact Onsager result in this case is

$$v_{\text{excl}}(\boldsymbol{a}_1, \boldsymbol{a}_2) = 8v_0\left\{\left(\frac{x}{\pi} + \frac{1}{4x}\right)[1 - (\boldsymbol{a}_1 \cdot \boldsymbol{a}_2)^2]^{1/2} + \frac{1}{4}(1 + |\boldsymbol{a}_1 \cdot \boldsymbol{a}_2|) + \frac{1}{\pi}E\left([1 - (\boldsymbol{a}_1 \cdot \boldsymbol{a}_2)^2]^{1/2}\right)\right\} , \tag{13.39}$$

where

$$E(\sin\phi) = \int_0^{\pi/2} (1 - \sin^2\phi \sin^2\psi)^{1/2} d\psi \quad .$$

A comparison between the approximate expression (13.38) and the exact expression (13.39) clearly shows that the approximation has the correct behaviour for all values of x. Within this framework it is relevant to remark that the exact expression (13.39) can also be approximated by resorting to an expansion in Legendre polynomials and truncating the expansion after a certain polynomial. However, such a procedure must be used carefully, because truncation after the second Legendre polynomial gives rise to partially incorrect approximations (Flapper and Vertogen 1981a).

The final step now concerns the calculation of (13.32) expressed in terms of (13.34):

$$Z_{\text{HR}} = (\lambda\tau)^{-3N} \int d\boldsymbol{a}_1 \ldots d\boldsymbol{a}_N Z(\boldsymbol{a}_1, \ldots, \boldsymbol{a}_N) \tag{13.40}$$

and using the expressions for the orientation-dependent excluded volume. This calculation is also hard to perform and approximation methods must be used. It should be noted that the integration over the spatial coordinates gives rise to many-particle interactions as far as the orientations are concerned. This follows immediately by rewriting (13.34) in the form

$$Z(\boldsymbol{a}_1, \ldots, \boldsymbol{a}_N) = \frac{\Omega^N}{N!} \exp\left\{\sum_{i=1}^{N} \ln\left[1 - \frac{1}{2\Omega} \sum_{j=1;j\neq i}^{N} v_{\text{excl}}(\boldsymbol{a}_i, \boldsymbol{a}_j)\right]\right\} \quad . \tag{13.41}$$

After substitution of (13.41) into (13.40) the Helmholtz free energy per rod, f_{HR}, is given by

$$\begin{aligned} f_{\text{HR}} &= -\frac{1}{\beta N} \ln Z_{\text{HR}} \\ &= \frac{1}{\beta}(3\ln\lambda + 3\ln\tau + \ln\varrho - 1) - \frac{1}{N}T(\Sigma_{\text{p}} + \Sigma_{\text{o}}) \end{aligned} \tag{13.42}$$

where $\Sigma_{\rm p}$ and $\Sigma_{\rm o}$ are respectively the packing entropy and the orientational or mixing entropy. These are given by

$$\Sigma_{\rm p} = k_{\rm B} \int d\boldsymbol{a}_1 \ldots d\boldsymbol{a}_N P(\boldsymbol{a}_1 \ldots \boldsymbol{a}_N) \times \sum_{i=1}^{N} \ln\left[1 - \frac{1}{2\Omega} \sum_{j=1;j\neq i}^{N} v_{\rm excl}(\boldsymbol{a}_i, \boldsymbol{a}_j)\right] \quad , \tag{13.43}$$

and

$$\Sigma_{\rm o} = -k_{\rm B} \int d\boldsymbol{a}_1 \ldots d\boldsymbol{a}_N P(\boldsymbol{a}_1 \ldots \boldsymbol{a}_N) \ln P(\boldsymbol{a}_1 \ldots \boldsymbol{a}_N) \quad , \tag{13.44}$$

where the probability $P(\boldsymbol{a}_1, \ldots, \boldsymbol{a}_N)$ for the appearance of a given orientational configuration is given by

$$P(\boldsymbol{a}_1, \ldots, \boldsymbol{a}_N) = \frac{Z(\boldsymbol{a}_1, \ldots, \boldsymbol{a}_N)}{\int d\boldsymbol{a}_1 \ldots d\boldsymbol{a}_N Z(\boldsymbol{a}_1, \ldots, \boldsymbol{a}_N)} \quad . \tag{13.45}$$

Next the mean field approximation is applied, i.e. the probability $P(\boldsymbol{a}_1, \ldots, \boldsymbol{a}_N)$ is approximated by a product of one-particle orientational distribution functions

$$P(\boldsymbol{a}_1, \ldots, \boldsymbol{a}_N) = \prod_{i=1}^{N} f(\boldsymbol{a}_i) \quad , \tag{13.46}$$

where the unknown distribution function $f(\boldsymbol{a})$ is determined by minimizing the free energy with respect to variations of $f(\boldsymbol{a})$. Finally the expression for the packing entropy is approximated still further by using

$$\Sigma_{\rm p} = k_{\rm B} \sum_{i=1}^{N} \ln\left[1 - \frac{1}{2\Omega} \sum_{j=1;j\neq i}^{N} \langle v_{\rm excl}(\boldsymbol{a}_i, \boldsymbol{a}_j)\rangle\right] \quad , \tag{13.47}$$

with

$$\langle v_{\rm excl}(\boldsymbol{a}_i, \boldsymbol{a}_j)\rangle = \int d\boldsymbol{a}_i d\boldsymbol{a}_j f(\boldsymbol{a}_i) f(\boldsymbol{a}_j) v_{\rm excl}(\boldsymbol{a}_i, \boldsymbol{a}_j) \quad . \tag{13.48}$$

The result of all these approximations is the following Van der Waals expression for the Helmholtz free energy per cylinder

$$\beta f_{\rm HR} = 3\ln\lambda + 3\ln\tau + \ln\varrho - 1 - \ln\left[1 - \tfrac{1}{2}\varrho\langle v_{\rm excl}(\boldsymbol{a}_1, \boldsymbol{a}_2)\rangle\right] + \int d\boldsymbol{a}\, f(\boldsymbol{a}) \ln f(\boldsymbol{a}) \quad , \tag{13.49}$$

where $f(\boldsymbol{a})$ is determined by minimizing the total Helmholtz free energy of the system with respect to variations of $f(\boldsymbol{a})$. This means that $f(\boldsymbol{a})$ is determined by minimizing (13.49) when considering a system of hard rods without attractive interactions.

It should be remarked here once more that the Van der Waals approximation can be only qualitatively correct because it overestimates, as is the case for hard spheres, the pressure and the reduction of the packing entropy. This directly follows from the overestimation of the third and higher virial coefficients. The enormous complexity of the problem, however, only allows qualitative descriptions and it is sufficient to resort to the simplest qualitative model. Painstaking calculations of the exact virial coefficients do not make sense in view of all other approximations involved. Even an exact calculation of the second virial coefficient, which appears in the Onsager theory, is superfluous, because the Onsager approach is quantitatively incorrect as soon as the system approaches the transition to the anisotropic state, as will be shown later.

(2) *The Thermal Average.* As stated before the thermal average is taken with respect to the hard-rod system and is given by

$$\left\langle \exp\left[\tfrac{1}{2}\beta \sum_{k\neq l} v(\boldsymbol{r}_{kl}, \boldsymbol{a}_k, \boldsymbol{a}_l)\right]\right\rangle_0 = \frac{\int \prod_{i=1}^{N} d\boldsymbol{a}_i d\boldsymbol{r}_i \exp\left(-\tfrac{1}{2}\beta \sum_{k\neq l} V_{kl}\right) \exp\left[\tfrac{1}{2}\beta \sum_{m\neq n} v(\boldsymbol{r}_{mn}, \boldsymbol{a}_m, \boldsymbol{a}_n)\right]}{\int \prod_{i=1}^{N} d\boldsymbol{a}_i d\boldsymbol{r}_i \exp\left(-\tfrac{1}{2}\beta \sum_{k\neq l} V_{kl}\right)} . \tag{13.50}$$

Clearly this thermal average is extremely hard to calculate and has to be approximated. An obvious approximation is

$$\left\langle \exp\left[\tfrac{1}{2}\beta \sum_{k\neq l} v(\boldsymbol{r}_{kl}, \boldsymbol{a}_k, \boldsymbol{a}_l)\right]\right\rangle_0 = \exp\left[\tfrac{1}{2}\beta \sum_{k\neq l} \langle v(\boldsymbol{r}_{kl}, \boldsymbol{a}_k, \boldsymbol{a}_l)\rangle_0\right] , \tag{13.51}$$

where

$$\langle v(\boldsymbol{r}_{kl}, \boldsymbol{a}_k, \boldsymbol{a}_l)\rangle_0 = \frac{1}{\Omega^2} \int d\boldsymbol{r}_k d\boldsymbol{r}_l d\boldsymbol{a}_k d\boldsymbol{a}_l \, g_{\mathrm{HR}}(\boldsymbol{r}_{kl}, \boldsymbol{a}_k, \boldsymbol{a}_l) v(\boldsymbol{r}_{kl}, \boldsymbol{a}_k, \boldsymbol{a}_l) \tag{13.52}$$

and $g_{\mathrm{HR}}(\boldsymbol{r}_{kl}, \boldsymbol{a}_k, \boldsymbol{a}_l)$ is the pair distribution function of the hard-rod model. It follows immediately from (13.52) that even isotropic attractive interactions can stabilize the anisotropic liquid due to the anisotropic nature of the pair distribution function!

The thermal average (13.51) can not be analyzed satisfactorily. This is entirely due to the fact that reliable information concerning the hard-rod model is not available. Consequently the influence of the attractive part on the thermodynamic properties cannot be estimated because of the unknown pair distribution function. Nevertheless it is quite clear from the present analysis that the pair distribution function itself may give rise to a considerable contribution to the anisotropic part of the free energy, even in those cases where the attractive part of the intermolecular interaction is only isotropic. Estimations (Gelbart and Gelbart 1977) indicate that this type of contribution can be as important as the contribution of the anisotropic part of the attractive interaction itself. For calculational purposes the thermal average must be approximated in a rather ad hoc way because of the absence of reliable information concerning the pair distribution function and the intermolecular interactions. A usual approximation is to write the thermal average as a product of two terms. The first term is an isotropic one analogous to (13.21). The second term takes the influence of the anisotropic interactions into account. For the sake of simplicity it is solely described in terms of the expectation value of the second Legendre polynomial. The anisotropic term is thought to arise both from the anisotropic part of the intermolecular attractions, and from the combination of the anisotropic part of the pair distribution function with the isotropic part of the attractive intermolecular interactions. In the spirit of the Van der Waals theory the attractive contribution to the Helmholtz free energy per cylinder is then given by

$$\begin{aligned}&\left\langle \exp\left[\frac{1}{2}\beta\sum_{k\neq l} v(\boldsymbol{r}_{kl},\boldsymbol{a}_k,\boldsymbol{a}_l)\right]\right\rangle_0 \\ &\quad = \exp\left[\frac{1}{2}N\beta\varrho J_0 + \frac{1}{2}N\beta\varrho J\int d\boldsymbol{a}_1 d\boldsymbol{a}_2 f(\boldsymbol{a}_1)f(\boldsymbol{a}_2)P_2(\boldsymbol{a}_1\cdot\boldsymbol{a}_2)\right] \quad . \qquad (13.53)\end{aligned}$$

The unknown orientational distribution function $f(\boldsymbol{a})$ is determined by the requirement that the total Helmholtz free energy must be minimal with respect to variations of $f(\boldsymbol{a})$.

(3) *The Equation of State.* Starting from the approximate expression (13.35) for the excluded volume of two cylinders, the expression (13.53) for the contribution of the attractive intermolecular interaction and the expression (13.49) for the hard-rod free energy, the Van der Waals expression for the Helmholtz free energy per molecule is given by

$$\begin{aligned}\beta f &= 3\ln\lambda + 3\ln\tau + \ln\varrho - 1 - \ln\left[1 - \tfrac{1}{2}\varrho(A_0 - A_1\langle P_2(\boldsymbol{a}_1\cdot\boldsymbol{a}_2)\rangle)\right] \\ &\quad - \tfrac{1}{2}\beta\varrho J_0 - \tfrac{1}{2}\beta\varrho J\langle P_2(\boldsymbol{a}_1\cdot\boldsymbol{a}_2)\rangle + \int d\boldsymbol{a}\, f(\boldsymbol{a})\ln f(\boldsymbol{a}) \quad , \qquad (13.54)\end{aligned}$$

where

$$\langle P_2(\boldsymbol{a}_1 \cdot \boldsymbol{a}_2)\rangle = \int d\boldsymbol{a}_1 d\boldsymbol{a}_2\, f(\boldsymbol{a}_1)\, f(\boldsymbol{a}_2)\, P_2(\boldsymbol{a}_1 \cdot \boldsymbol{a}_2) \quad .$$

The orientational distribution function follows from the minimization of (13.54) with respect to $f(\boldsymbol{a})$, i.e.

$$f(\boldsymbol{a}) = c \exp\left[b \int d\boldsymbol{a}_1\, f(\boldsymbol{a}_1) P_2(\boldsymbol{a}_1 \cdot \boldsymbol{a})\right] \quad \text{with} \tag{13.55}$$

$$b = \frac{\varrho A_1}{1 - \frac{1}{2}\varrho[A_0 - A_1\langle P_2(\boldsymbol{a}_1 \cdot \boldsymbol{a}_2)\rangle]} + \varrho\beta J \tag{13.56}$$

and c is a normalization constant such that $\int d\boldsymbol{a}\, f(\boldsymbol{a}) = 1$.

In order to facilitate the calculation as far as possible and to avoid unnecessary numerical effort, use will be made of the so-called spherical version of the present model. This means that the original unit vector $\boldsymbol{a}_i$, which describes the direction of the long molecular axis, is replaced by a vector $\boldsymbol{a}_i$ which is allowed to have any length with the restriction that

$$\sum_{i=1}^{N} \boldsymbol{a}_i^2 = N \quad . \tag{13.57}$$

The constraint (13.57) is known as the spherical constraint and taken into account by means of the method of Lagrange multipliers in the following way. The relevant integration must be performed taking the three-dimensional character of the vector $\boldsymbol{a}$ into account, i.e. $d\boldsymbol{a}$ must be replaced by d^3a. Accordingly the original partition function

$$Z = \int d\boldsymbol{a}_1 \ldots d\boldsymbol{a}_N \exp[-\beta H(\boldsymbol{a}_1, \boldsymbol{a}_2, \ldots, \boldsymbol{a}_N)] \tag{13.58}$$

changes into

$$Z = \int d^3a_1 \ldots d^3a_N \times \exp\left[-\mu\left(\sum_{i=1}^{N} \boldsymbol{a}_i^2 - N\right) - \beta H(\boldsymbol{a}_1, \boldsymbol{a}_2, \ldots, \boldsymbol{a}_N)\right] \quad , \tag{13.59}$$

where the Lagrange multiplier μ is determined by

$$\frac{\beta \partial f}{\partial \mu} = -\frac{1}{N}\frac{\partial \ln Z}{\partial \mu} = 0 \quad , \tag{13.60}$$

and f represents the Helmholtz free energy per molecule. In thermodynamic equilibrium it holds that

$$\langle a_i^2 \rangle = \langle a_j^2 \rangle \quad .$$

Consequently the spherical constraint implies

$$\langle a_i^2 \rangle = 1 \quad ,$$

i.e. the original constraint $\boldsymbol{a}_i^2 = 1$ is weakened to the constraint that the thermodynamic average of $\boldsymbol{a}_i^2$ must be one.

The spherical version of the present model gives rise to a Helmholtz free energy that can be obtained directly from (13.54). According to (13.59) the expression (13.54), which holds for the original model, is simply changed into

$$\begin{aligned}\beta f &= 3\ln\lambda + 3\ln\tau + \ln\varrho - 1 + \mu(\langle \boldsymbol{a}^2\rangle - 1) \\ &\quad - \ln\{1 - \tfrac{1}{2}\varrho[A_0 - A_1\langle P_2(\boldsymbol{a}_1\cdot\boldsymbol{a}_2)\rangle]\} \\ &\quad - \tfrac{1}{2}\beta\varrho J_0 - \tfrac{1}{2}\beta\varrho J\langle P_2(\boldsymbol{a}_1\cdot\boldsymbol{a}_2)\rangle + \int d^3a\, f(\boldsymbol{a})\ln f(\boldsymbol{a}) \quad ,\end{aligned} \tag{13.61}$$

where

$$\langle P_2(\boldsymbol{a}_1\cdot\boldsymbol{a}_2)\rangle = \int d^3a_1 d^3a_2 f(\boldsymbol{a}_1) f(\boldsymbol{a}_2) P_2(\boldsymbol{a}_1\cdot\boldsymbol{a}_2) \quad . \tag{13.62}$$

The unknown distribution function $f(\boldsymbol{a})$ satisfies the relation

$$\int d^3a\, f(\boldsymbol{a}) = 1 \tag{13.63}$$

and is determined by minimizing the Helmholtz free energy with respect to variations $\delta f(\boldsymbol{a})$ of $f(\boldsymbol{a})$. This procedure gives rise to

$$\begin{aligned}&\mu\int d^3a\, \boldsymbol{a}^2\delta f(\boldsymbol{a}) - \left\{\frac{\varrho A_1}{1 - \frac{1}{2}\varrho[A_0 - A_1\langle P_2(\boldsymbol{a}_1\cdot\boldsymbol{a}_2)\rangle]} + \varrho\beta J\right\} \\ &\quad\times \int d^3a \int d^3a_1 f(\boldsymbol{a}_1) P_2(\boldsymbol{a}\cdot\boldsymbol{a}_1)\,\delta f(\boldsymbol{a}) \\ &\quad + \int d^3a\,\delta f(\boldsymbol{a})\ln f(\boldsymbol{a}) + \int d^3a\,\delta f(\boldsymbol{a}) = 0 \quad .\end{aligned} \tag{13.64}$$

It follows directly from (13.64) that

$$f(\boldsymbol{a}) = C^{-1}\exp\left[-\mu\boldsymbol{a}^2 + B\int d^3a_1 f(\boldsymbol{a}_1) P_2(\boldsymbol{a}_1\cdot\boldsymbol{a})\right] \tag{13.65}$$

with

$$C = \int d^3a \exp\left[-\mu\boldsymbol{a}^2 + B\int d^3a_1 f(\boldsymbol{a}_1) P_2(\boldsymbol{a}_1\cdot\boldsymbol{a})\right] \tag{13.66}$$

and

$$B = \frac{\varrho A_1}{1 - \frac{1}{2}\varrho[(A_0 - A_1\langle P_2(\boldsymbol{a}_1 \cdot \boldsymbol{a}_2)\rangle]} + \varrho\beta J \quad . \tag{13.67}$$

Without loss of generality the anisotropic phase can be assumed to be oriented along the z axis. Then, assuming

$$f(\boldsymbol{a}) = g(a_x)g(a_y)g(a_z) \tag{13.68}$$

the distribution function $g(a_\alpha)$, $\alpha = x, y, z$, is given by

$$g(a_\alpha) = C_\alpha^{-1} \exp\left[-\left(\mu - \tfrac{3}{2}B\langle a_\alpha^2\rangle\right)a_\alpha^2\right] \quad , \tag{13.69}$$

and the normalization C_α is

$$C_\alpha = \int da_\alpha \exp\left[-\left(\mu - \frac{3}{2}B\langle a_\alpha^2\rangle\right)a_\alpha^2\right] = \left(\frac{\pi}{\mu - \frac{3}{2}B\langle a_\alpha^2\rangle}\right)^{1/2} . \tag{13.70}$$

The order parameter $\langle a_\alpha^2\rangle$ must be determined using the relation

$$\langle a_\alpha^2\rangle = \int da_\alpha a_\alpha^2 g(a_\alpha) = \tfrac{1}{2}\left(\mu - \tfrac{3}{2}B\langle a_\alpha^2\rangle\right)^{-1} \quad . \tag{13.71}$$

The Lagrange multiplier μ follows from (13.60), i.e. μ is determined from

$$\langle a_x^2\rangle + \langle a_y^2\rangle + \langle a_z^2\rangle = 1$$

or using (13.71)

$$\sum_\alpha \left(\mu - \frac{3}{2}B\langle a_\alpha^2\rangle\right)^{-1} = 2 \quad . \tag{13.72}$$

This relation expresses the Lagrange multiplier in terms of the quantities $\langle a_x^2\rangle$, $\langle a_y^2\rangle$ and $\langle a_z^2\rangle$.

The solution of Eqs. (13.71) and (13.72) is obtained in the following way. Because of the uniaxiality of the anisotropic phase the transition can be described in terms of an order parameter S, defined by

$$\langle a_x^2\rangle = \langle a_y^2\rangle = \tfrac{1}{3}(1 - S) \quad , \quad \langle a_z^2\rangle = \tfrac{1}{3}(1 + 2S) \quad . \tag{13.73}$$

Substitution of (13.73) into (13.72) yields μ in terms of S and B. It easily follows that

$$\mu = \tfrac{3}{4} + \tfrac{1}{2}B + \tfrac{1}{4}BS \pm \tfrac{3}{4}(B^2S^2 - \tfrac{2}{3}BS + 1)^{1/2} . \tag{13.74}$$

In order to satisfy the condition $\langle \boldsymbol{a}^2 \rangle = 1$ only the solution with the positive sign can be accepted, as the solution with the negative sign gives rise to divergent integrals.

The equation for the order parameter S follows directly from substituting (13.73) into (13.71) and eliminating μ from the two resulting equations. This leads to the following equation for S

$$BS = \frac{3S}{(1-S)(1+2S)} \quad . \tag{13.75}$$

It should be remarked here that in general B depends on S as well, namely

$$B = \frac{\varrho A_1}{1 - \frac{1}{2}\varrho(A_0 - A_1 S^2)} + \varrho\beta J \quad . \tag{13.76}$$

Consequently the order of the anisotropic state is obtained by solving a biquadratic equation. Equation (13.75) always contains the solution $S = 0$, the isotropic solution. The solution of (13.75) describing the ordered phase can easily be obtained in the Onsager and Maier-Saupe approaches.

The Helmholtz free energy now becomes

$$\begin{aligned} \beta f = {} & 3\ln\lambda + 3\ln\tau + \ln\varrho - \tfrac{7}{4} - \ln[1 - \tfrac{1}{2}\varrho(A_0 - A_1S^2)] \\ & - \tfrac{1}{2}\beta\varrho J_0 - \tfrac{1}{2}\beta\varrho J S^2 - \tfrac{1}{4}BS - \tfrac{3}{4}(B^2S^2 - \tfrac{2}{3}BS + 1)^{1/2} \\ & + BS^2 - \tfrac{2}{3}\ln\tfrac{2\pi}{3} - \ln(1-S) - \tfrac{1}{2}\ln(1+2S) \quad . \end{aligned} \tag{13.77}$$

The pressure p of the system is obtained by means of the relation

$$\frac{\beta p}{\varrho} = \beta\varrho\frac{\partial f}{\partial \varrho} \quad . \tag{13.78}$$

Consequently the equation of state is

$$p = \frac{\varrho k_{\mathrm{B}} T}{1 - \frac{1}{2}\varrho(A_0 - A_1S^2)} - \frac{1}{2}\varrho^2 J_0 - \frac{1}{2}\varrho^2 J S^2 \quad . \tag{13.79}$$

The Gibbs free energy per cylinder, g, is defined by

$$g = f + \frac{p}{\varrho} \quad . \tag{13.80}$$

Using (13.76), (13.77) and (13.79) and the expression for S as given by

(13.75) the Gibbs free energy can be explicitly expressed as a function of the pressure and the temperature. If the order parameter equation (13.75) allows several solutions, the thermodynamically stable solution is obtained by the requirement that it must give rise to the lowest Helmholtz free energy while keeping the volume constant, and the lowest Gibbs free energy at constant pressure.

Experimentally the most interesting thermodynamic data concern the density change and the value of the order parameter S_c at the order-disorder transition. These values are obtained by solving the equations

$$p_i = p_n \quad , \tag{13.81}$$

$$g_i = g_n \quad , \tag{13.82}$$

where p_i and p_n are the pressure and g_i and g_n the Gibbs free energy per cylinder of respectively the isotropic and anisotropic phase.

In general the set of Eqs. (13.81) and (13.82) must be solved numerically. Completely in accordance with expectations, the equation of state (13.79) is found to describe the nematic-isotropic transition rather well from a qualitative point of view. The experimentally observed behaviour of the order parameter and the density change at the transition can be reproduced fairly well by an appropriate choice of the parameters. The derived equation of state, however, is unable to describe the effects of different molecules in a physically satisfactory way, i.e. the required parameter values do not correlate with the molecular structure. This is clearly seen if an equation of state that describes the transition fairly well from a quantitative point of view is used. Such a phenomenological equation of state for nematics has been proposed by Flapper and Vertogen (1981b) using a straightforward generalization of the Carnahan-Starling equation of state for hard spheres. The three parameters J_0, J and x, i.e. the strengths of the isotropic and anisotropic attraction and the axial ratio, appear to suffice for a satisfactory description of the transition. The problems start as soon as the parameter values, which are obtained by a fitting procedure, are analyzed from a physical point of view. Comparing the parameter values of different nematic substances reveals conceptual difficulties (Flapper et al. 1981). Most notably the interpretation of the parameter x in terms of an axial ratio or even in terms of a monotonic increasing function of the axial ratio is highly problematical (see Table 13.1). These discrepancies are not at all surprising in view of the simplicity of the model. Some of the difficulties can be attributed to the rather poor approximation methods, which overestimate a number of quantities, e.g. the influence of the excluded volume. A significant improvement of these matters, however, seems to be quite hard to obtain. The second and probably most important source of the difficulties in

Table 13.1. A comparison between the axial ratio x used as a fit parameter (Flapper et al. 1981) and as calculated from molecular data (Leenhouts and Dekker 1981) for the homologous series APAPAn; x_A is the fitted value and x_B the calculated value

	x_A	x_B
APAPA	1.41	3.25
APAPA2	1.28	3.46
APAPA3	1.33	3.68
APAPA4	1.26	3.87
APAPA5	1.25	4.02
APAPA9	1.25	4.43

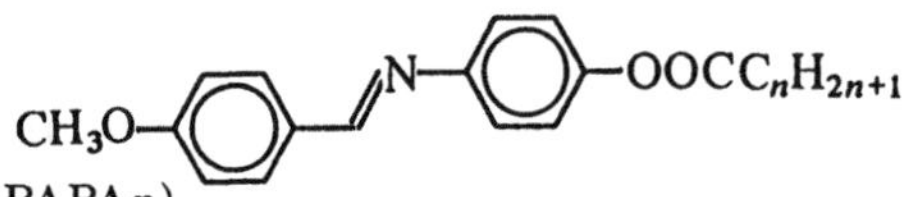

(APAPAn)

interpretation is the poor representation of the intermolecular interactions and the neglect of the molecular flexibility.

In order to obtain some insight in the effect of both an increasing axial ratio and an increasing anisotropic interaction strength two limiting cases will be considered. First the Onsager approach is treated, i.e. the effect of the attractive interactions is neglected. Secondly the Maier-Saupe approach is presented, i.e. the orienting effect of the repulsive interactions is not taken into account and the anisotropic phase is solely ascribed to the attractive anisotropic interaction. More information and references to other literature can be found in the review papers by Luckhurst (1979) and Gelbart (1982). For computer simulations of liquid crystals see, for example, the paper by Frenkel and Mulder (1985) and references therein.

13.3 The Onsager Approach

The first discussion of the isotropic-anisotropic transition has been given in terms of the steric hindering between hard rods with the shape of spherocylinders (Onsager 1949). This line of approach boils down to a discussion of the thermodynamic properties of a system of hard rods. According to the expression (13.36) for the excluded volume of two spherocylinders, the Helmholtz free energy (13.77) per spherocylinder, in the Van der Waals approximation, is given by

$$\begin{aligned}\beta f_{\mathrm{HR}} = {} & 3\ln\lambda + 3\ln\tau - \tfrac{4}{7} - \tfrac{3}{2}\ln\tfrac{2\pi}{3} + \ln\varrho \\ & - \ln\left[1 - 4\eta(1+\gamma-\gamma S^2)\right] - \tfrac{1}{4}BS - \tfrac{3}{4}\left(B^2S^2 - \tfrac{2}{3}BS + 1\right)^{1/2} \\ & + BS^2 - \ln(1-S) - \tfrac{1}{2}\ln(1+2S) \quad , \qquad (13.83)\end{aligned}$$

where $\eta = \varrho v_0$ is the packing fraction and B is given by

$$B = \frac{8\eta\gamma}{1 - 4\eta(1 + \gamma - \gamma S^2)} \quad . \tag{13.84}$$

Now Eq. (13.75) for the order parameter S can be solved analytically and gives

$$S_\pm = \frac{1}{7} \pm \frac{6}{7}\left[1 + \frac{7}{12\gamma}\left(1 - \frac{1}{4\eta}\right)\right]^{1/2} \quad . \tag{13.85}$$

Clearly a packing fraction $\eta > \frac{1}{4}$ does not make sense here because of the requirement $S \leq 1$.

The original Onsager approach applies to a low density expansion, (i.e. η is taken to be very small) and considers only the effect of the second virial coefficient. This means that the original approach of Onsager is obtained by setting

$$\ln\left[1 - 4\eta(1 + \gamma - \gamma S^2)\right] = -4\eta(1 + \gamma - \gamma S^2) \quad . \tag{13.86}$$

It should be remarked here that such a procedure is only valid for a low packing fraction η provided that $\gamma\eta$ is also small. The second condition does not appear in the case of spheres because $\gamma = 0$.

Following Onsager, the Helmholtz free energy per spherocylinder, using $B = 8\eta\gamma$, is

$$\begin{aligned}\beta f_{\mathrm{HR}} = {} & 3\ln\lambda + 3\ln\tau - \tfrac{7}{4} - \tfrac{3}{2}\ln\tfrac{2\pi}{3} + \ln\varrho + 4\eta(1 + \gamma + \gamma S^2) - 2\eta\gamma S \\ & - \tfrac{3}{4}(64\eta^2\gamma^2 S^2 - \tfrac{16}{3}\eta\gamma S + 1)^{1/2} \\ & - \ln(1 - S) - \tfrac{1}{2}\ln(1 + 2S) \quad ,\end{aligned} \tag{13.87}$$

where according to (13.75) the equation for the order parameter S is given by

$$8\eta\gamma S = \frac{3S}{(1 - S)(1 + 2S)} \quad . \tag{13.88}$$

This equation has three solutions, namely

(1) $S = 0$, the isotropic solution.
(2) Two solutions, describing an ordered state,

$$S_\pm = \frac{1}{4} \pm \frac{3}{4}\left(1 - \frac{1}{3\eta\gamma}\right)^{1/2} \quad . \tag{13.89}$$

The correct solution is selected by the criterion of the lowest free energy.

The equation of state follows directly from (13.79). Up to the second virial term the pressure is given by

$$p = \varrho k_{\mathrm{B}} T[1 + 4\eta(1 + \gamma - \gamma S^2)] \quad . \tag{13.90}$$

The result (13.89) is quite simple compared to the result of the original calculation by Onsager. This must be solely ascribed to the present simplified expression for the excluded volume and the use of the spherical constraint. Taking into account the full expression for the excluded volume of two spherocylinders leads to an integral equation that must be solved numerically. Onsager circumvented the numerical problem using a variational calculation based on one order parameter only. In principle, however, the full solution involves an infinite set of order parameters.

In order to describe the thermodynamic behaviour of the system at the phase transition equations (13.81) and (13.82) must be solved. Equation (13.81) gives rise to

$$\eta_{\mathrm{i}}[1 + 4\eta_{\mathrm{i}}(1 + \gamma)] = \eta_{\mathrm{n}}[1 + 4\eta_{\mathrm{n}}(1 + \gamma - \gamma S_{\mathrm{c}}^2)] \quad , \tag{13.91}$$

whereas (13.82) results in

$$\begin{aligned} \ln \eta_{\mathrm{i}} + 8\eta_{\mathrm{i}}(1 + \gamma) - \tfrac{3}{4} &= \ln \eta_{\mathrm{n}} + 8\eta_{\mathrm{n}}(1 + \gamma) - 2\eta_{\mathrm{n}}\gamma S_{\mathrm{c}} \\ &\quad - \tfrac{3}{4}(64\eta_{\mathrm{n}}^2\gamma^2 S_{\mathrm{c}}^2 - \tfrac{16}{3}\eta_{\mathrm{n}}\gamma S_{\mathrm{c}} + 1)^{1/2} \\ &\quad - \ln(1 - S_{\mathrm{c}}) - \tfrac{1}{2}\ln(1 + 2S_{\mathrm{c}}) \end{aligned} \tag{13.92}$$

with

$$S_{\mathrm{c}} = \frac{1}{4} + \frac{3}{4}\left(1 - \frac{1}{3\eta_{\mathrm{n}}\gamma}\right)^{1/2} \quad . \tag{13.93}$$

Expression (13.91) clearly shows that the Onsager approach only makes sense for a dilute system, since it requires

$$\gamma\eta_{\mathrm{n}} \geqq \tfrac{1}{3} \quad . \tag{13.94}$$

This relation immediately implies that the axial ratio of the molecules must exceed a critical value in order to obtain an anisotropic phase, as the packing fraction is smaller than one. This critical axial ratio must be larger than x_0, defined by

$$\gamma_{\mathrm{cr}} = \frac{2(x_0 - 1)^2}{\pi(3x_0 - 1)} = \frac{1}{3} \quad ,$$

or $x_0 = 3.08$. The order parameter S_c at the transition, corresponds to the S_+ solution found in (13.89). This S_+ solution gives rise to the lowest free energy and describes the anisotropic phase. This can also be seen at once from the fact that S_+ approaches one, i.e. perfect order, with increasing density.

Equations (13.91) and (13.92) still have to be solved numerically. Reasonable analytical approximations, however, can be obtained for relatively large values of the parameter γ, i.e. $\gamma \gg 1$. Introducing the dimensionless parameter y defined by

$$\eta_i = \eta_n(1-y) \quad , \tag{13.95}$$

and assuming γ to be large, (13.91) can be approximated by

$$(1-y)[1+4(1-y)\gamma\eta_n] = 1 + 4\gamma\eta_n(1-S_c^2) \tag{13.96}$$

or, assuming y to be small,

$$y = \frac{4\gamma\eta_n S_c^2}{8\gamma\eta_n + 1} \quad . \tag{13.97}$$

Similarly, using the approximation,

$$\ln(\eta_i/\eta_n) = -y$$

Eq. (13.92) becomes

$$\begin{aligned} y(1+8\eta_n\gamma) = & -\tfrac{3}{4} + \tfrac{3}{4}(64\gamma^2\eta_n^2 S_c^2 - \tfrac{16}{3}\gamma\eta_n S_c + 1)^{1/2} \\ & + 2\gamma\eta_n S_c + \ln(1-S_c) + \tfrac{1}{2}\ln(1+2S_c) \quad . \end{aligned} \tag{13.98}$$

Expressing y up to order S_c^2 leads to

$$y = \frac{(128\gamma^2\eta_n^2 - 9)S_c^2}{6(8\eta_n\gamma + 1)} \quad . \tag{13.99}$$

Then it holds, from (13.97) and (13.99), that

$$\gamma\eta_n = \tfrac{3}{8} \quad . \tag{13.100}$$

Thus according to (13.93) S_c is given by

$$S_c = 0.5 \quad . \tag{13.101}$$

The result of the corresponding Onsager calculation for large γ values is

$S_c = 0.8$. The relative change in the density at the transition is directly related to the parameter y using the relation

$$\Delta = \frac{\varrho_{\mathrm{n}} - \varrho_{\mathrm{i}}}{\varrho_{\mathrm{i}}} = \frac{y}{1-y} \quad . \tag{13.102}$$

Within the present approximation, using Eqs. (13.100) and (13.101) Δ is given by

$$\Delta = y = \tfrac{3}{32} \quad . \tag{13.103}$$

The relative density change in the full Onsager approach is about 25 %. Compared to the experimental results the relative density change is much too high, experimentally Δ is about 0.004. The jump in the order parameter is roughly 0.3 for most thermotropic nematics, much lower than the predicted value of S_c.

Clearly the Onsager approach only provides a qualitative insight into the effect of the steric interactions as far as real nematics are concerned. In fact the validity of the Onsager approach itself is highly questionable as soon as the system approaches the transition to the anisotropic state. This is due to the fact that the parameter $\gamma\eta$ is not small, see (13.100), meaning that the approximation (13.86) loses its quantitative validity, even for very dilute systems. This fact is emphasized further by comparing the expressions for the order parameter starting from the Onsager approach (13.89) and the Van der Waals approach (13.85). It is found that the Van der Waals expression does not contain the Onsager expression in the low packing fraction limit. For a further discussion of the validity of the Onsager approach the reader is also referred to Saupe (1979). Notwithstanding the correct criticism of the Onsager approach, the claim may be made that it gives a simple and qualitatively correct picture of the order-disorder transition caused by the shape of the molecules. The present spherical version shows in an analytical and transparent way the basic idea of Onsager, namely that the phase transition is brought about by the competition between the packing entropy, favouring an ordered state, and the orientational entropy, favouring a disordered state.

The Onsager approach also describes the ordering of disc-like molecules in a simple way, in which case expression (13.38) for the excluded volume of cylinders must be used. Compared with spherocylinders, the difference shows up at small axial ratios for which the parameter γ remains large, indicating ordering of the discs. The phase transition in a hard-disc system has also been described in terms of a phenomenological Carnahan-Starling equation (Flapper and Vertogen 1981a).

13.4 The Maier-Saupe Approach

The other extreme starting point is the idea that the existence of the nematic phase is caused by the anisotropic part of the dispersion interaction energy between the molecules. This energy originates from the intermolecular electrostatic interaction, which is taken into account by means of perturbation theory. The second-order perturbation term is called the dispersion energy. For the sake of simplicity Maier and Saupe (1959, 1960) approximated the electrostatic interaction by the first terms of its multipole expansion and they assumed that:

i) the influence of the permanent dipoles can be neglected as far as long-range nematic order is concerned;
ii) only the effect of the induced dipole-dipole interaction needs to be considered, because the higher order terms, are not thought to significantly affect the nematic order;
iii) in the first instance a molecule may be considered to be rotationally symmetric with respect to its long molecular axis described by the unit vector $\boldsymbol{a}$;
iv) with respect to a given molecule the distribution of the centres of mass of the remaining molecules may be taken to be spherically symmetric.

These assumptions result in the following anisotropic interaction V_{ij} between the molecules i and j

$$V_{ij} = -J_{ij}P_2(\boldsymbol{a}_i \cdot \boldsymbol{a}_j) \quad , \tag{13.104}$$

where the coupling constant J_{ij} depends on (1) the distance between the centres of mass of the molecules as r_{ij}^{-6} and (2) a molecular property, which is roughly the anisotropy of the polarizability. The proposed interaction (13.104), the famous Maier-Saupe (MS) model, appears to be quite successful in predicting a number of properties of the nematic phase and of the nematic-isotropic transition despite the, to say the least, highly questionable assumptions involved. For this reason the MS model is perhaps best considered as a phenomenological model, where its relevance is justified by its success. From this point of view the coefficient J_{ij} is just a coupling parameter which is not necessarily related to the original interpretation.

The thermodynamic properties of the MS model can in general not be calculated exactly. This means that approximation methods must be used that, in turn, increase the qualitative nature of the results. In order to keep the discussion as simple as possible the spherical version of the MS model is considered, which can be calculated in the mean field approximation with hardly any numerical effort (Vertogen and Van der Meer 1979b). Moreover, the phase transition is discussed at constant volume, i.e. only the Helmholtz free energy of the system needs to be considered.

The Helmholtz free energy of the MS model is obtained from (13.76) and (13.77) by putting $A_1 = 0$ (no orientation dependence of the excluded volume). The order parameter S of the MS model follows from (13.75) and is determined by the relation

$$\beta \varrho J S = \frac{3S}{(1-S)(1+2S)} \quad , \tag{13.105}$$

where

$$J = \int d\boldsymbol{r}\, J(r) g_{\mathrm{HC}}(r)$$

with $g_{\mathrm{HC}}(r)$ denoting the pair distribution of hard spheres because of assumption (iv). This means that the order of the nematic phase is described by the S_+ solution of

$$S_{\pm} = \frac{1}{4} \pm \frac{3}{4}\left(1 - \frac{8}{3\beta\varrho J}\right)^{1/2} \quad . \tag{13.106}$$

Just as in the Onsager approach the S_- solution gives rise to a higher free energy. In order to investigate the thermodynamic behaviour at the phase transition it suffices to study the difference Δf between the free energies of the ordered and isotropic phase:

$$\begin{aligned} \Delta f = {} & \frac{1}{2}\varrho J S^2 - \frac{1}{4}\varrho J S + \frac{3}{4\beta} - \frac{3}{4\beta}\left(\beta^2 \varrho^2 J^2 S^2 - \frac{2}{3}\beta\varrho J S + 1\right)^{1/2} \\ & - \frac{1}{\beta}\ln(1-S) - \frac{1}{2\beta}\ln(1+2S) \quad . \end{aligned} \tag{13.107}$$

This expression can easily be dealt with numerically. Graphs of Δf versus S for fixed β show curves similar to the ones already found in the discussion of the Landau-De Gennes theory (see Fig. 12.1). The thermodynamic properties near the phase transition are obtained by combining the possible solutions for the order parameter, i.e. $S = 0$ and expression (13.106), with the corresponding behaviour of the difference in free energy between the ordered and disordered phase. Four distinct temperature regions are then found.

(i) $T > T_c^\dagger$. Only the isotropic phase exists here. The solutions S_+ and S_- are both imaginary and consequently do not apply. The temperature $T_c^\dagger$ is determined by

$$\left[1 - \frac{8}{3\beta_c^\dagger \varrho J}\right]^{1/2} = 0 \quad .$$

This means that $T_c^\dagger$ and the corresponding value of the order parameter $S^\dagger = S(T_c^\dagger)$ are given by

$$T_c^\dagger = \tfrac{3}{8}\varrho J/k_B \quad , \quad S^\dagger = 0.25 \quad . \tag{13.108}$$

(ii) $T_c < T \leqq T_c^\dagger$. The minimum of Δf is still given by $S = 0$, i.e. the isotropic phase is the thermodynamically stable state. In addition a local minimum and local maximum are found for respectively $S = S_+$ and $S = S_-$. This means that there is a barrier in the free energy between the two minima $S = 0$ and $S = S_+$, that has a height $\Delta f(S_-) - \Delta f(S_+)$. Thus a metastable nematic state can occur in this temperature range. If $T = T_c^\dagger$ the height of the barrier is zero, since $S_+ = S_- = \frac{1}{4} = S^\dagger$, and the corresponding point in the free energy curve is a point of inflection.

(iii) $T_c^* \leqq T \leqq T_c$. Now the solution $S = S_+$ gives rise to the lowest free energy and $S = 0$ corresponds to a local minimum. The nematic-isotropic transition temperature follows from $\Delta f(S_+) = 0$. It is easily found that

$$T_c = 0.370\varrho J/k_B \quad , \quad S_c = 0.335 \quad . \tag{13.109}$$

The $S = S_-$ solution still corresponds to a local maximum. The height of the barrier in the free energy between the two minima is given by $\Delta f(S_-)$. As soon as $S_- = 0$ the height of the barrier vanishes, this corresponds to a temperature T_c^* determined by

$$\frac{1}{4} - \frac{3}{4}\left(1 - \frac{8}{3\beta_c^* \varrho J}\right)^{1/2} = 0 \quad , \quad \text{i.e.} \quad T_c^* = \tfrac{1}{3}\varrho J/k_B \quad .$$

Because of the existence of the barrier in the free energy between the S_+ and $S = 0$ solution, a metastable isotropic state can exist within this temperature range. This state can be supercooled down to the temperature T_c^*. It follows that

$$(T_c - T_c^*)/T_c = 0.1 \quad .$$

(iv) $T < T_c^*$. The nematic state is the thermodynamically stable state. Now the $S = S_-$ solution corresponds to a local minimum and the $S = 0$ solution to a local maximum. There is, however, no way to reach the S_-

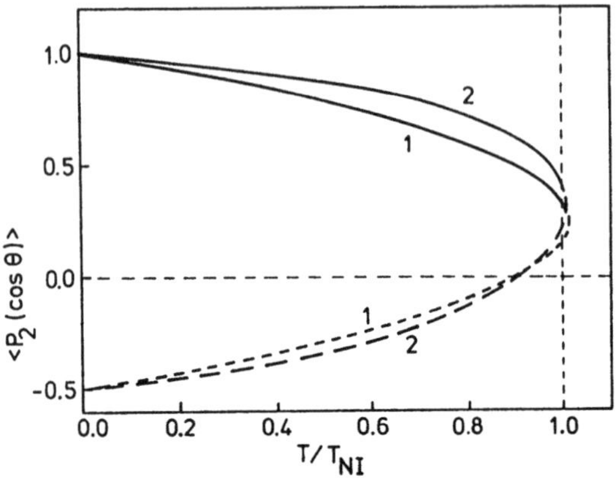

Fig. 13.4. Mean field solutions for the order parameter S as a function of the reduced temperature T/T_c for the Maier-Saupe model, where (*1*) relates to the present approach and (*2*) to the original MS approach. The solid lines give the solutions that correspond to the minimum of the free energy

state, which corresponds to a situation where the molecules tend to orient themselves perpendicular to the uniaxial axis.

All the solutions for the order parameter are once more illustrated in Fig. 13.4, both those obtained from (13.106) and the original MS solutions.

Finally the connection will be made between the molecular statistical theory and the phenomenological Landau theory by expanding (13.107) up to order S^4. In this way insight can be obtained into the qualitative merits of the Landau theory when dealing with first-order transitions in nematics.

According to the Landau-De Gennes theory the difference in free energy between the nematic and isotropic phase near the transition can be approximated by expanding (13.107) in the following way

$$\Delta f = \frac{1}{\beta}\left[\left(1-\frac{1}{3}\beta\varrho J\right)\left(\frac{3}{2}+\beta\varrho J\right)S^2 - \left(\frac{1}{9}\beta^3\varrho^3J^3+1\right)S^3 + \left(\frac{1}{27}\beta^4\varrho^4J^4+\frac{9}{4}\right)S^4\right] \quad . \qquad (13.110)$$

A comparison of expression (13.110) with the original Landau-De Gennes expression (12.13) then yields

$$a(T-T_c^*) = \frac{3}{\beta}\left(1-\frac{1}{3}\beta\varrho J\right)\left(\frac{3}{2}+\beta\varrho J\right) \quad ,$$

$$B = \frac{27}{2\beta}\left(\frac{1}{9}\beta^3\varrho^3 J^3 + 1\right) \quad ,$$

$$C = \frac{9}{\beta}\left(\frac{1}{27}\beta^4\varrho^4 J^4 + \frac{9}{4}\right) \quad .$$

Clearly a, B and C depend all three on the temperature, whereas T_c^* is determined by the relation

$$\tfrac{1}{3}\beta_c^*\varrho J = 1 \quad , \quad \text{i.e.} \quad T_c^* = \tfrac{1}{3}\varrho J/k_B \quad .$$

According to (12.19b) the expressions for S_c and T_c are given by

$$S_c = \frac{B}{3C} = \frac{6\beta_c^3\varrho^3 J^3 + 54}{4\beta_c^4\varrho^4 J^4 + 243} \quad ,$$

$$\frac{B^2}{27a(T_c - T_c^*)C} = \frac{2(\beta_c^3\varrho^3 J^3 + 9)^2}{(3 - \beta_c\varrho J)(3 + 2\beta_c\varrho J)(4\beta_c^4\varrho^4 J^4 + 243)} = 1 \quad ,$$

and it is found that

$$T_c = 0.385\varrho J/k_B \quad ; \quad S_c = 0.374 \quad . \tag{13.111}$$

The temperature $T_c^\dagger$, which determines the upper limit of the metastable nematic phase, is obtained from (12.20). It is easily found that $T_c^\dagger$, and the corresponding value of the order parameter $S^\dagger$ are given by

$$T_c^\dagger = 0.391\varrho J/k_B \quad ; \quad S^\dagger = \frac{B(T_c^\dagger)}{4C(T_c^\dagger)} = 0.280 \quad . \tag{13.112}$$

Now the Landau-De Gennes approach and the mean field approximation can be compared. It is found that the Landau-De Gennes expansion gives rise to an overestimation of $T_c^\dagger$ and T_c of the order of 4 %. The corresponding values of the order parameter $S^\dagger$ and S_c are overestimated by about 12 %. As expected, the temperature T_c^* is determined correctly. The ratio $(T_c - T_c^*)/T_c$, however, is overestimated by about 40 %.

13.5 Theory of the Chiral Nematic Phase

The appearance of the chiral nematic phase is due to the additional presence of so-called twist interactions originating from the molecular structure. The intermolecular interaction resulting from such a molecular structure can be represented in terms of tensors of an arbitrary rank. Evidently the question concerning the nature of the twist interactions reduces to a question of the required coupling between the different molecular tensors that produce twist. In general it can be stated that the intermolecular coupling between a tensor of odd rank and a tensor of even rank gives rise to a twist interaction because of the necessary appearance of an odd number of components of the intermolecular distance vector. Therefore a necessary condition for the existence of twist interactions is an intermolecular interaction that has to be described in terms of both odd and even rank tensors.

In order to describe the molecular structure, an orthogonal coordinate system with unit vectors $\boldsymbol{a}$, $\boldsymbol{b}$ and $\boldsymbol{c}$ is associated with a molecule, where

$$\boldsymbol{a} = \sin\theta\cos\phi\,\boldsymbol{e}_1 + \sin\theta\sin\phi\,\boldsymbol{e}_2 + \cos\theta\,\boldsymbol{e}_3 \quad , \tag{13.113a}$$

$$\begin{aligned}\boldsymbol{b} = {} & (\cos\theta\cos\phi\cos\psi - \sin\phi\sin\psi)\boldsymbol{e}_1 \\ & + (\cos\theta\sin\phi\cos\psi + \cos\phi\sin\psi)\boldsymbol{e}_2 - \sin\theta\cos\psi\,\boldsymbol{e}_3 \quad ,\end{aligned} \tag{13.113b}$$

$$\begin{aligned}\boldsymbol{c} = {} & -(\cos\theta\cos\phi\sin\psi + \sin\phi\cos\psi)\boldsymbol{e}_1 \\ & - (\cos\theta\sin\phi\sin\psi - \cos\phi\cos\psi)\boldsymbol{e}_2 + \sin\theta\sin\psi\,\boldsymbol{e}_3 \quad ,\end{aligned} \tag{13.113c}$$

and θ, ϕ, ψ are the Eulerian angles with respect to the laboratory coordinate system with unit vectors $\boldsymbol{e}_1$, $\boldsymbol{e}_2$ and $\boldsymbol{e}_3$ (see Fig. 5.1). The vector $\boldsymbol{a}$ is taken to be along the long molecular axis. The origin of this molecular coordinate system is situated at the centre of mass of the molecule. Starting with the nematic phase, the simplest intermolecular interaction that conforms with the macroscopic head-tail symmetry must be represented in terms of a second rand tensor

$$\Lambda_{\alpha\beta} = A a_\alpha a_\beta + B b_\alpha b_\beta + C c_\alpha c_\beta \quad , \tag{13.114}$$

where the axes $\boldsymbol{a}$, $\boldsymbol{b}$ and $\boldsymbol{c}$ are assumed to be principal axes. In order to simplify the discussion as far as possible the molecules are assumed to rotate independently of one another around the long molecular axis, i.e. the tensor $\Lambda_{\alpha\beta}$ can be averaged over the angle ψ. Averaging results in

$$\overline{\Lambda}_{\alpha\beta} = \int_0^{2\pi} d\psi\,\Lambda_{\alpha\beta} = \left(A - \frac{1}{2}B - \frac{1}{2}C\right) a_\alpha a_\beta + \frac{1}{2}(B + C)\delta_{\alpha\beta} \quad , \tag{13.115}$$

as can easily be derived from expressions (13.113). These immediately imply that

$$\overline{b_\alpha b_\beta} = \overline{c_\alpha c_\beta} = \tfrac{1}{2}\delta_{\alpha\beta} - \tfrac{1}{2}a_\alpha a_\beta \quad . \tag{13.116}$$

The interaction between the molecules i and j can then, for instance, be represented by

$$\begin{aligned} -J_{ij}^{(1)}\overline{\Lambda}_{\alpha\beta i}\overline{\Lambda}_{\alpha\beta j} = & -J_{ij}^{(1)}[(A - \tfrac{1}{2}B - \tfrac{1}{2}C)^2(\boldsymbol{a}_i \cdot \boldsymbol{a}_j)^2 \\ & + (A + B + C)(B + C)] \quad , \end{aligned} \tag{13.117}$$

where $J_{ij}^{(1)}$ depends on the distance r_{ij} between the centres of mass of the molecules i and j. Clearly the interaction (13.117) is nothing other than the interaction of the Maier-Saupe model. A second example of such a tensorial interaction is

$$\begin{aligned} & -J_{ij}^{(2)} u_{ij\alpha}u_{ij\beta}\overline{\Lambda}_{\alpha\gamma i}\overline{\Lambda}_{\beta\gamma j} \\ & \quad = -J_{ij}^{(2)}\{(A - \tfrac{1}{2}B - \tfrac{1}{2}C)^2(\boldsymbol{a}_i \cdot \boldsymbol{u}_{ij})(\boldsymbol{a}_j \cdot \boldsymbol{u}_{ij})(\boldsymbol{a}_i \cdot \boldsymbol{a}_j) \\ & \quad + \tfrac{1}{2}(A - \tfrac{1}{2}B - \tfrac{1}{2}C)(B + C)\,[(\boldsymbol{a}_i \cdot \boldsymbol{u}_{ij})^2 + (\boldsymbol{a}_j \cdot \boldsymbol{u}_{ij})^2] \\ & \quad + \tfrac{1}{4}(B + C)^2\} \quad , \end{aligned} \tag{13.118}$$

where $J_{ij}^{(2)}$ depends on r_{ij} while $\boldsymbol{u}_{ij} = \boldsymbol{r}_{ij}/r_{ij}$ and $\boldsymbol{r}_{ij}$ is the distance vector joining the centre of mass of molecule i to that of molecule j.

For the construction of twist terms, contractions between tensors of odd and even rank must be considered. The simplest twist interaction is obtained by a contraction between a tensor of the first rank, i.e. a vector, and a tensor of even rank, e.g. a tensor of the second rank. Because of the absence of any polar effects, however, tensors of the first rank cannot appear. Consequently the first possible candidate, which conforms to the symmetry, is a coupling between a third rank tensor and an even rank tensor. Such a third rank tensor is proportional for instance to

$$\Lambda_{\alpha\beta\gamma} = 2a_\alpha b_\beta c_\gamma \quad . \tag{13.119}$$

In contrast to a first rank tensor (vector) the mentioned third rank tensor does not vanish when averaged with respect to the up-down symmetry. In order to keep the discussion as simple as possible this tensor is averaged over the angle ψ as well. Because of the relation

$$b_\beta = \varepsilon_{\beta\mu\nu}c_\mu a_\nu$$

it follows, using (13.116), that the averaged tensor is given by

$$\overline{\Lambda}_{\alpha\beta\gamma} = 2a_\alpha a_\nu \varepsilon_{\beta\mu\nu}\overline{c_\mu c_\gamma} = a_\alpha \varepsilon_{\beta\gamma\nu} a_\nu \quad .$$

A well-known and simple example of a twist interaction between the molecules i and j is the coupling between the averaged third rank tensor and a second rank tensor:

$$\begin{aligned} -L_{ij}\overline{\Lambda}_{\alpha\beta\gamma i}\overline{\Lambda}_{\alpha\beta j} u_{ij\gamma} &= -\tfrac{1}{2}L_{ij}(A - \tfrac{1}{2}B - \tfrac{1}{2}C) a_{i\alpha} a_{j\alpha} u_{ij\gamma} \varepsilon_{\gamma\nu\beta} a_{i\nu} a_{j\beta} \\ &= -K_{ij}(\boldsymbol{a}_i \cdot \boldsymbol{a}_j)(\boldsymbol{a}_i \times \boldsymbol{a}_j \cdot \boldsymbol{u}_{ij}) \quad , \end{aligned} \tag{13.120}$$

where the coupling constants L_{ij} and $K_{ij} = \frac{1}{2}L_{ij}(A - \frac{1}{2}B - \frac{1}{2}C)$ depend on the intermolecular distance r_{ij}. This simple expression for the twist interaction was first proposed by Van der Meer et al. (1976). A specific case of this general expansion has also been given by Goossens (1971), who extended the original Maier-Saupe theory to the next higher order term in the multipole expansion. It should be noted here that notwithstanding the averaging procedure, the sense of a molecule, i.e. its property of being left-handed or right-handed, does not disappear but, on the contrary, is made much more explicit as is shown by the appearance of the Levi-Civita tensor.

The chiral nematic phase can be discussed in terms of the Maier-Saupe interaction energy

$$H_{\text{MS}} = -\frac{1}{2}\sum_{i,j} J(r_{ij})(\boldsymbol{a}_i \cdot \boldsymbol{a}_j)^2 \tag{13.121}$$

and the following twist interaction

$$H_{\text{twist}} = -\frac{1}{2}\sum_{i,j} K(r_{ij})(\boldsymbol{a}_i \cdot \boldsymbol{a}_j)(\boldsymbol{a}_i \times \boldsymbol{a}_j \cdot \boldsymbol{u}_{ij}) \quad . \tag{13.122}$$

The resulting model can easily be solved in the mean field approximation using the spherical constraint. For the sake of readability the discussion will be confined to the main characteristics and results of this calculation. The chiral nematic phase manifests itself in the dependence of the order parameters $\langle a_{i\alpha} a_{i\beta} \rangle$ on the position. Without loss of generality the direction of the helix axis can be identified with the macroscopic x axis. A given molecule will then experience the following mean field due to the presence of the other molecules

$$\begin{aligned} \overline{H} = &-J\langle a_x^2\rangle a_x^2 - \tfrac{1}{2}J(\langle a_y^2\rangle + \langle a_z^2\rangle)(a_y^2 + a_z^2) \\ &-\tfrac{1}{2}JA(q)(\langle a_y^2\rangle - \langle a_z^2\rangle)(a_y^2 - a_z^2) \end{aligned} \tag{13.123}$$

with

$$J = \sum_j J_{ij} \quad , \qquad \text{and} \tag{13.124a}$$

$$JA(q) = \sum_j [J_{ij} \cos(2qr_{ijx}) + K_{ij}u_{ijx} \sin(2qr_{ijx})] \quad , \tag{13.124b}$$

where q is the helical wave number of the chiral nematic. Now the free energy is calculated using the spherical constraint. The values of the order parameters and the helical wave number q follow from the requirement that they must minimize the free energy. Because of the constraint $\langle \boldsymbol{a}^2 \rangle = 1$ it is advantageous to introduce a new set of order parameters S and R defined by

$$\langle a_x^2 \rangle = \tfrac{1}{3}(1 - S + R) \quad , \tag{13.125a}$$

$$\langle a_y^2 \rangle = \tfrac{1}{3}(1 - S - R) \quad , \tag{13.125b}$$

$$\langle a_z^2 \rangle = \tfrac{1}{3}(1 + 2S) \quad . \tag{13.125c}$$

The order parameter S approaches the value of the nematic order parameter of the MS model in the limit $q \to 0$, in which case the biaxiality order parameter R will approach zero.

The equilibrium value of the helical wave number, q_0, is found to be independent of temperature. In practice q_0 is small, and can be approximated by (Scholte and Vertogen 1982)

$$q_0 = \frac{K}{2r_0} \quad \text{with} \tag{13.126}$$

$$r_0^2 = \frac{1}{J} \sum_j J_{ij} r_{ijx}^2 \quad ; \quad K = \frac{1}{Jr_0} \sum_j K_{ij} u_{ijx} r_{ijx} \quad . \tag{13.127}$$

The relevant quantities at the chiral nematic-isotropic transition calculated up to order $(q_0 r_0)^2$ are

$$T_c = 0.247[1 + \tfrac{3}{2}(q_0 r_0)^2] J/k_B \quad ,$$

$$S_c = 0.335 \quad ,$$

$$R_c = 0.333(q_0 r_0)^2 \quad .$$

Note that the chirality of the medium does not influence the value of the nematic order parameter S up to order $(q_0 r_0)^2$. Furthermore the chirality of the phase requires a second order parameter R, that describes the biaxiality of the system. In practice, where $|q_0 r_0| \approx 10^{-2}$, R is of the order of 10^{-5} and the system may be considered to be locally a nematic.

Experimentally a temperature-dependent pitch is observed. This means that the present model must be extended in order to describe this effect as well. Clearly such an extension does not present any problems because of the existence of an infinite variety of orientation-dependent interactions. The real problem concerns the relative relevance of all possible interactions, or to put it differently, the problem of the correlation between the molecular structure and the physical behaviour of the pitch is still largely unsolved.

13.6 On the Relevance of Molecular Models

In view of the preceding sections of this chapter it is not unlikely that the reader is inclined to question the relevance of the molecular models presented. Such an attitude may well be defended because the present theories are unable to describe the physical behaviour of nematics in a complete and satisfactory way. Moreover it seems quite obvious that the most interesting feature of a theory, namely the power to predict, is out of the question as far as the behaviour of specific molecules is concerned in view of the complicated intermolecular interactions. Even if the essentials of the intermolecular interaction were known, one might question whether a satisfactory molecular-statistical theory could be formulated because of the enormous calculational problems. Nevertheless, some arguments can be given that support the present molecular models. For this reason the merits and shortcomings of these models are summarized here.

First of all it must be emphasized that all these models are only of a qualitative value. This fact, and the qualitative nature of the calculations must be borne in mind in order to appreciate the theory correctly. The basic ingredients of the theory of nematics are the axial ratio x, the volume of a molecule v_0, a parameter J_0 related to the isotropic interaction between the molecules, and a parameter J describing the anisotropic interaction between the molecules. These parameters are assumed to be essential for a qualitative description of the physical behaviour of the nematic state. As discussed in Sect. 13.3 this model indeed gives a description of the NI transition but the relevant values of the parameters do not correlate, in general, with the molecular structure. Hence utmost care must be taken when correlating these parameters with molecular properties. Apart from the qualitative character of the calculations, which tend to overestimate considerably the influence of the parameters (e.g. the axial ratio), the molecular structure is so complex that it is not usually sufficient to describe such a structure in terms of for instance the isotropic and anisotropic part of the polarizability.

When formulating a molecular statistical theory of the nematic phase some form for the intermolecular interactions must be used in order to be explicit. Because of the symmetry of the nematic phase it is natural to start with molecules having a biaxial symmetry. The molecular structure is then represented mathematically in terms of tensors of that symmetry and only tensors of the lowest rank are taken into account. Unfortunately, however, such a choice cannot be justified at all on physical grounds since the electrostatic interaction cannot be approximated by the first few terms of the multipole expansion as assumed by Maier and Saupe (1959, 1960). Consequently, the relative success of this model does not imply that the dipole-dipole interaction is indeed of dominant importance. Similarly ascribing the existence of the chiral nematic state to the induced dipole-quadrupole interaction (Goossens 1971) lacks substantial physical justification. The chiral nematic state arises because of the molecular property "handedness", which shows up clearly in the optical activity of the molecule. The description of such a molecular structure involves odd-rank tensors resulting in an intermolecular twist interaction as was shown in Sect. 13.5. The induced dipole-quadrupole interaction is a simple example of such a twist interaction but not necessarily the dominant contribution.

The molecular models can be even further simplified by averaging the intermolecular interaction with respect to rotations around the long molecular axes. Such a procedure can only be justified experimentally, for instance by the smallness of the order parameter D (Sect. 9.3). Thus, effectively, molecules with uniaxial symmetry are obtained, which is a prerequisite for using the Maier-Saupe interaction and the simple twist interaction (13.122). This stresses once more the ad hoc nature of these models. They only represent some ideas about the nematic phase and at best indicate trends to be expected when varying simple molecular parameters like e.g. the axial ratio. At this stage it should be mentioned that one important molecular parameter has not been dealt with here, namely the flexibility of the tail, as no simple treatment of this effect can be given. For a discussion of the influence of the tail, e.g. the so-called odd-even effect, the reader is referred to Marcelja (1974). More recently the influence of the conformation of the alkyl chains in stabilizing the various liquid crystalline phases has been discussed by Dowell and Martire (1978) using lattice models.

Summarizing it can be concluded that the merits of the molecular statistical theory are on the conceptual level in the sense that some simple trends can be indicated. These trends, however, cannot be demonstrated experimentally in a clear way because it is impossible to vary one single molecular parameter. An explanation of the physical behaviour of the various nematics does not exist, e.g. their elastic constants cannot be satisfactorily correlated with their molecular structure (Flapper et al. 1981). In this respect it is worthwhile to mention that the MS model gives rise to elastic constants for splay, twist and bend, that are equal, whereas the induced dipole-dipole model gives rise to $K_1{:}K_2{:}K_3 = 5{:}11{:}5$ (see Sect. 5.4). Therefore the conclusion seems justified that, as far as theoretical physics is concerned, the design of nematics with specific elastic constants and viscosities should be considered to be an art rather than a science! The present remarks concerning the status of the molecular-statistical approach can also be applied to models for smectics which will be discussed in the next chapter.

14. Molecular Statistical Theory of the Smectic Phases

The topic of this chapter is the molecular theory of the smectic A and the smectic C phases. First we discuss the smectic A phase and the smectic A-nematic transition in terms of two models. The first model is very simple and describes a perfectly oriented smectic A phase. The second model deals with the effect of deviations from perfect orientational order, which can change the smectic A-nematic phase transition from second order to first order. Next we discuss the smectic C phase in terms of a model that ascribes the origin of the smectic C phase to the influence of the dipole-induced dipole interaction and we show that the smectic C-smectic A transition is a second-order transition. Finally we will deal with to the chiral smectic C phase.

14.1 Theory of the Smectic A-Nematic Transition

The transition of the nematic to the smectic phase involves a rearrangement of the centres of the molecules as is discussed in Chap. 3. The centres are disordered in the nematic state and ordered in the smectic state. The ordering in the smectic A (S_A) phase is such that the molecular centres are, on average, arranged in equidistant planes with interplanar spacing d. Within these layers the molecules move at random, with the restriction that the director remains perpendicular to the smectic layers. Consequently the density changes its behaviour at the S_AN transition from periodic to homogeneous. Thus the theory of the S_AN transition is a theory of melting and is closely related to the theory of fusion of crystals (Kirkwood 1951). Because the density is periodic in only one direction (perpendicular to the smectic planes) the S_AN transition can be considered to be a form of "one-dimensional" melting.

A molecular-statistical theory of smectic layering should be able to explain the layering and other basic observations regarding smectic behaviour as discussed in Sect. 1.2 and Chap. 3. In particular the following points must be taken care of:

i) the occurrence of smectic phases in a homologous series if the alkyl chain length is relatively long (see Fig. 1.5);

ii) the change of a second-order $S_A N$ phase transition into a first-order transition when T_{AN} approaches T_{NI}.

The second point has already been discussed within the context of the Landau theory (Sect. 12.4). However, such a theory does not allow for an interpretation in terms of intermolecular interactions. As mentioned in Sect. 1.2 the asymmetry of the molecules can also be important, and this could be added as a third point. However, in a first attempt to formulate a molecular-statistical theory of the smectic phase any molecular asymmetry will be disregarded. This has the advantage of simplifying the treatment, as interactions involving permanent dipole moments can be ignored. The results are still relevant, because homologous series of non-polar liquid crystals exist that show a phase behaviour similar to that shown in Fig. 1.5.

The first molecular statistical treatment of the smectic-nematic transition has been given by Kobayashi (1970, 1971) on the basis of a rather ad hoc model, namely the Maier-Saupe (MS) model with an additional isotropic intermolecular interaction. A similar but more explicit approach was followed by McMillan (1971), who discussed the $S_A N$ transition in terms of the MS model using, however, the additional constraint that the director is perpendicular to the smectic planes. This condition is necessary because the existence of the smectic phase requires a coupling of the orientations of the molecules with the orientations of the intermolecular distance vectors and evidently such a coupling is not present in the MS model. Before discussing the Kobayashi-McMillan approach, however, an even simpler treatment of the $S_A N$ transition due to Meyer and Lubensky (ML) (1976) will be mentioned. The ML model limits the discussion of the $S_A N$ transition to the case of ideal orientational order, i.e. the temperature dependence of the orientation of the molecules is neglected. Such an approach has the great advantage that attention is completely focused on the behaviour of the density, i.e. the melting phenomenon, and that the introduction of ad hoc anisotropic intermolecular interactions can be avoided as well.

14.1.1 The Meyer-Lubensky Model

In the ML model the molecules are assumed to be perfectly oriented in a given direction, say the z direction, whereas their centres of mass are situated on planes, parallel to the xy plane. The interaction between two arbitrary planes, situated at z and z', is assumed to have the form

$$-U(z-z') = -\int_{-\infty}^{+\infty} A(k) \cos[k(z-z')]\, dk \quad , \quad \text{where} \tag{14.1}$$

$$A(k) = \frac{1}{2\pi} \int_{-\infty}^{+\infty} U(z) \cos(kz)\, dz \quad . \tag{14.2}$$

Next the mean field approximation is applied for computational purposes, i.e. the interaction between the layer at position z and all other layers is replaced by the interaction with an effective field. This field, $M(z)$, is obtained by multiplying the interaction $-U(z-z')$ with the density $\varrho(z')$ at the site z' and integrating the resulting expression with respect to z'. Accordingly the effective field becomes

$$M(z) = -\int_{-\infty}^{+\infty} U(z-z')\varrho(z')dz' \quad . \tag{14.3}$$

The density $\varrho(z)$ is a periodic function with period d. Without loss of generality a smectic plane may be situated at $z = 0$. Consequently the density can be expanded in the following Fourier series

$$\varrho(z) = \varrho_0 + 2\varrho_0 \sum_{n=1}^{\infty} \varrho_n \cos(nqz) \quad , \tag{14.4}$$

where $q = 2\pi/d$ and

$$\varrho_0 = \frac{1}{d} \int_0^d \varrho(z)dz \quad , \tag{14.5a}$$

$$\varrho_n = \frac{1}{d\varrho_0} \int_0^d \varrho(z) \cos(nqz)\, dz \quad . \tag{14.5b}$$

Substitution of (14.4) into (14.3) gives the following expression for the effective field

$$M(z) = -\frac{1}{2}U_0 - \sum_{n=1}^{\infty} U_n \varrho_n \cos(nqz) \quad \text{with} \tag{14.6}$$

$$U_n = 4\pi\varrho_0 A(nq) = 2\varrho_0 \int_{-\infty}^{+\infty} U(z') \cos(nqz')\, dz' \quad . \tag{14.7}$$

It directly follows from (14.6) that the smectic phase appears as soon as the coefficients ϱ_n $(n \geq 1)$ are unequal to zero as an effective field results with period d. The average density of the system, ϱ_0, is known, whereas the order parameters ϱ_n $(n \geq 1)$, that describe the smectic phase, are determined in a self-consistent way using the definition of the density

$$\varrho(z) = \frac{\varrho_0}{Z} \exp[-\beta M(z)] \quad \text{where} \tag{14.8}$$

$$Z = \frac{1}{d} \int_0^d \exp[-\beta M(z)] dz \quad . \tag{14.9}$$

Equations (14.5b) and (14.8) immediately imply

$$\varrho_n = \frac{1}{Zd} \int_0^d \cos(nqz) \exp[-\beta M(z)] dz = \langle \cos(nqz) \rangle \quad . \tag{14.10}$$

The order parameters ϱ_n are the expectation values of $\cos(nqz)$. Within the framework of the ML model the nematic or isotropic phase is described by the uniform density ϱ_0 only. The smectic phase is described by the infinite set of order parameters $\{\varrho_n\}$.

The solution of the ML model can be simplified considerably by the truncation $U_n = 0$, $n > 2$. In general such a procedure is not justified as can be seen by considering a Gaussian potential

$$U(z - z') = A \exp\left[-\frac{(z - z')^2}{\lambda^2}\right] \tag{14.11}$$

with λ being a measure for the range of the potential. Then from (14.7) it follows that

$$U_n = 2\varrho_0 \lambda \pi^{1/2} A \exp[-(\tfrac{1}{2} nq\lambda)^2] \quad . \tag{14.12}$$

Consequently the truncation can be applied if λ is large, i.e. if the potential is flat. On the other hand a strongly peaked Gaussian potential, i.e. small λ, does not allow such a truncation.

Here only the truncated ML model will be considered which, apart from an irrelevant constant $-\frac{1}{2}U_0$, is defined by the effective field

$$M_t(\phi) = -\sum_{n=1}^{2} U_n \varrho_n \cos(n\phi) \quad , \tag{14.13}$$

where $\phi = qz$. The order parameter ϱ_1 and ϱ_2 are determined by the equations

$$\varrho_n = \langle \cos(n\phi) \rangle = \frac{1}{Z} \int_0^{2\pi} \cos(n\phi) \exp[-\beta M_t(\phi)] \, d\phi \quad , \quad n = 1, 2 \quad , \tag{14.14}$$

with

$$Z = \int_0^{2\pi} \exp[-\beta M_t(\phi)]d\phi \quad . \tag{14.15}$$

This formulation shows the analogy between the ML approach to the S_AN phase transition and the theory of magnetic transitions. The truncated ML model (14.13) is nothing but the mean field approximation of a Heisenberg model (Van Vleck 1932) of the form

$$H = -\frac{1}{2}\sum_{i,j} v_{ij}\boldsymbol{a}_i \cdot \boldsymbol{a}_j - \frac{1}{2}\sum_{i,j} w_{ij}[2(\boldsymbol{a}_i \cdot \boldsymbol{a}_j)^2 - 1] \quad , \tag{14.16}$$

where $\boldsymbol{a}_i$ and $\boldsymbol{a}_j$ are two-dimensional unit vectors and the coupling constants v_{ij} and w_{ij} are such that

$$\sum_j v_{ij} = U_1 \quad , \quad \sum_j w_{ij} = U_2 \quad . \tag{14.17}$$

Defining $a_x = \cos\phi$ and $a_y = \sin\phi$ the relation between the original order parameters ϱ_1 and ϱ_2 of the truncated ML model and the order parameters of the Heisenberg model $\langle a_x \rangle$, $\langle a_y \rangle$, $\langle a_x^2 \rangle$ and $\langle a_y^2 \rangle$ can be derived easily. It is immediately found that

$$\begin{aligned} &\langle a_x \rangle = \varrho_1 \quad , &&\langle a_y \rangle = 0 \\ &\langle a_x^2 \rangle = \tfrac{1}{2}(1+\varrho_2) \quad , &&\langle a_y^2 \rangle = \tfrac{1}{2}(1-\varrho_2) \quad . \end{aligned} \tag{14.18}$$

For simplicity the spherical version of the Heisenberg model (14.16) will be considered, i.e. the two-dimensional vector $\boldsymbol{a}_i$ is allowed to have any length but such that the spherical constraint is satisfied, meaning that

$$\sum_{i=1}^{N} \boldsymbol{a}_i^2 = N \quad . \tag{14.19}$$

In order to apply the mean field approximation the following identities are used, where the two components of $\boldsymbol{a}_i$ are denoted by a_{ix} and a_{iy}, respectively, and the Cartesian coordinate system has been chosen such that $\langle a_x a_y \rangle = 0$:

$$\begin{aligned} \boldsymbol{a}_i \cdot \boldsymbol{a}_j &= (a_{ix} - \langle a_x \rangle)(a_{jx} - \langle a_x \rangle) + (a_{iy} - \langle a_y \rangle)(a_{jy} - \langle a_y \rangle) \\ &\quad + \langle a_x \rangle (a_{ix} + a_{jx}) + \langle a_y \rangle (a_{iy} + a_{jy}) - \langle a_x \rangle^2 - \langle a_y \rangle^2 \quad , \end{aligned} \tag{14.20a}$$

$$(\boldsymbol{a}_i \cdot \boldsymbol{a}_j)^2 = (a_{ix}^2 - \langle a_x^2\rangle)(a_{jx}^2 - \langle a_x^2\rangle) + (a_{iy}^2 - \langle a_y^2\rangle)(a_{jy}^2 - \langle a_y^2\rangle)$$
$$+ 2a_{ix}a_{iy}a_{jx}a_{jy} + \langle a_x^2\rangle(a_{ix}^2 + a_{jx}^2)$$
$$+ \langle a_y^2\rangle(a_{iy}^2 + a_{jy}^2) - \langle a_x^2\rangle^2 - \langle a_y^2\rangle^2 \quad . \tag{14.20b}$$

These identities take into account that the expectation values $\langle a_\alpha\rangle$ and $\langle a_\alpha^2\rangle$ with $\alpha = x, y$, do not depend on the position. Next the fluctuations $a_{i\alpha} - \langle a_\alpha\rangle$, $a_{i\alpha}^2 - \langle a_\alpha^2\rangle$ and $a_{ix}a_{iy}$ are neglected. Now the partition function of the model (14.16) in the mean field approximation (see also Chap. 13) becomes

$$Z = \exp\left\{\mu N - \beta N\left[\frac{1}{2}U_2 + \sum_\alpha\left(\frac{1}{2}U_1\langle a_\alpha\rangle^2 + U_2\langle a_\alpha^2\rangle^2\right)\right]\right\}$$
$$\times \int d^2a_1 \ldots d^2a_N \exp\left\{\sum_{i=1}^N\sum_\alpha [(2\beta U_2\langle a_\alpha^2\rangle - \mu)a_{i\alpha}^2\right.$$
$$\left. + \beta U_1\langle a_\alpha\rangle a_{i\alpha}]\right\} \quad , \tag{14.21}$$

where the Lagrange multiplier μ has been introduced in order to take the spherical constraint (14.19) into account. This multiplier is determined by the condition $\partial f/\partial\mu = 0$ where f is the Helmholtz free energy per vector $\boldsymbol{a}$:

$$\beta f = -\mu + \frac{1}{2}\beta U_2 + \sum_\alpha\left[\frac{1}{2}\beta U_1\langle a_\alpha\rangle^2 + \beta U_2\langle a_\alpha^2\rangle^2\right.$$
$$\left. - \frac{1}{2}\ln\left(\frac{\pi}{\mu - 2\beta U_2\langle a_\alpha^2\rangle}\right) - \frac{\beta^2 U_1^2\langle a_\alpha\rangle^2}{4(\mu - 2\beta U_2\langle a_\alpha^2\rangle)}\right] \quad . \tag{14.22}$$

Consequently μ follows from solving the equation

$$\sum_\alpha\left[\frac{1}{2(\mu - 2\beta U_2\langle a_\alpha^2\rangle)} + \frac{\beta^2 U_1^2\langle a_\alpha\rangle^2}{4(\mu - 2\beta U_2\langle a_\alpha^2\rangle)^2}\right] = 1 \quad . \tag{14.23}$$

The order parameters $\langle a_\alpha\rangle$ and $\langle a_\alpha^2\rangle$ are determined in a self-consistent way by calculating the expectation values of a_α and a_α^2 using the distribution function that results from the mean field approximation. The equations for these order parameters also follow from the conditions $\partial f/\partial\langle a_\alpha\rangle = \partial f/\partial\langle a_\alpha^2\rangle = 0$:

$$\beta U_1\langle a_\alpha\rangle\left[1 - \frac{\beta U_1}{2(\mu - 2\beta_2\langle a_\alpha^2\rangle)}\right] = 0 \quad , \tag{14.24a}$$

$$\langle a_\alpha^2\rangle - \frac{1}{2(\mu - 2\beta U_2\langle a_\alpha^2\rangle)} - \frac{\beta^2 U_1^2\langle a_\alpha\rangle^2}{4(\mu - 2\beta U_2\langle a_\alpha^2\rangle)} = 0 \quad . \tag{14.24b}$$

In the following only the truncated Gaussian potential (14.11) is considered, i.e.

$$U_1 = 2\varrho_0 \lambda \pi^{1/2} \exp[-(\tfrac{1}{2}q\lambda)^2] \ , \quad U_2 = 2\varrho_0 \lambda \pi^{1/2} \exp[-(q\lambda)^2] \ . \tag{14.25}$$

The self-consistency relations (14.24) combined with equation (14.23) for μ can be solved in a straightforward way. The relevant physical order parameters ϱ_1 and ϱ_2 are obtained from the order parameters $\langle a_\alpha \rangle$ and $\langle a_\alpha^2 \rangle$ by using the expression (14.18). The following solutions with corresponding free energies are found:

(1) $\varrho_2 = 0 \ , \quad \langle a_x \rangle = \langle a_y \rangle = 0 \ , \quad \mu = 1 + \beta U_2 \ , \quad \varrho_1 = 0 \ ,$
$\beta f_1 = -1 - \ln \pi \ ;$

(2) $\varrho_2 = 0, \ \langle a_x \rangle^2 = \langle a_y \rangle^2 = \frac{1}{2} - \frac{1}{\beta U_1} \ , \quad \mu = \frac{1}{2}\beta U_1 + \beta U_2 \ ,$
no solution for ϱ_1,
$\beta f_2 = -\frac{1}{2}\beta U_1 - \ln\left(\frac{2\pi}{\beta U_1}\right) \ ;$

(3) $\varrho_2^2 = 1 - \frac{1}{\beta U_2} \ , \quad \langle a_x \rangle = \langle a_y \rangle = 0 \ , \quad \mu = 2\beta U_2 \ , \quad \varrho_1 = 0 \ ,$
$\beta f_3 = -\frac{1}{2} - \ln \pi - \frac{1}{2}\beta U_2 + \frac{1}{2}\ln(\beta U_2) \ ;$

(4) $\varrho_2 = \frac{1}{2} - \frac{U_1}{8U_2} \pm \frac{1}{2}\left[\left(\frac{U_1}{4U_2} + 1\right)^2 - \frac{2}{\beta U_2}\right]^{1/2} \ ,$
$\langle a_x \rangle = \varrho_1 \ , \quad \langle a_y \rangle = 0 \ ,$
$\varrho_1^2 = \frac{1}{2} + \frac{1}{2}\varrho_2 - \frac{1}{\beta U_1} \ , \quad \mu = \frac{1}{2}\beta U_1 + \beta U_2 + \beta U_2 \varrho_2 \ ,$
$$\beta f_4 = -\frac{1}{2}\beta U_1 - \beta U_2 \varrho_2 + \frac{1}{2}\beta U_2 \varrho_2^2 - \frac{1}{2}\ln\left(\frac{2\pi}{\beta U_1}\right) - \frac{1}{2}\ln\left(\frac{2\pi}{\beta U_1 + 4\beta U_2 \varrho_2}\right) \ .$$

Solution (1) always holds and represents the nematic or isotropic phase. The remaining solutions only hold for certain temperature regions. If several solutions exist at a given temperature the physically acceptable solution is obtained by means of the additional criterion that the free energy must be as low as possible.

The relevant parameter of the ML model is the ratio $x = U_2/U_1 = \exp[-\frac{3}{4}(q\lambda)^2]$. Depending on the value of x the S_AN transition is second or first order. Moreover, a second smectic A phase appears if x exceeds a critical value x_c. The behaviour of the ML model as a function of x and of temperature is shown in the phase diagram drawn in Fig. 14.1. Four relevant regions of the parameter x must be distinguished. These regions are discussed in the following.

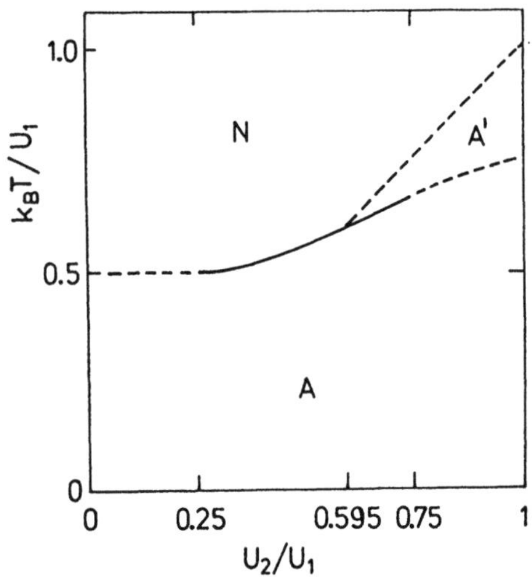

Fig. 14.1. The phase diagram of the truncated Meyer-Lubensky model. Four regions can be distinguished: $0 \leq U_2/U_1 \leq 1/4$, $1/4 < U_2/U_1 \leq 0.595$, $0.595 < U_2/U_1 \leq 3/4$ and $3/4 < U_2/U_1 < 1$. The solid line represents a first-order transition and the dashed lines correspond to second-order transitions. Three different phases appear: the nematic phase (N) having $\varrho_1 = \varrho_2 = 0$, the smectic A$'$ phase corresponding to $\varrho_1 = 0$, $\varrho_2 \neq 0$, and the smectic A phase where $\varrho_1 \neq 0$ and $\varrho_2 \neq 0$

(A) $0 \leq x \leq \frac{1}{4}$ or $\lambda/d \geq 0.22$. Here the range of the potential exceeds about a quarter of the interlayer distance d, which is roughly equal to the length of a molecule. The model describes a second-order S_AN transition at a temperature $T_c = \frac{1}{2}U_1/k_B$. Above T_c the nematic or isotropic phase described by solution (1) is found, whereas below T_c the S_A phase described by solution (4) is found. As an example the temperature dependence of the order parameters ϱ_1 and ϱ_2, that describe the S_A phase, is presented in Fig. 14.2a for the value $x = 3/16$ or $\lambda/d = 0.24$.

(B) $\frac{1}{4} < x < 0.595$ or $0.22 > \lambda/d > 0.13$. Now the S_AN transition is first order and the transition T_c must be determined numerically by solving $f_4 - f_1 = 0$. A tricritical point occurs at the ratio $x_c = \frac{1}{4}$, i.e. the nature of the transition depends on the sign of $x - x_c$. As an example Fig. 14.2b shows the temperature dependence of the order parameters ϱ_1 and ϱ_2 for $x = 7/16$ or $\lambda/d = 0.17$.

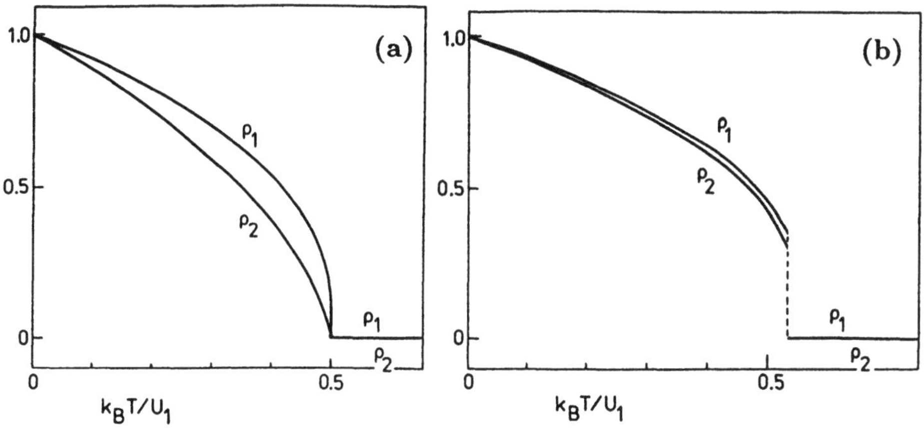

Fig. 14.2. (a) The temperature dependence of the order parameters ϱ_1 and ϱ_2 for $U_2/U_1 = 3/16$. (b) The temperature dependence of the order parameters ϱ_1 and ϱ_2 for $U_2/U_1 = 7/16$

(C) $0.595 \leq x < \frac{3}{4}$ or $0.13 \geq \lambda/d > 0.1$. For this range of the potential the ML model gives rise to an additional phase indicated by smectic A′. This smectic phase is characterized by solution (3) describing a periodic layered structure with period $d/2$. In addition the normal smectic A phase is found, characterized by solution (4) and a periodicity d. In order to understand such a behaviour a period $d/2$ instead of d must be associated with the molecular length, otherwise the absence of the order parameter ϱ_1 in the smectic A′ phase cannot be understood. The nematic or isotropic phase is stable at temperatures $T > T_1 = U_2/k_B$, the smectic A′ phase appears within the temperature region $T_2 < T < T_1$, where T_2 must be determined numerically from solving $f_4 - f_3 = 0$. The smectic A phase is found at temperatures $T < T_2$. The $S_A S_{A'}$ transition is a first-order transition. The triple point of the system occurs at the parameter ratio $x = 0.595$ and at a temperature $T_c = 0.595 U_1/k_B$. The occurrence of the additional smectic A′ phase should probably be considered as an artefact of the model that occurs for large values of U_2.

(D) $\frac{3}{4} \leq x \leq 1$ or $0.1 \geq \lambda/d \geq 0$. Now the $S_A S_{A'}$ transition is a second-order phase transition. The smectic A′ phase lies within the temperature region $T_2 < T < T_1$, where $T_1 = U_2/k_B$ and $T_2 = [1 - 1/(4x)]U_1/k_B$. The tricritical point for the $S_A S_{A'}$ transition occurs at $x = \frac{3}{4}$.

Summarizing, only three out of the four possible solutions of the combined set of order parameter equations (14.24) and the equation for the Lagrange multiplier (14.23) are relevant. As expected solution (2) never appears. Depending on the values of both x and the temperature, the system is in the nematic or isotropic phase, the simply layered smectic A phase with period d, or the smectic A$'$ phase with period $d/2$. The S_AN and $S_AS_{A'}$ transition are either first or second order. Finally it should be stressed that the connection between the molecular structure and the proposed potential is in fact unknown. Therefore the physical interpretation of the mathematical description is rather unclear. This is true in particular for the smectic A$'$ phase and the $S_AS_{A'}$ phase transition.

14.1.2 The Kobayashi-McMillan Model

The Kobayashi-McMillan (KM) description of the S_AN transition takes into account the effect of orientational order on the nature of the transition. Therefore an ad hoc orientational interaction must be introduced. As can be expected, the MS model is chosen as the starting point for the description. An additional constraint is added requiring that the average orientation of the molecules must be perpendicular to the smectic layers. Unfortunately the proposed model does not lend itself to a simplified treatment using the spherical constraint due to the simultaneous appearance of smectic and orientational order parameters. For this reason only the essentials of the KM treatment are discussed here.

In the KM theory the molecules are again assumed to be oriented along the z direction, but in contrast to the ML model the molecular orientational ordering does not have to be ideal. In the S_A phase the centres of mass will be situated on planes parallel to the xy plane. The starting point is the MS interaction. This means that the interaction (14.1) between two arbitrary planes, situated at z and z', is given by

$$-U(z-z')P_2(\cos\theta)P_2(\cos\theta') = -\int_{-\infty}^{+\infty} dk\, A(k)\cos[k(z-z')]P_2(\cos\theta)P_2(\cos\theta') \quad , \tag{14.26}$$

where θ and θ' are, respectively, the angles between the orientations of the smectic layers at z and z' and the direction of the z axis. The orientation-dependent part of the interaction, which has a broken symmetry, is also called the MS interaction, because it results from taking the following average of $P_2(\boldsymbol{a}_i\cdot\boldsymbol{a}_j)$. Defining

$$\boldsymbol{a} = (\sin\theta\cos\phi,\ \sin\theta\sin\phi,\ \cos\theta)$$

the breaking of the rotation symmetry of the interaction is obtained by averaging over ϕ_i and ϕ_j, as the following relation holds

$$\frac{1}{(2\pi)^2}\int_0^{2\pi} d\phi_i \int_0^{2\pi} d\phi_j P_2(\boldsymbol{a}_i \cdot \boldsymbol{a}_j) = P_2(\cos\,\theta_i)P_2(\cos\,\theta_j) \quad .$$

The next step concerns the replacement of the interaction of the layer at site z with all remaining layers by an interaction with an effective field. This field $M(z,\theta)$ is obtained by multiplying the interaction (14.26) with the density $\varrho(z',\theta')$ and subsequently integrating with respect to z' and $\cos\,\theta'$, i.e.

$$M(z,\theta) = -\int_{-\infty}^{+\infty} dz' \int_{-1}^{+1} d(\cos\,\theta')\varrho(z',\theta')U(z-z')P_2(\cos\,\theta)P_2(\cos\,\theta') \quad . \tag{14.27}$$

The density $\varrho(z,\theta)$, which describes the spatial distribution of molecules with a given orientation θ, is a periodic function with period d. This means that

$$\varrho(z,\theta) = \varrho_0 f(\theta) + 2\varrho_0 \sum_{n=1}^{\infty} \varrho_n(\theta)\,\cos(nqz) \quad , \tag{14.28}$$

where $q = 2\pi/d$ and

$$f(\theta) = \frac{1}{d\varrho_0}\int_0^d \varrho(z,\theta)dz \quad , \tag{14.29a}$$

$$\varrho_n(\theta) = \frac{1}{d\varrho_0}\int_0^d \varrho(z,\theta)\,\cos(nqz)\,dz \quad , \tag{14.29b}$$

with ϱ_0 representing the average density of the system. The orientational distribution function $f(\theta)$ satisfies the relation

$$\int_0^{\pi} f(\theta)\,\sin\,\theta\,d\theta = 1 \quad . \tag{14.30}$$

Substitution of (14.28) into (14.27) leads to the following expression for the effective field

$$M(z,\theta) = -\frac{1}{2}U_0 S P_2(\cos\,\theta) - \sum_{n=1}^{\infty} U_n\sigma_n\,\cos(nqz)P_2(\cos\,\theta) \tag{14.31}$$

with U_n defined by (14.7). The order parameters S and σ_n are given by

$$S = \int_0^\pi f(\theta') P_2(\cos\theta') \sin\theta' d\theta' \quad , \quad \text{and} \tag{14.32}$$

$$\sigma_n = \int_0^\pi \varrho_n(\theta') P_2(\cos\theta') \sin\theta' d\theta' \quad . \tag{14.33}$$

Now the density $\varrho(z,\theta)$ is defined by

$$\varrho(z,\theta) = \frac{\varrho_0}{Z} \exp[-\beta M(z,\theta)] \quad \text{with} \tag{14.34}$$

$$Z = \frac{1}{d} \int_0^d dz \int_0^\pi \exp[-\beta M(z,\theta)] \sin\theta \, d\theta \quad . \tag{14.35}$$

Consequently the distributions $f(\theta)$ and $\varrho_n(\theta)$ according to (14.29) and (14.34) become

$$f(\theta) = \frac{1}{Zd} \int_0^d \exp[-\beta M(z,\theta)] dz \quad , \tag{14.36a}$$

$$\varrho_n(\theta) = \frac{1}{Zd} \int_0^d \exp[-\beta M(z,\theta)] \cos(nqz) \, dz \quad . \tag{14.36b}$$

Substitution of (14.36) into (14.32) and (14.33) leads to an infinite set of coupled equations that determine the values of the order parameters S and σ_n. The KM approach limits the discussion of the $S_A N$ transition to a simplified version of the potential (14.31) by assuming $U_n = 0$, $n > 1$. Such a procedure can be justified if the range of the potential $U(z - z')$ is large as follows from (14.2). Consequently only two order parameters S and σ_1 appear. They are determined by the following two coupled equations

$$\begin{aligned} S &= \frac{1}{Zd} \int_0^d \int_0^\pi \exp[-\beta M(z,\theta)] \, P_2(\cos\theta) \sin\theta \, d\theta \, dz \\ &= \langle P_2(\cos\theta) \rangle \quad , \end{aligned} \tag{14.37}$$

$$\begin{aligned} \sigma_1 &= \frac{1}{Zd} \int_0^d \int_0^\pi \exp[-\beta M(z,\theta)] \cos(qz) P_2(\cos\theta) \sin\theta \, d\theta \, dz \\ &= \langle \cos(qz) P_2(\cos\theta) \rangle \quad \text{with} \end{aligned} \tag{14.38}$$

$$M(z,\theta) = -\tfrac{1}{2} U_0 S P_2(\cos\theta) - U_1 \sigma_1 \cos(qz) P_2(\cos\theta) \quad . \tag{14.39}$$

Equations (14.37) and (14.38) must be solved self-consistently. Depending on the values of the coupling parameters βU_0 and βU_1 three types of solution are possible:

i) $S = \sigma_1 = 0$, this solution is always found and describes the isotropic liquid phase;
ii) $S \neq 0$, $\sigma_1 = 0$, this solution describes the nematic phase in accordance with the Maier-Saupe theory;
iii) $S \neq 0$, $\sigma_1 \neq 0$, this solution describes the S_A phase in terms of the mixed orientational-translational order parameter σ_1.

If the self-consistency equations (14.37) and (14.38) allow more than one solution, the thermodynamically stable solution is selected by the criterion of the lowest free energy. According to (14.31) the internal energy per molecule, u, is given by

$$u = -\tfrac{1}{2}(\tfrac{1}{2}U_0 S^2 + U_1 \sigma_1^2), \tag{14.40}$$

whereas the entropy per molecule, Σ, is based on the probability distribution

$$P(z,\theta) = \frac{1}{Zd} \exp[-\beta M(z,\theta)] \quad , \tag{14.41}$$

i.e. $P(z,\theta)$ is a periodic function of z with periodicity d and

$$\int_0^d dz \int_0^\pi P(z,\theta) \sin\theta \, d\theta = 1 \quad .$$

The entropy Σ is given by

$$\Sigma = -k_B \int_0^d dz \int_0^\pi d\theta \sin\theta P(z,\theta) \ln P(z,\theta) \quad . \tag{14.42}$$

Substitution of (14.41) into (14.42) gives

$$\begin{aligned} \Sigma &= \frac{1}{T}\langle M(z,\theta)\rangle + k_B \ln(Zd) \\ &= -\frac{1}{2T}U_0 S^2 - \frac{U_1}{T}\sigma_1^2 + k_B \ln(Zd) \quad . \end{aligned} \tag{14.43}$$

Consequently the free energy per molecule is

$$f = u - T\Sigma = \tfrac{1}{4}U_0 S^2 + \tfrac{1}{2}U_1 \sigma_1^2 - k_B T \ln(Zd) \quad . \tag{14.44}$$

Clearly the requirement that f must be minimal with respect to S and σ_1 results in the self-consistency relations (14.37) and (14.38).

It is obvious that the self-consistency relations must be solved numerically. For a full discussion the reader is referred to the original paper of McMillan (1971). The $S_A N$ transition strongly depends on the ratio $\frac{1}{2}\alpha = U_1/U_0$ or, using the Gaussian potential (14.12) on the quantity $\alpha = 2\exp[-(\pi\lambda/d)^2]$. Three types of behaviour can be found:

(i) $\alpha < 0.7$ or $\lambda/d > 0.33$. The $S_A N$ transition is a second-order transition, i.e. the smectic order parameter σ_1 drops to zero continuously when approaching the critical temperature.

(ii) $0.7 < \alpha < 0.98$ or $0.27 < \lambda/d < 0.33$. The $S_A N$ transition is first order. The smectic order parameter σ_1 drops discontinuously to zero. Because of the coupling between σ_1 and S the nematic order parameter also changes discontinuously at the $S_A N$ transition.

(iii) $\alpha > 0.98$ or $\lambda/d < 0.27$. The nematic phase does not appear any more. Instead the S_A phase directly melts into the isotropic phase. The transition from the S_A phase to the isotropic phase is a first-order transition and both order parameters drop to zero discontinuously on reaching the isotropic phase.

This behaviour is represented in Fig. 14.3 for various values of α.

The main difference between the ML approach and the present approach concerns the nature of the smectic order parameter. The ML treatment is based upon a pure translational order parameter ϱ_1, whereas the McMillan model gives rise to a mixed order parameter σ_1 as the essential order parameter. Clearly σ_1 approaches ϱ_1 in the limit of increasing orientational order. The pure translational order parameter ϱ_1 can be incorporated into the McMillan model easily by adding an isotropic interaction between the molecules and subsequently expanding this interaction in a

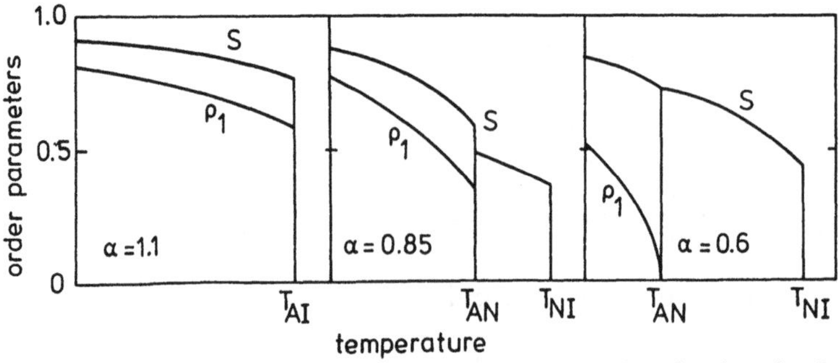

Fig. 14.3. Phase diagram concerning the $S_A N$ transition showing the dependence on the parameter α (or increasing chain length)

one-dimensional Fourier series with a period d along the z direction. For a further discussion of this approach, which presents an approximate description of the S_A phase in terms of three order parameters S, σ_1 and ϱ_1, the reader is referred to the literature (Kobayashi 1970; McMillan 1971; Lee et al. 1973).

Both the ML model and the KM model stress the relevance of the range of the intermolecular potential with respect to the molecular length as far as the nature of the S_AN transition is concerned. In addition, the KM model predicts that the nematic phase is no longer thermodynamically stable once the ratio between the molecular length and the range of the potential becomes too large. In this sense the KM model is qualitatively in agreement with the observed behaviour in homologous series if dispersion interactions between the central aromatic cores are assumed to be dominant. As the aromatic core remains the same in a homologous series the range λ of the intermolecular potential does not change along the series. Consequently the ratio λ/d decreases with increasing chain length, and the experimentally observed trend in the relative stability of smectic, nematic and isotropic phases is in agreement with the model predictions. It should be remarked here, however, that great care must be exercised in correlating the experimental data and in particular the molecular structure with the parameters of the ML model and KM model because of the ad hoc nature of both models and the impossibility to vary only one molecular parameter. Like in the previous section it must be stressed that, for example, the possible influence of the flexibility of the alkyl chains has not been considered (Dowell 1983).

Both in the ML model and the KM model a tricritical point is predicted where the S_AN phase transition changes its character from first order to second order. As in the Landau theory this is due to a coupling between ϱ_1 and either ϱ_2 (ML) or the orientational order parameter S (KM). In a sense these models give a specific interpretation to the Landau theory, and if desired the expansion coefficients of the latter theory can be expressed in the model parameters. In the KM model the tricritical point is found at $\lambda/d = 0.33$, which corresponds to $T_{AN}/T_{NI} = 0.88$. As mentioned at the end of Sect. 12.4 a tricritical point has been observed experimentally, but at values of T_{AN} much closer to T_{NI}. In this sense the predictions of the KM model are quantitatively incorrect. For a discussion of the general dependence of the tricritical point on both translational order parameters ϱ_1 and ϱ_2 and the orientational order parameter S the reader is referred to Longa (1986). In x-ray diffraction higher orders of the (001) reflection that correspond to the density wave, are often weak or even absent. This indicates that the density modulation is very weak and truncation of the Fourier expansion of the potential after the ϱ_1 term could be a good approximation.

However, smectics with a much stronger (002) reflection have also been found in which case the importance of ϱ_2 must be considered carefully.

As mentioned previously, molecular asymmetry may have an important influence on the occurrence of smectic phases. In particular the presence of strongly polar end groups like CN or NO_2, can lead to a variety of "exotic" smectic phases, that still have the overall symmetry of smectic A (see Sect. 3.2). The possibility of breaking the up-down symmetry of the mutual orientation of the molecules must now be included in the theory. This leads to long-range ordering of the polar heads within the layers, and thus to ferroelectric layers. These can in turn be combined to a ferroelectric S_{Af} phase or to an antiferroelectric S_{A2} phase with periodicity $2d$ (see Fig. 3.6). Longa and De Jeu (1983) have presented a simple molecular-statistical theory for the S_{A2} phase. According to their theory the S_{Af} phase is unstable, which is in agreement with experiments. It will be clear that theoretically the situation becomes rather complicated, as in addition to the dispersion interactions permanent dipoles also come into play. With so many partly competing effects it is not surprising that there are various possible ways to stabilize the different types of S_A phase. Even a qualitative molecular understanding of the S_A phase is obviously an extremely difficult task.

14.2 Theory of the Smectic C Phase

The main difference between the S_A phase and the S_C phase is the symmetry of the two phases as is discussed in Sect. 3.2. The S_A phase is optically uniaxial, whereas the S_C phase is optically biaxial. In order to account for the breaking of the uniaxial symmetry the S_C phase is thought to consist of smectic layers, where the director is no longer normal to the smectic planes. The angle between the normal and the director is called the tilt angle. Thus the difference between the S_C phase and the S_A phase is described in terms of a non-zero tilt angle. Because of the close relation between these two phases, it seems reasonable to explain the existence of the S_C phase in terms of the S_A phase, namely by taking the existence of the S_A phase for granted and only looking for additional intermolecular interactions that may give rise to a non-zero tilt angle. Several molecular models have been proposed in order to explain the appearance of the S_C phase. Experimentally, the permanent dipoles are found to play an essential part in bringing about the S_C phase (see Chap. 3). Consequently only the molecular models that are based upon dipole forces, will be mentioned here. These models are due to McMillan (1973), Cabib and Benguigui (1977) and Van der Meer and Vertogen (1979a). Both the model of McMillan and of Van der Meer and Vertogen (MV) account for the negligible biaxiality with respect to the director and stress the importance of transverse dipoles, i.e. dipoles that are directed more or less transverse to the long molecular axis. The first model considers the interaction between permanent dipole moments. Such an interaction, however, averages out because of the uniaxial rotations of the molecules around their long axes in the S_C phase. For this reason only the MV model is considered, which is based upon the dipole-induced dipole interaction that allows for such rotations.

The starting point of the considerations is the assumption that removal of the dipoles from the molecules gives rise to a S_A phase. For reasons of simplicity the molecules are assumed to be perfectly oriented in the S_A phase and their centres are taken to be situated on planes parallel to the xy plane. The orientation of the S_A phase is thus described by the director $\boldsymbol{n} = (0, 0, 1)$. Next, the S_A phase is changed into the S_C phase by only tilting the director without adding the dipoles, i.e. the orientation of the phase is described by the director $\boldsymbol{n} = (\sin\omega\cos\phi,\ \sin\omega\sin\phi,\ \cos\omega)$, where ω denotes the tilt angle. Clearly the S_A phase resists such a transformation into the S_C phase if the molecular structure remains unchanged. This resistance is expressed by the observation that the free energy density of the obtained S_C phase, f_C, exceeds the free energy density f_A of the S_A phase. Because of the symmetry of the S_C phase this difference in free energy density can be approximated by

$$\Delta f = f_C - f_A = v_1 \sin^2\omega + v_2 \sin^4\omega \quad , \tag{14.45}$$

where both v_1 and v_2 are positive quantities that depend on the temperature. This temperature dependence originates from the change in the contribution of the packing entropy to the free energy. Because of the tilted orientation of the molecules with respect to the layer their centres of mass have less free volume at their disposal for movement within the layer. This is due to an increased steric hinderance or excluded volume, as the intersection of a cylindrically shaped molecule with the plane is no longer a circle but an ellipse instead. The area of this ellipse depends on the tilt angle ω. The quantities v_1 and v_2 can be roughly approximated by

$$v_1 = a_1 + b_1 T \quad \text{and} \tag{14.46a}$$

$$v_2 = a_2 + b_2 T \quad , \tag{14.46b}$$

where the positive constants a_1 and a_2 are related to the increase of the internal energy. The terms $b_1 T$ and $b_2 T$ describe the decrease in packing entropy where b_1 and b_2 are positive quantities and T is the temperature. Clearly the S_C phase is unstable according to (14.45). The only way to obtain a stable tilted state is to change the sign of v_1, i.e. to change the molecular structure. The essence of the MV model is the idea that the incorporation of permanent dipoles within the molecular structure expresses itself mainly in terms of an intermolecular dipole-induced dipole interaction which can cause the required change of sign under suitable conditions.

In order to derive an explicit expression for the dipole-induced dipole interaction the MV model considers the interaction between a permanent dipole of one molecule and an isotropic polarizability in the centre of another molecule, separated by a distance $\boldsymbol{s}$ (see Fig. 14.4). The dipole moment operator is given by $\boldsymbol{p}$. For simplicity the isotropic polarizability is described in terms of an isotropic harmonic oscillator with a Hamiltonian

$$H = -\frac{\hbar^2}{2m}\nabla^2 + \frac{1}{2}m\omega^2\boldsymbol{r}^2 \quad . \tag{14.47}$$

The polarizability of this system, α, the oscillator strength $m\omega^2$ and the fluctuating electric charge are related according to

$$\alpha = \frac{e^2}{m\omega^2} \quad . \tag{14.48}$$

The interaction between the permanent dipole and the isotropic polarizability is given by

$$V = \frac{e}{s^3}[\boldsymbol{p}\cdot\boldsymbol{r} - 3(\boldsymbol{p}\cdot\hat{\boldsymbol{s}})(\boldsymbol{r}\cdot\hat{\boldsymbol{s}})] \quad , \tag{14.49}$$

where $e\boldsymbol{r}$ is the instantaneous dipole moment of the polarizability, $s = |\boldsymbol{s}|$ and $\hat{\boldsymbol{s}} = \boldsymbol{s}/s$. The interaction energy E between the isotropic polarizability

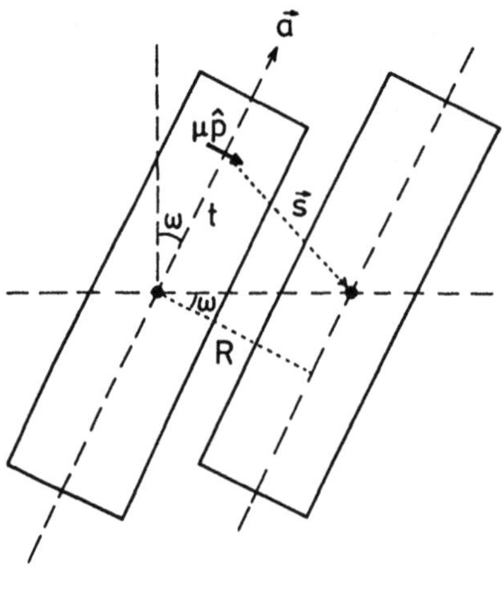

Fig. 14.4. The tilted situation occurring in the dipole-induced dipole model. Note that because of the tilt the distance between the two molecular centres changes from R to $R/\cos\omega$

and the permanent dipole is calculated using perturbation theory. Clearly the first-order contribution is zero since $\langle 0|\boldsymbol{p}\cdot\boldsymbol{r}|0\rangle = \langle 0|\hat{\boldsymbol{s}}\cdot\boldsymbol{r}|0\rangle = 0$, where $|0\rangle$ represents the ground state of the unperturbed system, i.e. the system without the interaction (14.49). The second-order contribution is

$$E = \sum_n{}' \frac{\langle 0|V|n\rangle\langle n|V|0\rangle}{E_0 - E_n} \quad , \tag{14.50}$$

where $|n\rangle$ represents an arbitrary excited state of the unperturbed system and the summation runs over all excited states. Using the well-known results for the matrix elements of the harmonic oscillator it follows that

$$E = \frac{-e^2\mu^2}{2m\omega^2 s^6}[1 + 3(\hat{\boldsymbol{p}}\cdot\hat{\boldsymbol{s}})^2] \quad , \tag{14.51}$$

where μ and $\hat{\boldsymbol{p}}$ are the magnitude and direction of the dipole, respectively. The only orientation-dependent term is the term proportional to $(\hat{\boldsymbol{p}}\cdot\hat{\boldsymbol{s}})^2$. Clearly the interaction energy is a minimum if the dipole is parallel to the line connecting its centre with that of the polarizability. The dipole $\mu\hat{\boldsymbol{p}}$ is caused by the molecular structure. In order to account for the negligible biaxiality of the phase with respect to the director, which coincides with the long molecular axis, the induction energy (14.51) is averaged over all possible orientations of the dipole-containing molecule for a fixed molecular orientation. For that purpose the direction of the permanent dipole is expressed as

$$\hat{\boldsymbol{p}} = p_1\boldsymbol{a} + p_2\boldsymbol{b} + p_3\boldsymbol{c} \quad , \tag{14.52}$$

where the unit vector $\boldsymbol{a}$ gives the orientation of the long molecular axis and the directions of the two remaining principal axes are denoted by $\boldsymbol{b}$ and $\boldsymbol{c}$. Using the relations (13.116) it follows that

$$\overline{(\hat{\boldsymbol{p}}\cdot\hat{\boldsymbol{s}})^2} = p_1^2(\boldsymbol{a}\cdot\hat{\boldsymbol{s}})^2 + \tfrac{1}{2}p_2^2[1 - (\boldsymbol{a}\cdot\hat{\boldsymbol{s}})^2] + \tfrac{1}{2}p_3^2[1 - (\boldsymbol{a}\cdot\hat{\boldsymbol{s}})^2] \quad . \tag{14.53}$$

Substitution of the relation

$$p_2^2 + p_3^2 = 1 - p_1^2 = 1 - (\hat{\boldsymbol{p}}\cdot\boldsymbol{a})^2$$

into expression (14.53) gives

$$\overline{(\hat{\boldsymbol{p}}\cdot\hat{\boldsymbol{s}})^2} = \tfrac{2}{3}P_2(\hat{\boldsymbol{p}}\cdot\boldsymbol{a})P_2(\hat{\boldsymbol{s}}\cdot\boldsymbol{a}) + \tfrac{1}{3} \quad . \tag{14.54}$$

Consequently the average induction energy becomes

$$E = -\frac{\alpha\mu^2}{s^6}[1 + P_2(\hat{\boldsymbol{p}}\cdot\boldsymbol{a})P_2(\hat{\boldsymbol{s}}\cdot\boldsymbol{a})] \quad . \tag{14.55}$$

Now the crucial step is to show that the average induction energy (14.55) indeed favours a tilted smectic phase. For that purpose two neighbouring molecules are considered with their centres of mass in the same smectic layer. The long molecular axes of the molecules are parallel to each other and are at an angle ω with respect to the normal of the smectic layer as shown in Fig. 14.4. The distance between their centres of mass is R in the S_A phase and $R/\cos\omega$ in the tilted phase, provided that the distance R between the long molecular axes remains unchanged in both phases. For simplicity only one of the two molecules carries a permanent dipole, which is situated on the long molecular axis $\boldsymbol{a}$ at a distance t from the centre of mass of the molecule in question. In order to calculate the dependence of the quantities s and $\hat{\boldsymbol{s}}\cdot\boldsymbol{a}$ on the tilt angle ω the unit vector $\boldsymbol{u}$ is introduced, which points from the centre of mass of the dipole-carrying molecule to the centre of mass of the other molecule. It follows easily that

$$t\boldsymbol{a} + s\hat{\boldsymbol{s}} = \frac{R}{\cos\omega}\boldsymbol{u} \quad . \tag{14.56}$$

This relation immediately implies that

$$\boldsymbol{a}\cdot\hat{\boldsymbol{s}} = -\frac{t}{s} + \frac{R}{s\cos\omega}\boldsymbol{a}\cdot\boldsymbol{u} \quad , \tag{14.57}$$

$$s^2 = t^2 + \frac{R^2}{\cos^2\omega} - 2\frac{Rt}{\cos\omega}\boldsymbol{a}\cdot\boldsymbol{u} \quad , \tag{14.58}$$

or using $\boldsymbol{a}\cdot\boldsymbol{u} = \sin\omega$ and defining $x = t/R$

$$\boldsymbol{a}\cdot\hat{\boldsymbol{s}} = -x\frac{R}{s} + \frac{R}{s}\tan\omega, \tag{14.59}$$

$$\left(\frac{s}{R}\right)^2 = x^2 + \frac{1}{\cos^2\omega} - 2x\tan\omega \quad . \tag{14.60}$$

Next the terms $(R/s)^2$ and $(\boldsymbol{a}\cdot\hat{\boldsymbol{s}})^2$ are expanded in terms of $\sin\omega$ up to order $\sin^2\omega$. Such an expansion results in the following expressions

$$\left(\frac{R}{s}\right)^2 = \frac{1}{1+x^2}\left[1 + \frac{2x}{1+x^2}\sin\omega + \frac{3x^2-1}{(1+x^2)^2}\sin^2\omega + \dots\right]$$

$$(\boldsymbol{a}\cdot\hat{\boldsymbol{s}})^2 = \frac{1}{1+x^2}\left[x^2 - \frac{2x}{1+x^2}\sin\omega - \frac{3x^2-1}{(1+x^2)^2}\sin^2\omega + \dots\right] \quad .$$

The average induction energy (14.55) up to order $\sin^2\omega$ now becomes

$$\begin{aligned} E = & -\frac{\alpha\mu^2}{R^6(1+x^2)^6}\left[(1+x^2)^3 + \left(x^6 + \frac{3}{2}x^4 - \frac{1}{2}\right)P_2(\hat{\boldsymbol{p}}\cdot\boldsymbol{a})\right] \\ & -\frac{6\alpha\mu^2 x}{R^6(1+x^2)^5}[1+x^2+(x^2-1)P_2(\hat{\boldsymbol{p}}\cdot\boldsymbol{a})]\sin\omega \\ & -\frac{3\alpha\mu^2}{R^6(1+x^2)^6}[7x^4+6x^2-1+(7x^4-12x^2+1)P_2(\hat{\boldsymbol{p}}\cdot\boldsymbol{a})] \\ & \times\sin^2\omega \quad . \end{aligned} \tag{14.61}$$

This expression can be simplified further by assuming no long range correlation between the permanent dipoles within the smectic layers, i.e. equal probabilities of occurrence exist for molecules with the permanent dipoles at $t\boldsymbol{a}$ or $-t\boldsymbol{a}$. Averaging over the situations $t\boldsymbol{a}$ and $-t\boldsymbol{a}$ results in the following induction energy

$$\begin{aligned} E = & -\frac{\alpha\mu^2}{R^6(1+x^2)^6}\left[(1+x^2)^3 + \left(x^6 + \frac{3}{2}x^4 - \frac{1}{2}\right)P_2(\hat{\boldsymbol{p}}\cdot\boldsymbol{a})\right] \\ & -\frac{3\alpha\mu^2}{R^6(1+x^2)^6}[7x^4+6x^2-1+(7x^4-12x^2+1)P_2(\hat{\boldsymbol{p}}\cdot\boldsymbol{a})] \\ & \times\sin^2\omega \quad . \end{aligned} \tag{14.62}$$

According to this expression an off-centre permanent dipole tends to tilt the long molecular axis $\boldsymbol{a}$ away from the normal of the smectic layer. Clearly the energy depends on the position of the permanent dipole as well as on the angle between the direction of the dipole and the long molecular axis.

The proposed interaction energy (14.62) should only be considered as a possible way of expressing the influence of the permanent dipoles. An alternative way of expressing their influence is to write down some tensorial interaction between the two molecules without making use of perturbation theory. The additional influence of the dipole-carrying molecule can then be described by making use of a tensorial interaction that must contain the tensor $p_\alpha p_\beta$. It is obvious that a tensorial interaction of the form

$$-f(s)p_\alpha p_\beta s_\alpha s_\beta \tag{14.63}$$

as appears in (14.51), is only one of the simplest types of interaction. The main message of the MV model therefore does not concern the exact form of the interaction but the statement that permanent dipoles induce a tendency to tilt the long molecular axis towards the centre of mass of a neighbouring molecule and that this tendency can be described by an energy term. The dependence on the tilt angle ω of this term can be approximated by

$$E = -w \sin^2 \omega \quad . \tag{14.64}$$

According to the MV model the appearance of the S_C state can be understood in the following way. The original difference in free energy density (14.45) between the S_C state and the S_A state changes into

$$\Delta f = (a_1 + b_1 T - w) \sin^2 \omega + (a_2 + b_2 T) \sin^4 \omega \quad , \tag{14.65}$$

where the constant w is due to the presence of permanent dipoles. It is clear that the S_C phase occurs as soon as

$$T < T_{CA} = \frac{w - a_1}{b_1} \quad . \tag{14.66}$$

This means that the S_C phase is stable provided that $w > a_1$. It is found that such a condition can be fulfilled for moderately strong and suitably located dipoles. The order parameter of the S_C phase is the tilt angle ω, written in Sect. 3.2 as $\chi = \omega \exp(\mathrm{i}\phi)$. From (14.65) it is easily found that the temperature dependence of this order parameter can be approximated by

$$\sin \omega = \pm \left[\frac{b_1 (T_{CA} - T)}{2(a_2 + b_2 T)} \right]^{1/2} \quad . \tag{14.67}$$

Clearly the $S_C S_A$ transition is second order.

Phase transitions from the S_C phase to the nematic phase occur as well. These S_CN transitions are found to be first order as might be expected from symmetry considerations since the nature of the S_CN transition inherently involves a finite jump of the tilt angle. This discontinuous behaviour signalizes a discontinuity in the internal energy at the transition, i.e. the presence of latent heat.

14.3 Theory of the Chiral Smectic C Phase

The S_C phase can be considered to consist of molecules with their centres of mass arranged in layers and with an average orientation described by a single unit vector, the director $\boldsymbol{n}$. Taking the smectic planes parallel to the xy plane, the ideal smectic C phase is described by $\boldsymbol{n} = (\sin \omega \cos \phi,$

sin ω sin ϕ, cos ω), where the tilt angle ω and the angle ϕ do not depend on the position z of the smectic planes. The projection of the director $\boldsymbol{n}$ on the smectic plane is called the director $\mathbf{c}$ (see Sect. 3.2). The components of that vector are given by $\boldsymbol{c} = (\cos\phi,\ \sin\phi,\ 0)$. Next, the symmetry of the molecular structure is changed such that intermolecular twist interactions appear, i.e. chiral molecules are considered. A chiral smectic C phase is found meaning that the direction of the C-director depends on the position of the smectic planes. In the following the influence of the chirality of the molecules will be discussed in terms of the simple intermolecular twist interaction (13.120) and the resulting dependence of the angle ϕ on the position z will be calculated (Van der Meer and Vertogen 1979b).

According to (13.120) the intermolecular twist interaction gives rise to the following interaction between two smectic C planes, situated at z and z' respectively,

$$
\begin{aligned}
H_{\text{twist}} = & -K(z-z')\boldsymbol{n}(z)\cdot\boldsymbol{n}(z')[\Theta(z-z')-\Theta(z'-z)] \\
& \times [n_x(z)n_y(z')-n_y(z)n_x(z')] \quad . \qquad (14.68)
\end{aligned}
$$

The interaction strength $K(z-z')$ only depends on the interlayer distance $|z-z'|$, but the sign of the interaction depends on the sign of $z-z'$, which is represented by the appearance of the step function $\Theta(z)$ defined by

$$
\Theta(z) = \begin{cases} 0 \ , & z<0 \\ 1 \ , & z>1 \end{cases} \quad .
$$

This twist interaction, in terms of the tilt angle ω and the position dependent angle ϕ, is given by

$$
\begin{aligned}
H_{\text{twist}} = & -K(z-z')[\theta(z-z')-\Theta(z'-z)] \\
& \times \{\cos^2\omega + \sin^2\omega\ \cos[\phi(z)-\phi(z')]\} \\
& \times \sin^2\omega\ \sin[\phi(z)-\phi(z')]
\end{aligned}
$$

or assuming $\phi(z)-\phi(z')$ to be small

$$
\begin{aligned}
H_{\text{twist}} = & -K(z-z')[\Theta(z-z')-\Theta(z'-z)] \\
& \times \sin^2\omega\,[\phi(z)-\phi(z')] \quad . \qquad (14.69)
\end{aligned}
$$

For simplicity it is also assumed that only neighbouring layers contribute to the energy. Consequently the twist distortion gives rise to an additional energy per layer of

$$H_{\rm twist} = K(d)\, d \sin^2 \omega \, \frac{d\phi}{dz} \quad , \tag{14.70}$$

where d is the interlayer distance.

The twist interaction is counteracted by the orientational interaction that favours the formation of the S_C phase. This means that a twist distortion increases the free energy of the non-chiral S_C state. In order to calculate the resulting increase, the intermolecular interactions must be known. Clearly this is out of the question. For this reason a simple interlayer interaction is postulated here that is based upon the KM model and the requirement that the difference between the free energy of a S_C layer and the free energy of a S_A layer is given by

$$\Delta f = -V_1 \sin^2 \omega + V_2 \sin^4 \omega \quad . \tag{14.71}$$

The postulated symmetry-breaking model interaction is given by

$$\begin{aligned} H = & - J(z-z')[\boldsymbol{n}(z)\boldsymbol{n}(z')]^2 \\ & + [V_1(z-z') - 2V_2(z-z')]\boldsymbol{n}(z)\cdot\boldsymbol{n}(z')n_z(z)n_z(z') \\ & + V_2(z-z')[\boldsymbol{n}(z)\cdot\boldsymbol{n}(z')n_z(z)n_z(z')]^2 \quad , \end{aligned} \tag{14.72}$$

where the first term (Maier-Saupe term) is supposed to be dominant. Taking only neighbouring layers into account, the contribution of the twist distortion to the layer free energy is found to be

$$\begin{aligned} H\left(\frac{d\phi}{dz}\right) = & - J(d)\Phi^2(d) + [V_1(d) - 2V_2(d)] \cos^2 \omega \, \Phi(d) \\ & + V_2(d) \cos^4 \omega \, \Phi^2(d) \end{aligned} \tag{14.73}$$

with

$$\Phi(d) = 1 - \frac{1}{2} d^2 \sin^2 \omega \left(\frac{d\phi}{dz}\right)^2 \quad , \tag{14.74}$$

where use has been made of the fact that $\phi(z) - \phi(z \pm d)$ is small and that the coupling constants $J(z)$, $V_1(z)$ and $V_2(z)$ are even functions. Consequently the increase of the twist suppressing term is up to order $(d\phi/dz)^2$

$$\begin{aligned} H\left(\frac{d\phi}{dz}\right) - H(0) = & \left[J(d) - \frac{1}{2}V_1(d) \cos^2 \omega + V_2(d) \cos^2 \omega \sin^2 \omega\right] \\ & \times d^2 \sin^2 \omega \left(\frac{d\phi}{dz}\right)^2 \quad . \end{aligned} \tag{14.75}$$

The combination of the expressions (14.70) and (14.75) describes the effect of the twist distortion on the free energy of the S_C state. The chirality introduces an additional term in the free energy of a chiral S_C layer of the form

$$\begin{aligned}\Delta f_{\text{twist}} &= K(d)\, d \sin^2\omega \left(\frac{d\phi}{dz}\right) \\ &+ \left[J(d) - \frac{1}{2}V_1(d)\cos^2\omega + V_2(d)\cos^2\omega\sin^2\omega\right] \\ &\times d^2 \sin^2\omega \left(\frac{d\phi}{dz}\right)^2 \quad , \end{aligned} \qquad (14.76)$$

where $\phi(z)$ must be determined by requiring that the free energy is minimal. The usual procedure results in

$$\phi = qz \quad , \qquad (14.77)$$

where the helical wave number q is given by

$$q = \frac{K(d)}{2d\left[J(d) - \frac{1}{2}V_1(d)\cos^2\omega + V_2(d)\cos^2\omega\sin^2\omega\right]} \quad . \qquad (14.78)$$

The observed temperature dependence of the pitch can be ascribed to (1) the temperature dependent behaviour of the tilt angle ω and (2) the contribution of the packing entropy to the free energy of the layer resulting in temperature dependent coupling constants as indicated for instance in (14.46). According to the present model the helical wave number q jumps to zero at the S_CS_A transition.

Finally it should be stressed here that this analysis of the chiral S_C state does not pretend to give any quantitative statements because of the complete ignorance of the relevant intermolecular interactions. The only purpose of the present model is to describe in mathematical terms the general ideas concerning the origin of the chiral S_C phase and the temperature dependent behaviour of such a state.

References

Abragam, A. (1961): *The Principles of Nuclear Magnetism* (Oxford Univ. Press, Oxford)
Als-Nielsen, J., Litster, J.D., Birgeneau, R.J., Kaplan, M., Safinya, C.R., Lindegaard-Andersen, A., Mathiesen, B. (1980): Phys. Rev. B **22**, 312
Arnold, H. (1964): Z. Phys. Chem. (Leipzig) **226**, 146

Baessler, H., Malya, P.A.G., Nes, W.R., Labes, M.M. (1970): Mol. Cryst. Liq. Cryst. **6**, 329
Bahadur, B. (1976): J. Chim. Phys. **73**, 255
Barker, J.A., Henderson, D. (1976): Rev. Mod. Phys. **48**, 587
Barrall, E.M., Johnson, J.F. (1974): in *Liquid Crystals and Plastic Crystals*, ed. by G.W. Gray, P.A. Winsor (Ellis Horwood, Chichester) Vol. 2, Chap. 10
Beens, W.W., Jeu, W.H. de (1983): J. de Phys. **44**, 129
Beens, W.W., Jeu, W.H. de (1985): J. Chem. Phys. **82**, 3841
Benattar, J.J., Doucet, J., Lambert, M., Levelut, A.M. (1979): Phys. Rev. A **20**, 2505
Benattar, J.J., Moussa, F., Lambert, M., Germain, C. (1981): J. de Phys. Lett. **42**, L67
Benguigui, L. (1984): Phys. Rev. A **29**, 2968
Berreman, D.W., Scheffer, T.J. (1970): Mol. Cryst. Liq. Cryst. **11**, 395
Biering, A., Demus, D., Richter, L., Sackmann, H., Wiegeleben, A., Zaschke, H. (1980): Mol. Cryst. Liq. Cryst. **62**, 1
Birgeneau, R.J., Litster, J.D. (1978): J. de Phys. Lett. **39**, L399
Blinc, R., Pirš, Zupančič, I. (1973): Phys. Rev. Lett. **30**, 546
Blinov, L.M. (1979): J. de Phys. Colloq. **40**, C3-247
Blumstein, A. (ed.) (1980): *Liquid Crystalline Order in Polymers* (Academic Press, New York)
Boden, N., Clark, L.D., Bushby, R.J., Emsley, J.W., Luckhurst, G.R., Stockley, C.P. (1981): Mol. Phys. **42**, 565
Böttcher, C.J.F. (1973): *Theory of Electric Polarization*, Vol. I (Elsevier, Amsterdam)
Böttcher, C.J.F., Bordewijk, P. (1978): *Theory of Electric Polarization*, Vol. II (Elsevier, Amsterdam)
Bouligand, Y. (1972): J. de Phys. **33**, 715
Bouligand, Y. (1978): in *Liquid Crystals*, ed. by L. Liébert, Solid State Phys. Suppl. 14 (Academic Press, New York) p. 259
Bradshaw, M.J., Raynes, E.P., Fedak, I., Leadbetter, A.J. (1984): J. de Phys. **45**, 157
Brochard, F. (1972): J. de Phys. **33**, 607
Brownsey, G.J., Leadbetter, A.J. (1980): Phys. Rev. Lett. **44**, 1608
Bunning, J.D., Crellin, D.A., Faber, T.E. (1986): Liq. Cryst. **1**, 37

Cabib, D., Benguigui, L. (1977): J. de Phys. **38**, 419
Cagnon, M., Durand, G. (1980): Phys. Rev. Lett. **45**, 1418
Caillé, A. (1972): C.R. Acad. Sci. Paris **274 B**, 891
Carnahan, N.F., Starling, K.E. (1969): J. Chem. Phys. **51**, 635
Chandrasekhar, S., Sadashiva, B.K., Suresh, K.A., Mathusudana, N.V., Kumar, S., Shashidhar, R., Venkatesh, G. (1979): J. de Phys. Colloq. **40**, C3-120

Chapmann, D. (1978): in *Liquid Crystals and Plastic Crystals,* ed. by G.W. Gray, P.A. Winsor (Ellis Horwood, Chichester) Vol. 1, Chap. 6
Charvolin, J., Deloche, B. (1979): in *The Molecular Physics of Liquid Crystals,* ed. by G.R. Luckhurst, G.W. Gray (Academic Press, London) Chap. 15
Charvolin, J., Tardieu, A. (1978): in *Liquid Crystals,* ed. by L. Liébert, Solid State Phys. Suppl. 14 (Academic Press, New York) p. 209
Chatelain, P. (1951): Acta Crystallogr. **4**, 453
Chatelain, P., Germain, M. (1964): C.R. Acad. Sci. Paris **259**, 127
Chu, K.-S., Moroi, D.S. (1975): J. de Phys. Colloq. **36**, C1-99
Ciferri, A., Krigbaum, W.R., Meyer, R.B. (eds.) (1982): *Polymer Liquid Crystals* (Academic Press, New York)
Cladis, P.E. (1972): Phys. Rev. Lett. **28**, 1629
Clark, N.A., Lagerwall, S.T. (1984): Ferroelectrics **58**, 389
Coates, D., Gray, G.W. (1973): Mol. Cryst. Liq. Cryst. **24**, 163
Cognard, J. (1982): Mol. Cryst. Liq. Cryst. Suppl. **1**, 1
Cotter, M.A. (1977): J. Chem. Phys. **66**, 1098

Dalmolen, L.G.P., Jeu, W.H. de (1983): J. Chem. Phys. **78**, 7353
Dalmolen, L.G.P., Picken, S.J., Jong, A.F. de, Jeu, W.H. de (1985): J. de Phys. **46**, 1443
Davies, M., Moutran, R., Price, A.H., Beevers, M.S., Williams, G. (1976): J. Chem. Soc. Faraday Trans. II **72**, 1447
Demus, D., Demus, H., Zaschke, H. (1983): *Flüssige Kristalle in Tabellen,* 2nd ed. (VEB Verlag, Leipzig)
Demus, D., Gloza, A., Hartung, H., Hauser, A., Rapthel, I., Wiegeleben, A. (1981): Cryst. Res. Techn. **16**, 1445
Demus, D., Marzotko, D., Sharma, N.K., Wiegeleben, A. (1980): Cryst. Res. Techn. **15**, 331
Demus, D., Richter, L. (1978): *Textures of Liquid Crystals* (Verlag Chemie, Weinheim)
Destrade, C., Gasparoux, H., Foucher, P., Nguyen Huu Tinh, Malthête, J., Jacques, J. (1983): J. Chim. Phys. **80**, 137
Deuling, H.J. (1978): in *Liquid Crystals,* ed. by L. Liébert, Solid State Phys. Suppl. 14 (Academic Press, New York) p. 77
Dianoux, A.J., Volino, F. (1979): J. de Phys. **40**, 181
Dijk, J.W. van, Beens, W.W., Jeu, W.H. de (1983): J. Chem. Phys. **79**, 3888
Diogo, A.C., Martins, A.F. (1981): Mol. Cryst. Liq. Cryst. **66**, 133
Dong, R.Y., Samulski, E.T. (1982): Mol. Cryst. Liq. Cryst. Lett. **82**, 73
Doucet, J. (1979): J. de Phys. Lett. **40**, L185
Doucet, J., Levelut, A.M., Lambert, M. (1973): Mol. Cryst. Liq. Cryst. **24**, 317
Doucet, J., Levelut, A.M., Lambert, M. (1974): Phys. Rev. Lett. **32**, 301
Doucet, J., Levelut, A.M., Lambert, M., Liébert, L., Strzelecki, L. (1975): J. de Phys. Colloq. **36**, C1-13
Dowell, F. (1983): Phys. Rev. A **28**, 3520
Dowell, F., Martire, D.E. (1978): J. Chem. Phys. **68**, 1094
Dubois, J.C., Billard, J. (1984): in *Liquid Crystals and Ordered Fluids,* ed. by A.C. Griffin, J.F. Johnson (Plenum, New York) p. 1043
Dubois-Violette, E., Durand, G., Guyon, E., Manneville, P., Pieranski, P. (1978): in *Liquid Crystals,* ed. by L. Liébert, Solid State Phys. Suppl. 14 (Academic Press, New York) p. 147
Dubois-Violette, E., Gennes, P.G. de, Parodi, O. (1971): J. de Phys. **32**, 305
Durand, G., Léger, L., Rondelez, F., Veyssié, M. (1969): Phys. Rev. Lett. **22**, 227
Dvolaitsky, M., Poldy, F., Taupin, C. (1973): Phys. Lett. **45A**, 454
Dzyaloshinskii, I.E. (1970): Sov. Phys.-JETP **31**, 773

Emsley, J.W. (ed.) (1985): *Nuclear Magnetic Resonance of Liquid Crystals* (Reidel, Dordrecht)

Emsley, J.W., Luckhurst, G.R. (1980): Mol. Phys. **41**, 19
Emsley, J.W., Luckhurst, G.R., Gray, G.W., Mosley, A. (1978): Mol. Phys. **35**, 1499
Emsley, J.W., Luckhurst, G.R., Stockley, C.P. (1982): Proc. Roy. Soc. London A **381**, 117
Etherington, G., Leadbetter, A.J., Wang, X.J., Gray, G.W., Tajbakhsh, T. (1986): Liq. Cryst. **1**, 209

Feldkamp. G.E., Handschy, M.A., Clark, N.A. (1981): Phys. Lett. A **85**, 359
Finkelman, H. (1980): in *Liquid Crystals of One- and Two-Dimensional Order*, ed. by W. Helfrich, G. Heppke (Springer, Berlin, Heidelberg) p. 238
Finkelman, H. (1983): Phil. Trans. Roy. Soc. London A **309**, 105
Flapper, S.D.P., Vertogen, G. (1981a): Phys. Rev. A **24**, 2089
Flapper, S.D.P., Vertogen, G. (1981b): J. Chem. Phys. **75**, 3599
Flapper, S.D.P., Vertogen, G., Leenhouts, F. (1981): J. de Phys. **42**, 1647
Fontell, K. (1981): Mol. Cryst. Liq. Cryst. **63**, 59
Frank, F.C. (1958): Discuss. Faraday Soc. **25**, 19
Frank, F.C., Chandrasekhar, S. (1980): J. de Phys. **41**, 1285
Frenkel, D., Mulder, B.M. (1985): Mol. Phys. **55**, 1171

Gane, P.A.C., Leadbetter, A.J., Benattar, J.J., Moussa, F., Lambert, M. (1981a): Phys. Rev. A **24**, 2694
Gane, P.A.C., Leadbetter, A.J., Wrighton, P.G. (1981b): Mol. Cryst. Liq. Cryst. **66**, 247
Gasparoux, H., Prost, J. (1971): J. de Phys. **32**, 953
Gelbart, W.M. (1982): J. Phys. Chem. **86**, 4298
Gelbart, W.M., Gelbart, A. (1977): Mol. Phys. **23**, 1387
Gennes, P.G. de (1968a): Solid State Commun. **6**, 163
Gennes, P.G. de (1968b): C.R. Acad. Sci. Paris B **226**, 571
Gennes, P.G. de (1971): Mol. Cryst. Liq. Cryst. **12**, 193
Gennes, P.G. de (1972): Solid State Commun. **10**, 753
Gennes, P.G. de (1973): Mol. Cryst. Liq. Cryst. **21**, 49
Gennes, P.G. de (1974): *The Physics of Liquid Crystals* (Clarendon, Oxford)
Geurst, J.A., Spruijt, A.M.J., Gerritsma, C.J. (1975): J. de Phys. **36**, 653
Goodby, J.W. (1981): Mol. Cryst. Liq. Cryst. Lett. **72**, 95
Goossens, W.J.A. (1971): Mol. Cryst. Liq. Cryst. **12**, 237
Govers, E., Vertogen, G. (1984): Z. Naturforsch. **39a**, 537
Graf, V., Noack, F., Stohrer, M. (1977): Z. Naturforsch. **32a**, 61
Gramsbergen, E.F., Longa, L., Jeu, W.H. de (1986): Phys. Rep. **135**, 195

Hakemi, H., Labes, M.M. (1974): J. Chem. Phys. **61**, 4020
Hanson, H., Dekker, A.J., Woude, F. van der (1975): J. Chem. Phys. **62**, 1941
Hardouin, F., Levelut, A.M., Achard, M.F., Sigaud, G. (1983): J. Chim. Phys. **80**, 53
Haven, T., Armitage, D., Saupe, A. (1981): J. Chem. Phys. **75**, 352
Helfrich, W. (1969): J. Chem. Phys. **51**, 4092
Helfrich, W. (1970): J. Chem. Phys. **53**, 2267
Heppke, G., Schneider, F. (1974): Ber. Bunsenges. Phys. Chem. **78**, 981
Heppke, G., Schneider, F., Sterzl, A. (1976): Z. Naturforsch. **31a**, 1700
Hérino, R. (1981): J. Chem. Phys. **74**, 3016
Hervet, H., Dianoux, A.J., Lechner, R.E., Volino, F. (1976): J. de Phys. **37**, 587
Höhener, A. (1978): Chem. Phys. Lett. **53**, 97
Höhener, A., Müller, E., Ernst, R.R. (1979): Mol. Phys. **38**, 909
Horn, R.G., Faber, T.E. (1979): Proc. Roy. Soc. London A **368**, 199
Hornreich, R.M., Shtrikman, S. (1980): J. de Phys. **41**, 335
Huang, C.C., Viner, J.M. (1982): Phys. Rev. A **25**, 3385

Jackson, J.D. (1975): *Classical Electrodynamics* (Wiley, London)
Jackson, W.J., Kuhfuss, H.F. (1976): J. Polym. Sci. **14**, 2043
Jähnig, F. (1979): J. Chem. Phys. **70**, 3279
Jen, S., Clark, N.A., Pershan, P.S., Priestly, E.B. (1977): J. Chem. Phys. **66**, 4635
Jeu, W.H. de (1977): J. de Phys. **38**, 1265
Jeu, W.H. de (1978): in *Liquid Crystals*, ed. by L. Liébert, Solid State Phys. Suppl. 14 (Academic Press, New York) p. 109
Jeu, W.H. de (1980): *Physical Properties of Liquid Crystalline Materials* (Gordon and Breach, New York)
Jeu, W.H. de (1981): Mol. Cryst. Liq. Cryst. **63**, 83
Jeu, W.H. de (1983): Phil. Trans. Roy. Soc. London A **309**, 217
Jeu, W.H. de, Bordewijk, P. (1978): J. Chem. Phys. **68**, 109
Jeu, W.H. de, Claassen, W.A.P. (1977): J. Chem. Phys. **67**, 3705
Jeu, W.H. de, Eidenschink, R. (1983): J. Chem. Phys. **78**, 4637
Jeu, W.H. de, Goossens, W.J.A., Bordewijk, P. (1974): J. Chem. Phys. **61**, 1985)
Jost, W., Hauffe, K. (1972): *Diffusion* (Steinkopff, Darmstadt)

Kats, E.I. (1978): Sov. Phys.-JETP **48**, 916
Kirkwood, J.G. (1951): in *Phase Transitions in Solids*, ed. by R. Smoluchowski (Wiley, New York) p. 67
Kléman, M. (1983): *Points, Lines and Walls* (Wiley, Chichester)
Kneppe, H., Schneider, F. (1981): Mol. Cryst. Liq. Cryst. **65**, 23
Kneppe, H., Schneider, F. (1983): J. Physics E **16**, 512
Kneppe, H., Schneider, F., Sharma, N.K. (1981): Ber. Bunsenges. Phys. Chem. **85**, 784
Kobayashi, K.K. (1970): J. Phys. Soc. Japan **29**, 101
Kobayashi, K.K. (1971): Mol. Cryst. Liq. Cryst. **13**, 137
Kohli, M., Otnes, K., Pynn, R., Riste, T. (1976): Z. Phys. B **24**, 147
Kolbe, A., Demus, D. (1968): Z. Naturforsch. **23a**, 1237
Kreibig, U., Wetter, C. (1980): Z. Naturforsch. **35c**, 750
Kresse, H. (1982): Fortschritte der Physik **30**, 507
Krüger, G.J. (1982): Phys. Rep. **82**, 229
Krüger, G.J., Spiesecke, H. (1973): Z. Naturforsch. **28a**, 964

Landau, L.D., Lifshitz, E.M. (1959a): *Theory of Elasticity* (Pergamon, Oxford)
Landau, L.D., Lifshitz, E.M. (1959b): *Fluid Dynamics* (Pergamon, Oxford)
Landau, L.D., Lifshitz, E.M. (1980): *Statistical Physics*, Vol. 1, 3rd ed. (Pergamon, Oxford)
Lawson, K.D., Flautt, T.J. (1968): J. Phys. Chem. **72**, 2066
Leadbetter, A.J. (1979): in *The Molecular Physics of Liquid Crystals*, ed. by G.R. Luckhurst, G.W. Gray (Academic Press, London) Chap. 13
Leadbetter, A.J., Frost, J.C., Mazid, M.A. (1979a): J. de Phys. Lett. **40**, L325
Leadbetter, A.J., Mazid, M.A., Kelly, B.A., Goodby, J.W., Gray G.W. (1979b): Phys. Rev. Lett. **43**, 630
Leadbetter, A.J., Norris, E.K. (1979): Mol. Phys. **38**, 669
Leadbetter, A.J., Richardson, R.M. (1979): in *The Molecular Physics of Liquid Crystals*, ed. by G.R. Luckhurst, G.W. Gray (Academic Press, London) Chap. 20
Leadbetter, A.J., Richardson, R.M., Colling, C.N. (1975): J. de Phys. Colloq. **36**, C1-37
Lee, F.T., Tan, H.T., Shih, Y.M., Woo, C.-W. (1973): Phys. Rev. Lett. **31**, 1117
Leenhouts, F., Dekker, A.J. (1981): J. Chem. Phys. **74**, 1956
Leenhouts, F., Jeu, W.H. de, Dekker, A.J. (1979): J. de Phys. **40**, 989
Leslie, F.M. (1970): Mol. Cryst. Liq. Cryst. **12**, 57
Leslie, F.M. (1979): in *Advances in Liquid Crystals*, ed. by G.H. Brown (Academic Press, New York) p. 1
Levelut, A.M. (1983): J. Chim. Phys. **80**, 149

Levelut, A.M., Malthête, J., Collet, A. (1986): J. de Phys. **47**, 351
Levelut, A.M., Tarento, R.J., Hardouin, F., Achard, M.F., Sigaud, G. (1981): Phys. Rev. A **24**, 2180
Levi-Civita, T. (1977): *The Absolute Differential Calculus* (Dover, New York)
Limmer, St., Findeisen, M., Schmiedel, H., Hillner, B. (1981): J. de Phys. **42**, 1665
Lippens, D., Parneix, J.P., Chapoton, A. (1977): J. de Phys. **38**, 1465
Litster, J.D. (1980): in *Liquid Crystals of One- and Two-Dimensional Order*, ed. by W. Helfrich, G. Heppke (Springer, Berlin, Heidelberg) p. 65
Long, D.A. (1978): *Raman Spectroscopy* (McGraw-Hill, New York)
Longa, L. (1986): J. Chem. Phys. **85, 2974**
Longa, L., Jeu, W.H. de (1983): Phys. Rev. A **28**, 2380
Luckhurst, G.R. (1979): in *The Molecular Physics of Liquid Crystals*, ed. by G.R. Luckhurst, G.W. Gray (Academic Press, London) Chap. 4

Maier, W., Meier, G. (1961a): Z. Naturforsch. **16a**, 262
Maier, W., Meier, G. (1961b): Z. Naturforsch. **16a**, 470
Maier, W., Saupe, A. (1959): Z. Naturforsch. **14a**, 882
Maier, W., Saupe, A. (1960): Z. Naturforsch. **15a**, 287
Malthête, J., Destrade, C. Nguyen Huu Tinh, Jacques, J. (1981): Mol. Cryst. Liq. Cryst. Lett. **64**, 233
Malthête, J., Levelut, A.M., Nguyen Huu Tinh (1985): J. de Phys. Lett. **46**, L875
Manneville, P. (1981): Mol. Cryst. Liq. Cryst. **70**, 223
Marčelja, S. (1974): J. Chem. Phys. **60**, 3599
Marcerou, J.P., Prost, J. (1980): Mol. Cryst. Liq. Cryst. **58**, 259
Martin, A.J., Meier, G., Saupe, A. (1971): Symp. Faraday Soc. **5**, 119
Martinan, J.L., Durand, G. (1972): Solid State Commun. **10**, 815
McMillan, W.L. (1971): Phys. Rev. A **4**, 1238
McMillan, W.L. (1972): Phys. Rev. A **6**, 936
McMillan, W.L. (1973): Phys. Rev. A **8**, 1921
Meer, B.W. van der, Vertogen, G. (1979a): J. de Phys. Colloq. **40**, C3-222
Meer, B.W. van der, Vertogen, G. (1979b): Phys. Lett. **74A**, 239
Meer, B.W. van der, Vertogen, G., Dekker, A.J., Ypma, J.G.J. (1976): J. Chem. Phys. **65**, 3935
Meiboom, S., Sethna, J.P., Anderson, P.W., Brinkman, W.F. (1981): Phys. Rev. Lett. **46**, 1216
Meulen, J.P. van der, Zijlstra, R.J.J. (1984): J. de Phys. **45**, 1627
Meyer, R.B. (1969): Phys. Rev. Lett. **22**, 918
Meyer, R.B. (1977): Mol. Cryst. Liq. Cryst. **40**, 38
Meyer, R.B., Lubensky, T.G. (1976): Phys. Rev. A **14**, 2307
Meyerhofer, D. (1975): in *Introduction to Liquid Crystals*, ed. by E.B. Priestly, P.J. Wojtowicz, Ping Sheng (Plenum, New York) p. 129
Miesowicz, M (1936): Bull. Intern. Acad. Polon. Ser. A, 228
Mircéa-Roussel, A., Léger, L., Rondelez, F., Jeu, W.H. de (1975): J. de Phys. Colloq. **36**, C1-93
Mittal, K.L. (ed.) (1977): *Micellization, Solubilization and Microemulsions* (Plenum, New York)
Mourey, B., Perbet, J.N., Hareng, M., Le Berre, S. (1982): Mol. Cryst. Liq. Cryst. **84**, 193

Nehring, J., Saupe, A. (1972): J. Chem. Phys. **56**, 5527
Nguyen Huu Tinh, Destrade, C., Gasparoux, H. (1979): Phys. Lett. **75A**, 251
Nguyen Huu Tinh, Malthête, J., Destrade, C. (1981): J. de Phys. Lett. **42**, L417
Noack, F. (1984): Mol. Cryst. Liq. Cryst. **113**, 247

Onsager, L. (1949): Ann. N.Y. Acad. Sci. **51**, 627
Orsay Liquid Crystal Group (1969): J. Chem. Phys. **51**, 816
Orsay Liquid Crystal Group (1972): Phys. Lett. **39A**, 181
Oseen, C.W. (1933): Trans. Faraday Soc. **29**, 883
Otani, S. (1981): Mol. Cryst. Liq. Cryst. **63**, 249

Parodi, O. (1970): J. de Phys. **31**, 581
Pelzl, G., Sharma, N.K., Richter, L., Wiegeleben, A., Schröder, G., Diele, S., Demus, D. (1981): Z. Phys. Chem. (Leipzig) **262**, 815
Penz, P.A. (1970): Phys. Rev. Lett. **24**, 1405
Penz, P.A. (1974): Phys. Rev. A **10**, 1300
Pieranski, P., Brochard, F., Guyon, E. (1973): J. de Phys. **34**, 35
Pincus, P. (1969): Solid State Commun. **7**, 415
Pindak, R., Moncton, D.E., Davey, S.C., Goodby, J.W. (1981): Phys. Rev. Lett. **46**, 1135
Pindak, R., Sprenger, W.O., Bishop, D.J., Osheroff, D.D., Goodby, J.W. (1982): Phys. Rev. Lett. **48**, 173
Pines, A., Ruben, D.J., Allison, S. (1974): Phys. Rev. Lett. **33**, 1002
Prost, J., Marcerou, J.P. (1977): J. de Phys. **38**, 315

Ratna, B.R., Vijaya, M.S., Shashidar, R., Sadashiva, B.K. (1973): Pramana Suppl. I, 69
Raynes, E.P. (1983): Phil. Trans. Roy. Soc. London A **309**, 167
Richardson, R.M., Leadbetter, A.J., Bonsor, D.H., Krüger, G.J. (1980): Mol. Phys. **40**, 741
Richter, L., Demus, D., Sackmann, H. (1981): Mol. Cryst. Liq. Cryst. **71**, 269
Rondelez, F. (1974): Solid State Commun. **14**, 815
Rose, M.E. (1957): *Elementary Theory of Angular Momentum* (Wiley, New York)
Rutar, V., Blinc, R., Vilfan, M., Zann, A., Dubois, J.C. (1982): J. de Phys. **43**, 761

Samulski, E.T. (1978): in *Liquid Crystalline Order in Polymers,* ed. by A. Blumstein (Academic Press, New York)
Samulski, E.T. (1980): Ferroelectrics **30**, 83
Samulski, E.T., Dong, R.Y. (1982): J. Chem. Phys. **77**, 5090
Saupe, A. (1979): J. de Phys. Colloq. **40**, C3-207
Schadt, M., Gerber, P.R. (1982): Z. Naturforsch. **37a**, 165
Schadt, M., Helfrich, W. (1971): Appl. Phys. Lett. **18**, 127
Scholte, P.M.L.O., Vertogen, G. (1982): Physica **113A**, 587
Schulze, H., Schumann, G. (1978): Z. Phys. Chem. (Leipzig) **6**, 1037
Skoulios, A. (1978): Ann. Phys. **3**, 421
Stegemeyer, H., Blümel, Th., Hiltrop, K., Onusseit, H., Porsch, F. (1986): Liq. Cryst. **1**, 3
Straley, J.P. (1976): Phys. Rev. A **14**, 1835
Stubbs, G., Warren, S., Holmes, K. (1977): Nature **267**, 216
Svedberg, T. (1914): Ann. Physik **44**, 1121
Svedberg, T. (1915): Jahrb. Radioakt. Electr. **12**, 129

Thoen, J., Marynissen, H., Dael, W. van (1982): Phys. Rev. A **26**, 2886
Thoen, J., Marynissen, H., Dael, W. van (1984): Phys. Rev. Lett. **52**, 204
Tiddy, G.J.T. (1980): Phys. Rep. **57**, 1

Ukleja, P., Pirš, J., Doane, J.W. (1976): Phys. Rev. A **14**, 414

Veen, J. van der, Grobben, A.H. (1971): Mol. Cryst. Liq. Cryst. **15**, 239
Veen, J. van der, Jeu, W.H. de, Wanninkhof, M.W.M., Tienhoven, C.A.M. (1973): J. Phys. Chem. **77**, 2153
Vertogen, G., Meer, B.W. van der (1979a): Phys. Rev. A **19**, 370
Vertogen, G., Meer, B.W. van der (1979b): Physica **99A**, 237
Vilfan, M., Zumer, S. (1980): Phys. Rev. A **21**, 672
Viner, J.M., Lamey, D., Huang, C.C., Pindak, R., Goodby, J.W. (1983): Phys. Rev. A **28**, 2433
Vleck, J.H. van (1932): *The Theory of Electric and Magnetic Susceptibilities* (Oxford University Press, London and New York)
Vries, A. de (1977): Mol. Cryst. Liq. Cryst. Lett. **41**, 27
Vries, H. de (1951): Acta Crystallogr. **4**, 219
Vuks, M.F. (1966): Optics Spectr. **20**, 361

Wade, C.G. (1977): Ann. Rev. Phys. Chem. **28**, 47
Wahlstrom, E.E. (1979): *Optical Crystallography*, 2nd ed. (Wiley, New York)
Weyl, H. (1918): *Raum-Zeit-Materie* (Springer, Berlin). This citation was translated in English as follows (Dover publications, 1952): "The study of tensor calculus is, without doubt, attended by conceptual difficulties – over and above the apprehension inspired by indices – which must be overcome. From the formal aspect, however, the method of reckoning used is of extreme simplicity, it is much easier than, e.g., the apparatus of elementary vector calculus."
Wiegeleben, A., Richter, L., Deresch, J., Demus, D. (1980): Mol. Cryst. Liq. Cryst. **59**, 329

Ypma, J.G.J., Vertogen, G. (1978): Phys. Rev. A **17**, 1490
Yu, L.J., Saupe, A. (1980a): J. Am. Chem. Soc. **102**, 4879
Yu, L.J., Saupe, A. (1980b): Phys. Rev. Lett. **45**, 1000

Zannoni, C. (1979): in *The Molecular Physics of Liquid Crystals*, ed. by G.R. Luckhurst, G.W. Gray (Academic Press, London) Chap. 3
Zeller, H.R. (1982a): Phys. Rev. A **26**, 1785
Zeller, H.R. (1982b): Phys. Rev. Lett. **48**, 334
Zimmer, J.E., White, J.L. (1982): in *Advances in Liquid Crystals*, ed. by G.H. Brown (Academic Press, New York) p. 157
Zink, H., Jeu, W.H. de (1985): Mol. Cryst. Liq. Cryst. **124**, 287
Zocher, H. (1933): Trans. Faraday Soc. **29**, 945
Zumer, S., Vilfan, M. (1978): Phys. Rev. A **17**, 424
Zupančič, I. Pirš, J., Luzar, M., Blinc, R., Doane, J.W. (1974): Solid State Commun. **15**, 227

Index of Compounds

Subject Index

MIX
Papier aus verantwortungsvollen Quellen
Paper from responsible sources
FSC® C105338

If you have any concerns about our products,
you can contact us on
ProductSafety@springernature.com

In case Publisher is established outside the EU,
the EU authorized representative is:
Springer Nature Customer Service Center GmbH
Europaplatz 3, 69115 Heidelberg, Germany

Printed by Libri Plureos GmbH
in Hamburg, Germany